Essentials of Modern Optical Fiber Communication

Reinhold Noé

Essentials of Modern Optical Fiber Communication

Prof. Dr.-Ing. Reinhold Noé
Universität Paderborn
Fakultät für Elektrotechnik,
Informatik und Mathematik
Institut für Elektrotechnik und Informationstechnik
Optische Nachrichtentechnik und Hochfrequenztechnik
Warburger Str. 100
33098 Paderborn
Germany
E-mail: noe@upb.de

ISBN 978-3-642-04871-5 e-ISBN 978-3-642-04872-2

DOI 10.1007/978-3-642-04872-2

Library of Congress Control Number: 2009943433

Typesetting: Data supplied by the authors

Production: Scientific Publishing Services Pvt. Ltd., Chennai, India

Cover Design: WMX Design, Heidelberg, Germany

Printed in acid-free paper

9 8 7 6 5 4 3 2 1

springer.com

Preface

This book covers important aspects of modern optical communication. It is intended to serve both students and professionals. Consequently, a solid coverage of the necessary fundamentals is combined with an in-depth discussion of recent relevant research results.

The book has grown from lecture notes over the years, starting 1992. It accompanies my present lectures Optical Communication A (Fundamentals), B (Mode Coupling), C (Modulation Formats) and D (Selected Topics) at the University of Paderborn, Germany.

I gratefully acknowledge contributions to this book from Dr. Timo Pfau, Dr. David Sandel, Dr. Sebastian Hoffmann and Mohamed El-Darawy.

Contents

Chapter 1
Introduction

At the end of the 1970ies, telecom carriers started to lay optical fiber between telecom exchange offices, and coaxial cable for electrical data communication was no longer deployed. The performance of optical fiber communication has since then grown exponentially, very much like Moore's law for the complexity of electronic circuits. In the electronic domain, rising clock speeds, miniaturization of feature sizes and chip size increase along two, maybe soon along the third dimension, are contributing to this truly impressive growth. The performance of optical communication is determined by clock speed as offered by a state-of-the-art electronic technology, availability of several or if needed many fibers in one cable, multiple optical channels carried on a single optical fiber by means of wavelength division multiplex, and recently the transmission of several bit per symbol.

The economic and societal impact is dramatic: Optical fiber communication is a key enabler of the worldwide web, of e-mail and of all but local telephone connections. The technically exploitable fiber bandwidth is roughly 10 THz, orders of magnitude higher than in other media. Fiber attenuation is extremely small: After 100 km of fiber there is still about 1% of the input power left. Optical amplifiers, with 4 THz bandwidth or more, overcome fiber loss so that transoceanic transmission is possible without intermediate signal regeneration.

Around the year 2000, in the so-called dotcom era, growth rates of information exchange of about one order of magnitude per year were forecast. This triggered massive investments and resulted in the foundation of many new companies in a short time. A significant part of that investment was lost, while achieved technical progress remains available at large. The telecom industry has consolidated since then because investments make sense only if customers pay them back. Of course, customers don't want to spend a significant part of their household budget for communication, even though available bandwidth has grown by more than two orders of magnitude thanks to DSL technology. In the years 2002 to 2004 there was even a telecom recession due to the earlier overspending.

But today's communication does indeed grow by a factor of 1.5 per year or so. Private communication such as music downloading, video portals, personal websites and of course also the ever more complex and video-laden media and enterprise websites are responsible for this, along with video telephony services, drastically increasing usage of the internet in developing countries, and so on. As a consequence there is healthy business. In contrast, revenues increase only on a single-digit percent scale annually. The quantitative growth is entertained by the technical and productivity progress.

With the rather conservative spending pattern of end users in mind, telecom carriers want to preserve their enormous investments in fiber infrastructure, and to use newly deployed fiber most economically. Multilevel modulation schemes, including the use of two orthogonal polarization modes, are needed to exploit fibers optimally. At the same time, phase modulation increases noise tolerance. Recent research and development places special emphasis on these issues, and so does this book.

Understanding fibers requires a knowledge of dielectric waveguides and their modes, including polarizations. Chapter 2 is therefore devoted to wave propagation in ideal and nonideal optical waveguides, also exhibiting polarization mode dispersion and polarization-dependent loss, to mode coupling, electrooptic components and nonlinear effects in silica fibers. Most optical components and transmission effects are based on these features.

Chapter 3 discusses optical transmission systems of all kinds. The simplest are standard intensity-modulated direct-detection systems. Their reach can be dramatically extended by optical amplifiers, the theory of which is thoroughly described. Performance is enhanced by binary and quadrature phase shift keying with interferometric detection. The same is possible also with coherent optical systems. But these can as well detect signal synchronously, which again increases performance. The principle is that the received signal and the unmodulated signal of a local laser are superimposed. The power fluctuations resulting from this interference are detected. Several signal superpositions and detectors allow to obtain an electrical replica of the optical field vector. Coherent optical transmission systems can therefore electronically compensate all linear distortions suffered during transmission. Signal processing and control algorithms for high-performance digital synchronous coherent optical receivers conclude the book.

Fiber-to-the-home services can increase customer data rates by several more orders of magnitude and make it likely that the pressure for increased capacity at moderate cost in metropolitan area and long haul communication will continue.

Chapter 2
Optical Waves in Fibers and Components

2.1 Electromagnetic Fundamentals

2.1.1 Maxwell's Equations

Electromagnetic radiations obeys Maxwell's equations

$$\operatorname{curl}\mathbf{H} = \frac{\partial \mathbf{D}}{\partial t} + \mathbf{J} \qquad \textit{Ampere's law,} \tag{2.1}$$

$$\operatorname{curl}\mathbf{E} = -\frac{\partial \mathbf{B}}{\partial t} \qquad \textit{Maxwell-Faraday equation,} \tag{2.2}$$

$$\operatorname{div}\mathbf{D} - \rho \qquad \textit{Gauß's law,} \tag{2.3}$$

$$\operatorname{div}\mathbf{B} = 0 \qquad \textit{Gauß's law for magnetism.} \tag{2.4}$$

We take the divergence of (2.1) and obtain with $\operatorname{div}\operatorname{curl}\mathbf{A} = 0$ the

$$\operatorname{div}\mathbf{J} = -\frac{\partial \rho}{\partial t} \qquad \textit{continuity equation.} \tag{2.5}$$

It says that the current drained from the surface of a differential volume element equals the reduction of charge per time interval (preservation of charge). The equations can be brought into integral form, using the integral theorems of Gauß and Stokes,

$$\oint \mathbf{H}\cdot d\mathbf{s} = \iint\left(\frac{\partial \mathbf{D}}{\partial t} + \mathbf{J}\right)\cdot d\mathbf{a} = \frac{\partial \Psi_e}{\partial t} + I \tag{2.6}$$

($I = \iint \mathbf{J}\cdot d\mathbf{a}$: enclosed current; $\Psi_e = \iint \mathbf{D}\cdot d\mathbf{a}$: electric flux),

$$U_{ind} = \oint \mathbf{E}\cdot d\mathbf{s} = -\frac{\partial}{\partial t}\left(\iint \mathbf{B}\cdot d\mathbf{a}\right) = -\frac{\partial \Psi_m}{\partial t} \tag{2.7}$$

(U_{ind} : induced voltage; $\Psi_m = \iint \mathbf{B}\cdot d\mathbf{a}$: magnetic flux),

$$\oiint \mathbf{D}\cdot d\mathbf{a} = \iiint \rho\, dV = Q \qquad (Q = \iiint \rho\, dV \text{ : enclosed charge}), \tag{2.8}$$

$$\oiint \mathbf{B} \cdot d\mathbf{a} = 0\,, \tag{2.9}$$

$$I = \oiint \mathbf{J} \cdot d\mathbf{a} = -\frac{\partial}{\partial t}\left(\iiint \rho\, dV\right) = -\frac{\partial Q}{\partial t}\,. \tag{2.10}$$

The relations between fields and flux densities are given by the material equations

$$\mathbf{D} = \varepsilon_0 \mathbf{E} + \mathbf{P}\,, \tag{2.11}$$

$$\mathbf{B} = \mu_0 (\mathbf{H} + \mathbf{M})\,. \tag{2.12}$$

In isotropic media electric (**P**) and magnetic (**M**) dipole moment have the same direction as the corresponding field. Therefore the material equations simplify into

$$\mathbf{D} = \varepsilon \mathbf{E} = \varepsilon_0 \varepsilon_\mathrm{r} \mathbf{E} = \varepsilon_0 (1+\chi) \mathbf{E} \qquad (\chi:\ \text{susceptibility}), \tag{2.13}$$

$$\mathbf{B} = \mu \mathbf{H} = \mu_0 \mu_\mathrm{r} \mathbf{H}\,. \tag{2.14}$$

But the same equations can also be applied for anisotropic media if ε (and χ) and μ are not defined as scalars but as rank-2 tensors (matrices),

$$\mathbf{D} = \boldsymbol{\varepsilon} \mathbf{E} = \varepsilon_0 \boldsymbol{\varepsilon}_r \mathbf{E} = \varepsilon_0 (\mathbf{1} + \boldsymbol{\chi}) \mathbf{E} \qquad \mathbf{B} = \boldsymbol{\mu} \mathbf{H}\,. \tag{2.15}$$

The material tensors are quadratic 3×3 matrices. In non-magnetic media, as employed in optics, it holds $\mu_r = 1$, $\mu = \mu_0$. All the same we will occasionally set μ instead of μ_0 in order to emphasize the analogy of treatment of magnetic and elektric field or to show that equations can be used also outside the optical domain. The relative dielectricity constant serves to define the refractive index n through $\varepsilon_r = n^2$. In vacuum it holds $\mu_r = 1$, $\varepsilon_r = n^2 = 1$.

Ohm's law

$$\mathbf{J} = \sigma \mathbf{E}\,, \tag{2.16}$$

which is another material equation, relates current density and electric field.

Time-dependent signals can be expressed by summation of Fourier componets with different frequencies in the frequency domain. Therefore a complex separation ansatz of space and time dependence such as $\underline{\mathbf{H}}(\mathbf{r},t) = \underline{\mathbf{H}}(\mathbf{r}) e^{j\omega t}$ is particularly apt to solve Maxwell's equations. The physical, scalar or vectorial amplitude quantity is simply the real part of the corresponding complex quantity. If one replaces $\partial/\partial t$ by $j\omega$ then (2.1) becomes

$$\operatorname{curl} \underline{\mathbf{H}} = j\omega \underline{\mathbf{D}} + \underline{\mathbf{J}} = j\omega\varepsilon \underline{\mathbf{E}} + \underline{\mathbf{J}} \qquad \text{with } \underline{\mathbf{D}} = \varepsilon \underline{\mathbf{E}}\,, \tag{2.17}$$

where we have assumed time-invariance of ε. Losses are taken into account in the current density $\underline{\mathbf{J}}$, which facilitates the interpretation of Poynting's vector. But in optics it is often more convenient to take losses into account in a complex dielectricity constant

$$\underline{\varepsilon} = \varepsilon_0 \underline{\varepsilon}_r = \varepsilon_0(\varepsilon_r - j\varepsilon_{ri}) = \varepsilon_0\varepsilon_r - j\sigma/\omega = \varepsilon - j\sigma/\omega, \quad (2.18)$$

here defined for isotropic media. This results in a re-defined complex flux density

$$\underline{\mathbf{D}} = \underline{\varepsilon}\underline{\mathbf{E}} = \varepsilon\underline{\mathbf{E}} - j\frac{\sigma}{\omega}\underline{\mathbf{E}} = \varepsilon\underline{\mathbf{E}} - j\frac{1}{\omega}\underline{\mathbf{J}}, \quad (2.19)$$

$$\operatorname{div}\underline{\mathbf{D}} = \operatorname{div}\varepsilon\underline{\mathbf{E}} + \frac{1}{j\omega}\operatorname{div}\underline{\mathbf{J}} = \underline{\rho} - \frac{1}{j\omega}\frac{\partial\underline{\rho}}{dt} = 0. \quad (2.20)$$

Here (2.5) has been inserted. In (2.17) the term $j\omega\underline{\mathbf{D}} + \underline{\mathbf{J}}$ is replaced by the re-defined (by (2.19)) $j\omega\underline{\mathbf{D}}$. One obtains

$$\operatorname{curl}\underline{\mathbf{H}} = \frac{\partial\underline{\underline{\mathbf{D}}}}{\partial t} = \underline{\varepsilon}\frac{\partial\underline{\underline{\mathbf{E}}}}{\partial t} = j\omega\underline{\mathbf{D}} = j\omega\underline{\varepsilon}\underline{\mathbf{E}} \qquad \text{with } \underline{\mathbf{D}} = \underline{\varepsilon}\underline{\mathbf{E}}. \quad (2.21)$$

Note that the effects of current density are duly taken into account, like in (2.17).

If there are pure ohmic losses then σ is frequency-independent. Generally it depends on frequency. Losses are characterized by $\sigma > 0$, $\varepsilon_{ri} > 0$. In lasers and optical amplifiers one utilizes media which amplify electromagnetic radation in the optical domain, where $\sigma < 0$, $\varepsilon_{ri} < 0$ is valid.

The two definitions of $\underline{\mathbf{D}}$ are based on two different usages in the literature. While (2.21) is formally (2.17) in contradiction with (2.1) the current density is correctly taken into account by the complex dielectricity constant $\underline{\varepsilon}$.

In an analog fashion magnetic losses can be expressed by a complex permeability constant

$$\underline{\mu} = \mu_0\underline{\mu}_r = \mu_0(\mu_r - j\mu_{ri}). \quad (2.22)$$

For anisotropic media one uses complex material tensors $\underline{\varepsilon}$, $\underline{\mu}$.

We obtain Maxwell's equation in complex notation

$$\operatorname{curl}\underline{\mathbf{H}} = j\omega\underline{\mathbf{D}} = j\omega\underline{\varepsilon}\underline{\mathbf{E}}, \quad (2.21)$$

$$\operatorname{curl}\underline{\mathbf{E}} = -j\omega\underline{\mathbf{B}} = -j\omega\underline{\mu}\underline{\mathbf{H}}, \quad (2.23)$$

$$\operatorname{div}\underline{\mathbf{D}} = \operatorname{div}(\underline{\varepsilon}\underline{\mathbf{E}}) = 0, \quad (2.20)$$

$$\operatorname{div}\underline{\mathbf{B}} = \operatorname{div}(\underline{\mu}\underline{\mathbf{H}}) = 0. \quad (2.24)$$

With real dielectricity constant when using the other definition of the eletric flux density (dielectric displacement) it holds instead

$$\operatorname{curl}\underline{\mathbf{H}} = j\omega\underline{\mathbf{D}} + \underline{\mathbf{J}} = j\omega\varepsilon\underline{\mathbf{E}} + \underline{\mathbf{J}}, \quad (2.17)$$

$$\operatorname{div}\underline{\mathbf{D}} = \operatorname{div}(\varepsilon\underline{\mathbf{E}}) = \underline{\rho}, \ \operatorname{div}\underline{\mathbf{J}} = \operatorname{div}(\sigma\underline{\mathbf{E}}) = -j\omega\underline{\rho}. \quad (2.25)$$

2.1.2 Boundary Conditions

Normally the medium of wave propagation is not homogeneous and infinite in space. For example, between air (refractive index $n = \sqrt{\varepsilon_r} \approx 1$) and silica ($n \approx 1{,}46$) there is a refractive index difference which must be taken into account in the calculations. The most effective way to do this is to solve the wave propagation equations at both sides of the boundary and to equate the two solutions with free parameters, using the boundary conditions.

We first determine the *normal boundary conditions* for electric and magnetic flux densities perpendicular to the boundary. In Fig. 2.1a the boundary region between two media with different material properties is sketched. Let the two media 1, 2 be homogeneous and isotropic. Bottom and lid of a shallow cylinder, both with area *F*, lie in media 1 and. 2 with material constants μ_1, ε_1 and μ_2, ε_2, respectively. The unit vector **n** is perpendicular to the boundary plane. Let the cylinder height *h* approach zero, so that its surface can be neglected. Gauß's law for magnetism in integral form (2.9) yields

$$0 = \oiint \mathbf{B} \cdot d\mathbf{a} = (\mathbf{B}_2 - \mathbf{B}_1) \cdot \mathbf{n}\, F \quad \Rightarrow \quad B_{2n} - B_{1n} = 0\,. \tag{2.26}$$

The normal components B_{1n}, B_{2n} of the magnetic flux density in direction of the normal vector **n** are identical on both sides of the boundary. In other words, it must be continuous while passing the boundary. Gauß's law in integral form (2.8) yields the enclosed charge. Assuming an area charge density ρ_A, which in the boundary itself corresponds to an infinite space charge density, the enlosed charge equals $Q = \rho_A F$. In optics it usually holds $\rho_A = 0$. In summary it holds for the normal components D_{1n}, D_{2n} of the electric flux density

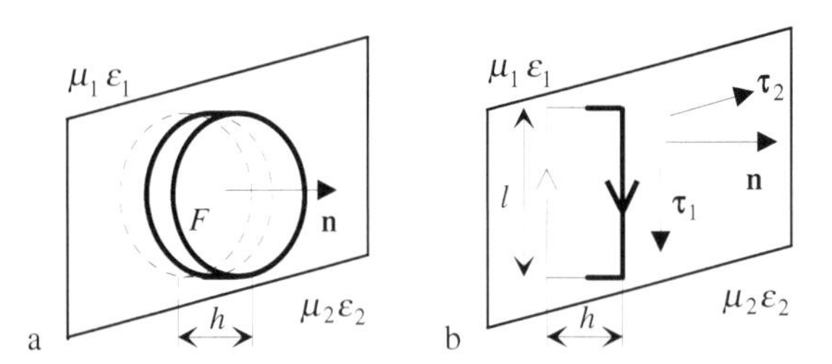

Fig. 2.1 Derivation of normal **a** and tangential **b** boundary conditions

$$Q = \iiint \rho dV = \oiint \mathbf{D} \cdot d\mathbf{a} = (\mathbf{D}_2 - \mathbf{D}_1) \cdot \mathbf{n}\, F \quad \Rightarrow \quad D_{2n} - D_{1n} = \rho_A\,. \tag{2.27}$$

For deduction of the *tangential boundary conditions* we look at Fig. 2.1b. A rectangular area element have length *l* and height *h*, which tends again toward zero. The closed-loop integral of the magnetic field is

$$\oint \mathbf{H} \cdot d\mathbf{s} = l\left(\mathbf{H}_2 - \mathbf{H}_1\right) \cdot \boldsymbol{\tau}_1 , \tag{2.28}$$

where $\boldsymbol{\tau}_1$ is the unit vector in the tangential plane parallel to a side of the rectangle. For finite temporal changes of electric flux and current densities the right-hand side of Ampere's law in integral form (2.6), applied to the area element, equals zero, since height h approaches zero. But if the boundary conductivity is infinite then there can be an area current density $\mathbf{J}_A$ with

$$\lim_{h \to 0} \int_{-h/2}^{h/2} \mathbf{J} dn = \mathbf{J}_A . \tag{2.29}$$

Ampere's law then yields

$$\left(\mathbf{H}_2 - \mathbf{H}_1\right) \cdot \boldsymbol{\tau}_1 = \mathbf{J}_A \cdot \boldsymbol{\tau}_2 , \tag{2.30}$$

where $\boldsymbol{\tau}_2$ is the unit vector in the tangential plane that is perpedicular to $\boldsymbol{\tau}_1$. If one replaces $\boldsymbol{\tau}_1$ by $\boldsymbol{\tau}_2 \times \mathbf{n}$ one obtains on the left-hand side a spade product of three vectors, which may be cyclically exchanged according to $\mathbf{u} \cdot \left(\mathbf{v} \times \mathbf{w}\right) = \mathbf{v} \cdot \left(\mathbf{w} \times \mathbf{u}\right)$,

$$\boldsymbol{\tau}_2 \cdot \left[\mathbf{n} \times \left(\mathbf{H}_2 - \mathbf{H}_1\right)\right] = \boldsymbol{\tau}_2 \cdot \mathbf{J}_A . \tag{2.31}$$

Since the direction of $\boldsymbol{\tau}_2$ in the tangential plane can be chosen at will, and $\mathbf{J}_A$ and $\mathbf{n} \times \left(\mathbf{H}_2 - \mathbf{H}_1\right)$ lie in the tangential plane, we may write

$$\mathbf{n} \times \left(\mathbf{H}_2 - \mathbf{H}_1\right) = \mathbf{J}_A . \tag{2.32}$$

Infinite conductivity excluded the tangential components (index t) of the magnetic field are continuous when passing the boundary,

$$\mathbf{H}_{2t} = \mathbf{H}_{1t} \qquad \text{for} \qquad \mathbf{J}_A = 0 . \tag{2.33}$$

The Maxwell-Faraday equation in integral form (2.7) allows to deduce in analog fashion the continuity of the tangential electric field components in the boundary,

$$\mathbf{E}_{2t} = \mathbf{E}_{1t} . \tag{2.34}$$

In (2.26), (2.27) we have deduced the conditions for the normal components of the flux densities. The corresponding fields are found using the material equations. Similarly, the tangential components of the flux densities can be found from the material equations once (2.32), (2.33) and (2.34) have specified the tangential field components.

Tangential and normal bounary conditions are interrelated. To show this one bends the area element of Fig. 2.1b to a complete cylinder wall of Fig. 2.1a. This way the continuity of the tangential electric (magnetic) field becomes equivalent to the continuity of the normal magnetic (electric) flux density. It is therefore sufficient to fulfill either

- the tangential or
- the normal boundary conditions or
- the normal boundary condition for the electric flux density and the tangential one for the electric field or
- the normal boundary condition for the magnetic flux density and the tangential one for the magnetic field.

The other boundary conditions are then automatically fulfilled.

The homogeneous region may be limited to the immediate surroundings of the boundary.

2.1.3 Wave Equation

We use complex notation and take losses into account in the imaginary parts of complex material parameters $\underline{\varepsilon}$, $\underline{\mu}$. The medium be isotropic so that $\underline{\varepsilon}$, $\underline{\mu}$ are scalars. We take the curl operator on both sides of Maxwell-Faraday equation (2.23) and apply on the right-hand side the general relation $\operatorname{curl}(\underline{F}\underline{\mathbf{a}}) = \underline{F}\operatorname{curl}\underline{\mathbf{a}} - \underline{\mathbf{a}}\times\operatorname{grad}\underline{F}$,

$$\operatorname{curl}(\operatorname{curl}\underline{\mathbf{E}}) = -j\omega\operatorname{curl}(\underline{\mu}\underline{\mathbf{H}}) = j\omega(-\underline{\mu}\operatorname{curl}\underline{\mathbf{H}} + \underline{\mathbf{H}}\times\operatorname{grad}\underline{\mu}). \tag{2.35}$$

Ampere's law (2.21) is inserted into first term, while (2.23) is again inserted into the second term on the right-hand side,

$$\operatorname{curl}(\operatorname{curl}\underline{\mathbf{E}}) = \omega^2\underline{\mu}\underline{\varepsilon}\underline{\mathbf{E}} - \frac{1}{\underline{\mu}}\operatorname{curl}\underline{\mathbf{E}}\times\operatorname{grad}\underline{\mu}. \tag{2.36}$$

The second term on the right-hand side is roughly zero if $\underline{\mu}$ changes only little within one optical wavelength. This is quite common. In optics it even holds $\underline{\mu} = \mu_0$ so that $\operatorname{grad}\underline{\mu} = 0$ holds. As a result we obtain

$$\operatorname{curl}(\operatorname{curl}\underline{\mathbf{E}}) = \omega^2\underline{\mu}\underline{\varepsilon}\underline{\mathbf{E}}. \tag{2.37}$$

According to (2.20) and with $\operatorname{div}(\underline{F}\underline{\mathbf{A}}) = \underline{F}\operatorname{div}\underline{\mathbf{A}} + \underline{\mathbf{A}}\cdot\operatorname{grad}\underline{F}$ we can write

$$0 = \operatorname{div}(\underline{\mathbf{D}}) = \underline{\varepsilon}\operatorname{div}\underline{\mathbf{E}} + \underline{\mathbf{E}}\cdot\operatorname{grad}\underline{\varepsilon}. \tag{2.38}$$

We insert into (2.37) the Laplace operator $\Delta\underline{\mathbf{A}} = \operatorname{grad}(\operatorname{div}\underline{\mathbf{A}}) - \operatorname{curl}(\operatorname{curl}\underline{\mathbf{A}})$ and (2.38),

$$\Delta\underline{\mathbf{E}} - \operatorname{grad}(\operatorname{div}\underline{\mathbf{E}}) = \Delta\underline{\mathbf{E}} + \operatorname{grad}\left(\frac{1}{\underline{\varepsilon}}\underline{\mathbf{E}}\cdot\operatorname{grad}\underline{\varepsilon}\right) = -\omega^2\underline{\mu}\underline{\varepsilon}\underline{\mathbf{E}}. \tag{2.39}$$

In vacuum and other homogeneous media (i.e., $\underline{\varepsilon}$, $\underline{\mu}$ are position-independent), but with sufficient accuracy also in slightly inhomogeneous media it holds $(\text{grad}\,\underline{\varepsilon})/\underline{\varepsilon} = 0$. This results in a simplified wave equation for the electric field,

$$\Delta\underline{\mathbf{E}} + \omega^2 \underline{\mu}\underline{\varepsilon}\underline{\mathbf{E}} = 0 . \tag{2.40}$$

Due to the symmetry of Maxwell's equations one can derive in analog fashion for the magnetic field

$$\Delta\underline{\mathbf{H}} + \omega^2 \underline{\mu}\underline{\varepsilon}\underline{\mathbf{H}} = 0 . \tag{2.41}$$

The vectorial wave equation (2.40) tells only the relation between space and time-dependence of the wave amplitude. However, the direction of the field vector is yet unclear. Once (2.40) is solved one may choose a tentative arbitrary vector direction $\underline{\mathbf{E}}$. Then one calculates $\underline{\mathbf{H}}$ through the Maxwell-Faraday equation. Finally $\underline{\mathbf{H}}$ is inserted into Ampere's law and one obtains a usually modified $\underline{\mathbf{E}}$ which is the correct solution. Instead of this complicated procedure one may start with certain assumptions (subchapter 2.1.4) or may eliminate a degree of freedom of the field vector (subchapter 2.3.2). The same holds for solutions of (2.41). An elegant possibility for "direct" solution of Maxwell's equations are electromagnetic potentials.

We assume now a nonmagnetic medium ($\mu = \mu_0$), insert the

$$c = \frac{1}{\sqrt{\varepsilon_0 \mu_0}} \qquad \text{speed of light in vacuum} \tag{2.42}$$

and the definition of the refractive index

$$n = \sqrt{\varepsilon_r}\,, \qquad \underline{n} = (\underline{\varepsilon}_r)^{1/2} \tag{2.43}$$

and write as an example (2.40) as

$$\Delta\underline{\mathbf{E}} = \frac{\underline{n}^2}{c^2}\frac{\partial^2 \underline{\mathbf{E}}}{\partial t^2} = -\omega^2 \frac{\underline{n}^2}{c^2}\underline{\mathbf{E}} . \tag{2.44}$$

The wave equation is often solved numerically. Knowledge of the position-dependence of $\underline{\varepsilon}$ or $\underline{n}$ and, if applicable, of an incident field $\underline{\mathbf{E}}$, is required.

2.1.4 Homogeneous Plane Wave in Isotropic Homogeneous Medium

We investigate wave propagation in an isotropic, homogeneous medium and write Ampere's law (2.21) and the Maxwell-Faraday equation (2.23) in cartesian coordinates

$$\frac{\partial \underline{H}_z}{\partial y} - \frac{\partial \underline{H}_y}{\partial z} = \underline{\varepsilon} \frac{\partial \underline{E}_x}{\partial t}, \tag{2.45}$$

$$\frac{\partial \underline{H}_x}{\partial z} - \frac{\partial \underline{H}_z}{\partial x} = \underline{\varepsilon} \frac{\partial \underline{E}_y}{\partial t}, \tag{2.46}$$

$$\frac{\partial \underline{H}_y}{\partial x} - \frac{\partial \underline{H}_x}{\partial y} = \underline{\varepsilon} \frac{\partial \underline{E}_z}{\partial t}, \tag{2.47}$$

$$\frac{\partial \underline{E}_z}{\partial y} - \frac{\partial \underline{E}_y}{\partial z} = -\underline{\mu} \frac{\partial \underline{H}_x}{\partial t}, \tag{2.48}$$

$$\frac{\partial \underline{E}_x}{\partial z} - \frac{\partial \underline{E}_z}{\partial x} = -\underline{\mu} \frac{\partial \underline{H}_y}{\partial t}, \tag{2.49}$$

$$\frac{\partial \underline{E}_y}{\partial x} - \frac{\partial \underline{E}_x}{\partial y} = -\underline{\mu} \frac{\partial \underline{H}_z}{\partial t}. \tag{2.50}$$

Without loss of generality we initially choose the z axis as the propagation direction. We furthermore assume a plane wave (in the x-y plane). This means that the derivaties $\partial / \partial x$ and $\partial / \partial y$ of the fields are zero. Note that this assumption is not generally valid!

As a consequence the temporal derivatives of $\underline{E}_z$ and $\underline{H}_z$ vanish. We are not interested in static fields, which can not propagate a wave. Therefore it holds $\underline{E}_z = \underline{H}_z = 0$. This plane wave has only transversal, no longitudinal field components. If one takes the derivative $\partial / \partial z$ of (2.49) and inserts into the derivative $\partial / \partial t$ of (2.45) one obtains the plane wave equation

$$\frac{\partial^2 \underline{E}_x}{\partial z^2} = \underline{\varepsilon\mu} \frac{\partial^2 \underline{E}_x}{\partial t^2}. \tag{2.51}$$

The same equation is obtained if one sets $\partial / \partial x = \partial / \partial y = 0$ in (2.44) and considers only the x-component of the field. The general solution of (2.51) is

$$\underline{E}_x(z,t) = \underline{E}_{x0}^{+} e^{j(\omega t - \underline{k}z)} + \underline{E}_{x0}^{-} e^{j(\omega t + \underline{k}z)}. \tag{2.52}$$

$\underline{k}$ is the *wave number* or *propagation constant*) of the wave in z-direction with

$$\begin{matrix} \underline{k}^2 = \omega^2 \underline{\varepsilon\mu} = \omega^2 \underline{n}^2 / c^2 & (\underline{n}: \text{only for } \underline{\mu} = \mu_0) & \underline{k} = \omega \underline{n}/c = 2\pi \underline{n}/\lambda_0 \\ j\underline{k} = \alpha + j\beta = \underline{\gamma} & \underline{k} = \beta - j\alpha = -j\underline{\gamma} & k_0 = \beta_0 = 2\pi/\lambda_0 \end{matrix}. \tag{2.53}$$

For description of transversal electromagnetic (TEM) lines one sometimes uses $\underline{\gamma}$ instead of $j\underline{k}$. Quantity λ_0 is the wavelength in vacuum. The solution with exponent $\omega t - \underline{k}z$ corresponds to wave propagation in positive z-direction; the solution with $\omega t + \underline{k}z$ propagates in negative z-direction. Wavenumber $\underline{k}$ is, like dielectricity constant $\underline{\varepsilon}$ or refractive index $\underline{n}$ complex in general. β is the *phase constant,* α the *attenuation constant* of the wave amplitude. Quantity $k_0 = \beta_0$ is the wave number and phase constant of vacuum.

If one sets, for example for the wave propagating in positive z-direction, $\omega t - \beta z = \text{const.}$, then one obtains the temporal derivative $\omega - \beta\, \partial z/\partial t = 0$. This allows to define generally

$$v_{ph} = \partial z/\partial t = \omega/\beta = c/\mathrm{Re}(\underline{n}) \qquad \textit{phase velocity,} \tag{2.54}$$

$$v_g = \partial\omega/\partial\beta \qquad \textit{group velocity,} \tag{2.55}$$

$$\frac{1}{v_g} = \frac{\partial\beta}{\partial\omega} = \frac{n_g}{c} \qquad n_g = n + \omega\frac{\partial n}{\partial\omega} \qquad \textit{group refractive index.} \tag{2.56}$$

The sinusoidal, single-frequency or monochromatic ansatz (2.52) is called a harmonic electromagnetic wave. Since any wave can be expressed, by Fourier transformation, as a linear combination of sinusoidal waves with various frequencies and since (2.51) is a linear differential equation the general solution consists in a linear combination of terms of type (2.52) with various propagation directions, angular frequencies ω, propagation constants $\underline{k}$ and phases. (The phase is contained above in the complex field amplitude, for example $\underline{E}_{x0}^{+}$.) The same holds for the magnetic field (which coexists with the electric field). Insertion of (2.52) into (2.49) yields

$$\underline{H}_y(z,t) = \underline{H}_{y0}^{+}\, e^{j(\omega t - \underline{k}z)} + \underline{H}_{y0}^{-}\, e^{j(\omega t + \underline{k}z)} \tag{2.57}$$

with

$$\underline{E}_{x0}^{\pm} = \pm\underline{Z}_F \cdot \underline{H}_{y0}^{\pm}, \tag{2.58}$$

$$\underline{Z}_F = \left(\frac{\underline{\mu}}{\underline{\varepsilon}}\right)^{\frac{1}{2}} = Z_{F0}\left(\frac{\underline{\mu}_r}{\underline{\varepsilon}_r}\right)^{\frac{1}{2}} \text{(generally)},\quad Z_{F0} = \sqrt{\frac{\mu_0}{\varepsilon_0}} \approx 377\,\Omega \text{ (vacuum).} \tag{2.59}$$

$\underline{Z}_F$, Z_F is the *field characteristic impedance*; Z_{F0} is the field characteristic impedance of vacuum. In lossless media the complex $\underline{Z}_F$ becomes the real Z_F. In special cases it holds

$$Z_F = \sqrt{\frac{\mu}{\varepsilon}} = Z_{F0}\sqrt{\frac{\mu_r}{\varepsilon_r}} \text{ (real } \varepsilon, \mu\text{)},\ \underline{Z}_F = \frac{Z_{F0}}{\underline{n}} \text{ (optics, } \mu = \mu_0\text{).} \tag{2.60}$$

While we have considered so far only $\underline{E}_x$, $\underline{H}_y$ there exist similar, indepenent solutions also for $\underline{E}_y$, $\underline{H}_x$. Like in (2.58) one finds

$$\underline{E}_{y0}^{\pm} = \mp \underline{Z}_F \cdot \underline{H}_{x0}^{\pm} . \tag{2.61}$$

Note the opposite signs compared to (2.58). Since phase and amplitude of forward and backward traveling homogeneous plane waves (2.52) depend in space only on z one may replace the product $\underline{k}z$ by the scalar product $\underline{\mathbf{k}} \cdot z\mathbf{e}_z$. For propagation in any direction $\underline{k}z$ must be replaced by the scalar product $\underline{\mathbf{k}} \cdot \mathbf{r}$, where $\mathbf{r} = [x, y, z]^T = x\mathbf{e}_x + y\mathbf{e}_y + z\mathbf{e}_z$ is the position vector.

We generalize our findings. Contained in the

$$\underline{\mathbf{k}} = \boldsymbol{\beta} - j\boldsymbol{\alpha} \qquad \textit{wave vector} \tag{2.62}$$

there is the *phase vector* $\boldsymbol{\beta}$ and the *amplitude vector* $\boldsymbol{\alpha}$. For *homogeneous* waves these two have the same direction. Eqn. (2.52) is a homogeneous wave due to $\underline{k}z = \underline{k}\mathbf{e}_z \cdot \mathbf{r}$. For a general plane but not necessarily homogeneous wave it holds

$$\underline{\mathbf{E}}(\mathbf{r},t) = \underline{\mathbf{E}}_0 e^{j(\omega t - \underline{\mathbf{k}} \cdot \mathbf{r})} = \underline{\mathbf{E}}_0 e^{-\boldsymbol{\alpha} \cdot \mathbf{r}} e^{j(\omega t - \boldsymbol{\beta} \cdot \mathbf{r})} . \tag{2.63}$$

Let $\underline{\mathbf{E}}_0$ and wave vector be position-independent. Inseration into (2.44) yields

$$\underline{k}_x^2 + \underline{k}_y^2 + \underline{k}_z^2 = \omega^2 \frac{\underline{n}^2}{c^2} . \tag{2.64}$$

The phase gradient is $-\boldsymbol{\beta}$, the amplitude gradient is $-\boldsymbol{\alpha}$. For a position-independent phase vector we have a *plane wave*. Its direction is the *propagation direction*. We write

$$\boldsymbol{\beta} = \beta \mathbf{s} . \tag{2.65}$$

The unit vector $\mathbf{s}$ in propagation direction is also called *wave normal vector* because it is perpendicular to the equiphase planes. For plane waves in lossless media amplitude and phase vectors are perpendicular to each other because an amplitude vector component in the direction of the phase vector would mean an attenuation or amplification of the wave along its propagation path. A homogeneous plane wave in a lossless medium has the amplitude vector $\boldsymbol{\alpha} = \mathbf{0}$, because it is both parallel and perpendicular to the phase vector.

In inhomogeneous media $\underline{n}$ and $\underline{\mathbf{k}}$ are position-dependent. In this context β, the length of phase vector $\boldsymbol{\beta} = \beta \mathbf{s}$, is not a phase *constant*. But in sufficiently small areas of inhomogeneous media waves usually can be considered as plane waves.

2.1.5 Power and Energy

We assume scalar material constants, insert (2.13) and (2.14) into Ampere's law (2.1) and the Maxwell-Faraday equation (2.2),

$$\text{curl}\,\mathbf{H} = \frac{\partial}{\partial t}(\varepsilon\mathbf{E}) + \mathbf{J}\,, \tag{2.66}$$

$$\text{curl}\,\mathbf{E} = -\frac{\partial}{\partial t}(\mu\mathbf{H}), \tag{2.67}$$

and take the scalar product of $\mathbf{E}$ with (2.66) and of $\mathbf{H}$ with (2.67). The right-hand sides can be manipulated, using the product rule of differentiation (for example $\mathbf{E}\cdot\partial\mathbf{E}/\partial t = (1/2)\,\partial|\mathbf{E}|^2/\partial t$). One thereby obtains

$$\mathbf{E}\cdot\text{curl}\,\mathbf{H} = \frac{\varepsilon}{2}\frac{\partial}{\partial t}|\mathbf{E}|^2 + \mathbf{E}\cdot\mathbf{J}\,, \tag{2.68}$$

$$\mathbf{H}\cdot\text{curl}\,\mathbf{E} = -\frac{\mu}{2}\frac{\partial}{\partial t}|\mathbf{H}|^2. \tag{2.69}$$

The real electromagnetic power density through a differential area element is given by the

$$\mathbf{S} = \mathbf{E}\times\mathbf{H} \qquad \text{Poynting vector} \qquad \text{with} \qquad P = \oint\!\!\oint \mathbf{S}\cdot d\mathbf{a}\,. \tag{2.70}$$

Here $\mathbf{E}$ and $\mathbf{H}$ must be due to the same source. P is the power which is emitted by a volume with a known surface. If one is interested in the power through a certain area one integrates only over this area. If one subtracts (2.69) from (2.68) and applies the general rule $\text{div}(\mathbf{A}\times\mathbf{B}) = \mathbf{B}\cdot\text{curl}\,\mathbf{A} - \mathbf{A}\cdot\text{curl}\,\mathbf{B}$ one obtains the differential form of Poynting's theorem

$$-\text{div}\,\mathbf{S} = \mathbf{E}\cdot\mathbf{J} + \frac{\partial}{\partial t}\left(\frac{\varepsilon}{2}|\mathbf{E}|^2 + \frac{\mu}{2}|\mathbf{H}|^2\right), \tag{2.71}$$

which allows to show the conservation of energy. Energies are transferred from their original type into another type. This is because power densities are temporal derivatives of energy densities. In the medium we find

$$\begin{aligned} p_{\text{ve}} &= \mathbf{E}\cdot\mathbf{J} && \text{Ohmic electric loss density,} \\ w_{\text{e}} &= \frac{\varepsilon}{2}|\mathbf{E}|^2 && \text{stored electric and} \\ w_{\text{m}} &= \frac{\mu}{2}|\mathbf{H}|^2 && \text{magnetic energy density.} \end{aligned} \tag{2.72}$$

Magnetic losses are not taken into account in the above. Quantity $-\text{div}\,\mathbf{S}$ is the power density of electromagnetic radation flowing into the differential volume element. Integration over the volume yields the integral form of Poynting's theorem,

$$-\oiint \mathbf{S}\cdot d\mathbf{a} = \iiint \left(\mathbf{E}\cdot\mathbf{J} + \frac{\partial}{\partial t}\left(\frac{\varepsilon}{2}|\mathbf{E}|^2 + \frac{\mu}{2}|\mathbf{H}|^2 \right) \right) dV \,. \tag{2.73}$$

At the left-hand side the volume integral of $\mathrm{div}\,\mathbf{S}$ has been replaced by a surface intergral of $\mathbf{S}$ according to the integral theorem of Gauß.

We define the complex Poynting vector

$$\underline{\mathbf{T}} = \frac{1}{2}\left(\underline{\mathbf{E}} \times \underline{\mathbf{H}}^* \right). \tag{2.74}$$

$\underline{\mathbf{T}}$ is *not* simply the complex form of $\mathbf{S}$, because complex notation of real quanties is not possible in *products* of complex quantities. After insertion of the complex monochromatic expressions

$$\begin{aligned} \mathbf{E} &= \mathrm{Re}\left(\underline{\mathbf{E}}_0 e^{j\omega t} \right) = \frac{1}{2}\left(\underline{\mathbf{E}}_0 e^{j\omega t} + \underline{\mathbf{E}}_0^* e^{-j\omega t} \right) \\ \mathbf{H} &= \mathrm{Re}\left(\underline{\mathbf{H}}_0 e^{j\omega t} \right) = \frac{1}{2}\left(\underline{\mathbf{H}}_0 e^{j\omega t} + \underline{\mathbf{H}}_0^* e^{-j\omega t} \right) \end{aligned} \tag{2.75}$$

into (2.70) we find

$$\mathbf{S} = \frac{1}{2}\mathrm{Re}\left(\underline{\mathbf{E}}_0 \times \underline{\mathbf{H}}_0^* \right) + \frac{1}{2}\mathrm{Re}\left(\underline{\mathbf{E}}_0 \times \underline{\mathbf{H}}_0 e^{j2\omega t} \right). \tag{2.76}$$

The second term on the right-hand side is an alternating signal at twice the frequency of the fields, the temporal average of which equals zero. The first tem is constant and therefore is the temporal average $\bar{\mathbf{S}}$ of the (real) Poynting vector. Comparison with (2.74) yields

$$\bar{\mathbf{S}} = \mathrm{Re}\left(\underline{\mathbf{T}} \right) = \frac{1}{2}\mathrm{Re}\left(\underline{\mathbf{E}} \times \underline{\mathbf{H}}^* \right). \tag{2.77}$$

To compute $\underline{\mathbf{T}}$ we must take the scalar products of $\underline{\mathbf{E}}$ with the complex conjugate of (2.17), (2.21) and of $\underline{\mathbf{H}}^*$ with (2.23),

$$\begin{aligned} \underline{\mathbf{E}}\cdot\mathrm{curl}\,\underline{\mathbf{H}}^* &= -j\omega\varepsilon|\underline{\mathbf{E}}|^2 + \underline{\mathbf{E}}\cdot\underline{\mathbf{J}}^* = -j\omega\underline{\varepsilon}^*|\underline{\mathbf{E}}|^2 \\ \underline{\mathbf{H}}^*\cdot\mathrm{curl}\,\underline{\mathbf{E}} &= -j\omega\underline{\mu}|\underline{\mathbf{H}}|^2 \end{aligned} \,. \tag{2.78}$$

Subtraction and division by 2 yields, in analogous fashion to the above,

$$-\mathrm{div}\,\underline{\mathbf{T}} = \frac{1}{2}\underline{\mathbf{E}}\cdot\underline{\mathbf{J}}^* + j\omega\left(\frac{\mu}{2}|\underline{\mathbf{H}}|^2 - \frac{\varepsilon}{2}|\underline{\mathbf{E}}|^2 \right) = j\omega\left(\frac{\underline{\mu}}{2}|\underline{\mathbf{H}}|^2 - \frac{\underline{\varepsilon}^*}{2}|\underline{\mathbf{E}}|^2 \right). \tag{2.79}$$

After integration we get

$$\begin{aligned} -\oiint \underline{\mathbf{T}} \cdot d\mathbf{a} &= \iiint \left(\frac{1}{2} \underline{\mathbf{E}} \cdot \underline{\mathbf{J}}^* + j\omega \left(\frac{\underline{\mu}}{2} |\underline{\mathbf{H}}|^2 - \frac{\underline{\varepsilon}}{2} |\underline{\mathbf{E}}|^2 \right) \right) dV \\ &= \iiint j\omega \left(\frac{\underline{\mu}}{2} |\underline{\mathbf{H}}|^2 - \frac{\underline{\varepsilon}^*}{2} |\underline{\mathbf{E}}|^2 \right) dV \end{aligned} . \tag{2.80}$$

The mean values of

$$\begin{aligned} &\overline{p_{\mathrm{ve}}} = \frac{\sigma}{2} |\underline{\mathbf{E}}|^2 = \frac{\omega \varepsilon_0 \varepsilon_{\mathrm{ri}}}{2} |\underline{\mathbf{E}}|^2 && \text{electric,} \\ &\overline{p_{\mathrm{vm}}} = \frac{\omega \mu_0 \mu_{\mathrm{ri}}}{2} |\underline{\mathbf{H}}|^2 && \text{magnetic and} \\ &\overline{p_{\mathrm{v}}} = \overline{p_{\mathrm{ve}}} + \overline{p_{\mathrm{vm}}} && \text{total loss power density,} \\ &\overline{w_{\mathrm{e}}} = \frac{\varepsilon}{4} |\underline{\mathbf{E}}|^2 = \frac{\mathrm{Re}(\underline{\varepsilon})}{4} |\underline{\mathbf{E}}|^2 && \text{stored electric and} \\ &\overline{w_{\mathrm{m}}} = \frac{\mu}{4} |\underline{\mathbf{H}}|^2 = \frac{\mathrm{Re}(\underline{\mu})}{4} |\underline{\mathbf{H}}|^2 && \text{magnetic energy density} \end{aligned} \tag{2.81}$$

allow to write

$$\begin{aligned} -\operatorname{div} \underline{\mathbf{T}} &= \overline{p_{\mathrm{v}}} + 2j\omega \left(\overline{w_{\mathrm{m}}} - \overline{w_{\mathrm{e}}} \right) \\ -\oiint \underline{\mathbf{T}} \cdot d\mathbf{a} &= \iiint \left(\overline{p_{\mathrm{v}}} + 2j\omega \left(\overline{w_{\mathrm{m}}} - \overline{w_{\mathrm{e}}} \right) \right) dV \end{aligned} . \tag{2.82}$$

$\mathrm{Im}(\underline{\mathbf{T}})$ gives the reactive power density through a differential area element. The mean active power through an area is

$$\overline{P} = \iint \frac{1}{2} \mathrm{Re}\left(\underline{\mathbf{E}} \times \underline{\mathbf{H}}^* \right) \cdot d\mathbf{a} = \iint \mathrm{Re}(\underline{\mathbf{T}}) \cdot d\mathbf{a} = \iint \overline{\mathbf{S}} \cdot d\mathbf{a} \quad \left(= \iint \overline{S}_z da \right), \tag{2.83}$$

where the rightmost expression in parentheses holds only for integration over the plane $z = \text{const}$.

For a plane wave in z-direction there is only a z component of the Poynting vector, due to $\underline{E}_z = \underline{H}_z = 0$. Using (2.52), (2.57) we get for a lossless medium

$$\begin{aligned} \overline{S}_z &= \mathrm{Re}(\underline{T}_z) = \frac{1}{2} \mathrm{Re}\left(\underline{E}_x \underline{H}_y^* - \underline{E}_y \underline{H}_x^* \right) \\ &= \frac{1}{2 Z_F} \left(\left| \underline{E}_{x0}^+ \right|^2 + \left| \underline{E}_{y0}^+ \right|^2 - \left| \underline{E}_{x0}^- \right|^2 - \left| \underline{E}_{y0}^- \right|^2 \right) \end{aligned} . \tag{2.84}$$

As expected the total power density is the difference of power densities flowign in positive and negative z-directions.

Let the attenuation constant of fields and – if defined – amplitudes be α. Since Poynting vector and power are proportional to squares of fields the attenuation constant of Poynting vector and power is 2α,

$$U, I, \mathbf{E}, \mathbf{H} \sim e^{-\alpha z} \qquad \Leftrightarrow \qquad P, \mathbf{S}, \underline{\mathbf{T}} \sim e^{-2\alpha z}. \tag{2.85}$$

Problem: An electric field $E_{in}(t) = a_{in}(t) E_0 e^{j\omega_0 t}$ (simplified as scalar) has an angular carrier frequency ω_0 and a Gaussian envelope

$$a_{in}(t) = e^{-2(1+j\alpha_H)(t/t_0)^2} \tag{2.86}$$

of pulse width t_0. Which is the temporal width t_w where power decays to $1/e$ times the maximum power? There is also a quadratic phase modulation; constant α_H is the (real-valued) chirp factor or Henry factor. The field propagates through a lossless, isotropic medium of length L with phase constant

$$\beta(\omega) = \beta(\omega_0) + (\omega - \omega_0)\beta'(\omega_0) + (1/2)(\omega - \omega_0)^2 \beta''(\omega_0). \tag{2.87}$$

Here $\beta''(\omega_0)$ is responsible for chromatic dispersion. Calculate the output field $E_{out}(t)$, determine its pulse width t_0' and chirp factor α_H'. Use the normalized chromatic dispersion constant

$$\Gamma = 4\beta''(\omega_0)L/t_0^2. \tag{2.88}$$

Which input α_H is needed to make $t_0' = t_0$? Which output chirp factor α_H' results from this? Which range of Γ is suitable for this? Find an approximation for $|\Gamma| << 1$.

Solution: With $P \sim |E|^2$ we set $|a_{in}(t_w/2)|^2 = e^{-1}$ and find $-4(t_w/2)^2 = -1$, $t_w = t_0$. A Fourier tranform allows to calculate the input spectrum $E_{in}(\omega) = \int_{-\infty}^{\infty} E_{in}(t) e^{-j\omega t} dt = E_0 \int_{-\infty}^{\infty} e^{-2(1+j\alpha_H)(t/t_0)^2 - j(\omega-\omega_0)t} dt$. We postulate $-2(1+j\alpha_H)(t/t_0)^2 - j(\omega - \omega_0)t \overset{!}{=} -u^2 + C = -(At+B)^2 + C = -A^2t^2 - 2ABt + (C - B^2)$. A comparison of coefficients yields $u = At + B$, $du/A = dt$, $A = t_0^{-1}(2(1+j\alpha_H))^{1/2}$, $B = \dfrac{-j(\omega-\omega_0)}{-2A} = \dfrac{j(\omega-\omega_T)t_0}{2(2(1+j\alpha_H))^{1/2}}$, $C = B^2 = -\dfrac{(\omega-\omega_0)^2 t_0^2}{8(1+\alpha_H^2)}(1 - j\alpha_H)$. The input spectrum becomes

$$E_{in}(\omega) = E_0 \int_{-\infty}^{\infty} e^{-(At+B)^2 + C} dt = \frac{E_0}{A} e^C \int_{-\infty}^{\infty} e^{-u^2} du = \frac{E_0}{A} e^C \sqrt{\pi}. \tag{2.89}$$

Note that the spectrum is also Gaussian with chirp, but as a function of $\omega - \omega_0$, not t. The optical transfer function is $H(\omega) = e^{-j\beta(\omega)L}$, the output spectrum is $E_{out}(\omega) = H(\omega)E_{in}(\omega)$.

The output field is the inverse Fourier tranform $E_{out}(t)=\frac{1}{2\pi}\int_{-\infty}^{\infty} H(\omega)E_{in}(\omega)e^{j\omega t}d\omega$,

$E_{out}(t)=\frac{E_0}{A}\frac{\sqrt{\pi}}{2\pi}\int_{-\infty}^{\infty} e^{-j\left(\beta(\omega_0)+(\omega-\omega_0)\beta'(\omega_0)+(1/2)(\omega-\omega_0)^2\beta''(\omega_0)\right)L+C+j\omega t}d\omega$. Similar to above, we postulate

$$
\begin{aligned}
&-j\left(\beta(\omega_0)+(\omega-\omega_0)\beta'(\omega_0)+(1/2)(\omega-\omega_0)^2\beta''(\omega_0)\right)L-\frac{(\omega-\omega_0)^2 t_0^2}{8(1+j\alpha_H)}+j(\omega-\omega_0)t+j\omega_0 t \\
&\overset{!}{=}-v^2+C'=-\left(A'(\omega-\omega_0)+B'\right)^2+C'=-A'^2(\omega-\omega_0)^2-2A'B'(\omega-\omega_0)+\left(C'-B'^2\right)
\end{aligned}
$$

and find $v=A'(\omega-\omega_0)+B'$, $dv/A'=d\omega$,

$$
\begin{aligned}
A' &= \frac{t_0}{2\sqrt{2}}\left(\frac{1-\alpha_H\Gamma+j\Gamma}{1+j\alpha_H}\right)^{1/2}, \\
B' &= -j\sqrt{2}\left(\left(t-\beta'(\omega_0)L\right)/t_0\right)\left(\frac{1+j\alpha_H}{1-\alpha_H\Gamma+j\Gamma}\right)^{1/2}, \\
C' &= j\left(\omega_0 t-\beta(\omega_0)L\right)-2\left(\left(t-\beta'(\omega_0)L\right)/t_0'\right)^2\left(1+j\alpha_H'\right) \text{ with}
\end{aligned}
$$

$$
t_0' = t_0\sqrt{(1-\alpha_H\Gamma)^2+\Gamma^2}\,, \quad \alpha_H' = \alpha_H-\left(1+\alpha_H^2\right)\Gamma\,. \tag{2.90}
$$

This results in

$$
E_{out}(t)=\frac{E_0}{A}\frac{\sqrt{\pi}}{2\pi}\int_{-\infty}^{\infty} e^{-\left(A'(\omega-\omega_0)+B'\right)^2+C'}d\omega=\frac{E_0}{AA'}e^{C'}\frac{\sqrt{\pi}}{2\pi}\int_{-\infty}^{\infty} e^{-v^2}dv=\frac{E_0}{2AA'}e^{C'},
$$

$$
E_{out}(t) = E_0\left(1-\alpha_H\Gamma+j\Gamma\right)^{-1/2}e^{-2\left(\left(t-\beta'(\omega_0)L\right)/t_0'\right)^2\left(1+j\alpha_H'\right)}e^{j\left(\omega_0 t-\beta(\omega_0)L\right)}. \tag{2.91}
$$

The carrier delay time is $\beta(\omega_0)/\omega_0\, L$, the (envelope) group delay time is $\beta'(\omega_0)L$, and output pulse width and chirp factor are t_0', α_H' as per (2.90). For $|\Gamma|>>1$ it holds $t_0'=t_0|\Gamma|\sqrt{1+\alpha_H^2}$. According to (2.88), chromatic dispersion aggravates proportional to L and to the inverse square of t_0 in that case!

Unchanged pulse width $t_0'=t_0$ requires $(1-\alpha_H\Gamma)^2+\Gamma^2=1$, $\alpha_H=(1/\Gamma)\left(1\pm\sqrt{1-\Gamma^2}\right)$. A real α_H requires $|\Gamma|\le 1$. We keep $\alpha_H=(1/\Gamma)\left(1-\sqrt{1-\Gamma^2}\right)$ to obtain the smallest possible $|\alpha_H|$. The output chirp factor becomes now $\alpha_H'=-\alpha_H$. For $|\Gamma|<<1$ we find $\alpha_H\approx\Gamma/2$ as the condition needed for $t_0'=t_0$.

In practice, optical pulses are not Gaussian. Yet, for $\beta'(\omega_0)<0$, which is the case in standard optical fibers at 1.55 μm wavelength, a negative α_H can reduce pulse broadening.

2.2 Dielectric Waveguides

A homogeneous plane wave can, strictly speaking, only exist in laterally infinite space. Therefore waveguides are of advantage. We cover the relatively simple dielectric slab waveguide to show the principle and the cylindrical dielectric waveguide, commonly called glass fiber.

2.2.1 Dielectric Slab Waveguide

We investigate wave propagation in a lossless symmetric dielectric slab waveguide [1]. It provides fundamental insight. However, most practical dielectric waveguides have two-dimensional rather than one-dimensional confinement. Table 2.1 categorizes the modes (field types) of waveguides.

Table 2.1 Modes (field types) in waveguides

Type	Properties	Occurrence
L or TEM modes (Lecher-modes, transversal electromagnetic modes)	$E_z = 0$, $H_z = 0$	Homogeneous plane wave, electrical wires
H or TE modes (transversal electric modes)	$E_z = 0$, $H_z \neq 0$	dielectric waveguides, hollow waveguides
E or TM modes (transversal magnetic modes)	$E_z \neq 0$, $H_z = 0$	dielectric waveguides, hollow waveguides
Hybrid modes (HE or EH)	$E_z \neq 0$, $H_z \neq 0$	2-dimensionally confined dielectric waveguide

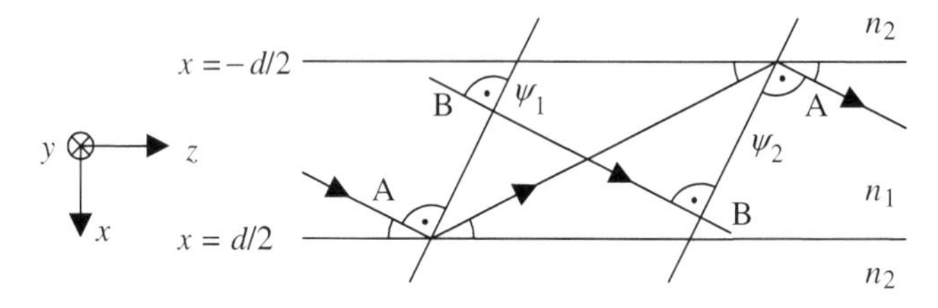

Fig. 2.2 Symmetric dielectric slab waveguide. All marked angles (except the right angles) are ϑ .

In Fig 2.2 the slab $x \in [-d/2, d/2]$ consists of a dielectric with refractive index n_1 whereas the refractive index is n_2 with $n_2 < n_1$ everywhere else. Two wavefronts are also shown. In the plane ψ_1 both rays *A* and *B* an incident plane wave have the same phase ψ_1. In the plane ψ_2 they must also have the same phase ψ_2. In spite of the unequal distances this is possible, because ray *A* is twice totally reflected between ψ_1 and ψ_2. The condition is only fulfilled for certain incidence angles $\pm\vartheta$, which correspond to different modes.

We write Maxwell's equations (2.21), (2.23) in cartesian coordinates. We assume fields which are proportional to $e^{j(\omega t-\beta z)}$, which means $\partial/\partial t = j\omega$, $\partial/\partial z = -j\beta$. With respect to the geometry of Fig 2.2 it holds also $\partial/\partial y = 0$,

$$j\beta H_y = j\omega\varepsilon E_x, \tag{2.92}$$

$$-j\beta H_x - \frac{\partial H_z}{\partial x} = j\omega\varepsilon E_y, \tag{2.93}$$

$$\frac{\partial H_y}{\partial x} = j\omega\varepsilon E_z, \tag{2.94}$$

$$j\beta E_y = -j\omega\mu_0 H_x, \tag{2.95}$$

$$-j\beta E_x - \frac{\partial E_z}{\partial x} = -j\omega\mu_0 H_y, \tag{2.96}$$

$$\frac{\partial E_y}{\partial x} = -j\omega\mu_0 H_z. \tag{2.97}$$

One family of solutions contains E_y, H_x, H_z. Here the electric field is purely transversal, and a longitudinal component exists for the magnetic field. These are the TE or H modes. We insert

$$H_x = -\frac{\beta}{\omega\mu_0}E_y, \qquad H_z = \frac{j}{\omega\mu_0}\frac{\partial E_y}{\partial x} \tag{2.98}$$

from (2.95), (2.97) into (2.93) and find

$$\beta^2 E_y - \frac{\partial^2 E_y}{\partial x^2} = \omega^2\varepsilon\mu_0 E_y. \tag{2.99}$$

A solution is possible for fields proportional to $e^{-jk_x x}$, which leads to $\partial^2/\partial x^2 = -k_x^2$ and $\beta^2 + k_x^2 = \omega^2\varepsilon\mu_0$. The material propagation constants inside and outside the waveguide are $\beta_{1,2} = \omega\sqrt{\varepsilon_{1,2}\mu_0} = \frac{\omega n_{1,2}}{c}$. Being tangential, the field E_y is identical at both sides of the boundaries. For this reason the effective phase constant $\beta \equiv \beta_{eff} = \frac{\omega n_{eff}}{c}$ is the same everywhere. A solution is possible for $\beta_2 < \beta < \beta_1$ with $k_{1x} = \pm\beta_{1x}$ and $k_{2x} = \pm j\alpha_{2x}$, because this results in $\beta_{1x}^2 = \beta_1^2 - \beta^2 > 0$, $\alpha_{2x}^2 = \beta^2 - \beta_2^2 > 0$. We define

$$u = \frac{d\beta_{1x}}{2} = \frac{d}{2}\sqrt{\beta_1^2 - \beta^2} \quad \textit{normalized transversal phase constant,} \tag{2.100}$$

$$v = \frac{d\alpha_{2x}}{2} = \frac{d}{2}\sqrt{\beta^2 - \beta_2^2} \quad \textit{normalized transversal attenuation,} \tag{2.101}$$

$$V = \sqrt{u^2 + v^2} = \frac{d}{2}\sqrt{\beta_1^2 - \beta_2^2} = \frac{d\omega}{2c}\sqrt{n_1^2 - n_2^2} \quad \textit{normalized frequency.} \tag{2.102}$$

In the upper and lower cladding the fields must decay for $|x| \to \infty$; otherwise there would be no waveguiding. This results in

$$E_y = e^{-j\beta z}\begin{cases} Ae^{-\alpha_{2x}x} & x > d/2 \\ Be^{j\beta_{1x}x} + Ce^{-j\beta_{1x}x} & |x| \le d/2 \\ De^{\alpha_{2x}x} & x < -d/2 \end{cases}. \tag{2.103}$$

Together with $\beta^2 + \beta_{1x}^2 = \beta_1^2$, $\cos\vartheta = \beta/\beta_1$, $\sin\vartheta = \beta_{1x}/\beta_1$ this shows that in the center there are two superimposed homogeneous plane waves propagating with angles $\pm\vartheta$ off the z axis. The longitudinal magnetic field is

$$H_z = \frac{j}{\omega\mu_0}\frac{\partial E_y}{\partial x} = \frac{j}{\omega\mu_0}e^{-j\beta z}\begin{cases} -\alpha_{2x}Ae^{-\alpha_{2x}x} & x > d/2 \\ j\beta_{1x}\left(Be^{j\beta_{1x}x} - Ce^{-j\beta_{1x}x}\right) & |x| \le d/2 \\ \alpha_{2x}De^{\alpha_{2x}x} & x < -d/2 \end{cases}. \tag{2.104}$$

Both must be continuous at the boundaries $x = \pm d/2$. Using (2.100), (2.101), this means

$$Ae^{-v} = Be^{ju} + Ce^{-ju}, \tag{2.105}$$

$$De^{-v} = Be^{-ju} + Ce^{ju}, \tag{2.106}$$

$$-vAe^{-v} = ju\left(Be^{ju} - Ce^{-ju}\right), \tag{2.107}$$

$$vDe^{-v} = ju\left(Be^{-ju} - Ce^{ju}\right). \tag{2.108}$$

Problem: Solve this for A/D, B/C, $v = f(u)$, $B/A = g(u,v)$.

Solution: (2.105) times (2.108) yields $vADe^{-2v} = ju\left(B^2 + BCe^{-j2u} - BCe^{j2u} - C^2\right)$. (2.106) times the negative of (2.107) yields $vADe^{-2v} = ju\left(-B^2 + BCe^{-j2u} - BCe^{j2u} + C^2\right)$. The two right hand sides must be equal, which results in $2B^2 = 2C^2$, $B/C = \pm 1$.

Case $B=C$: (2.105), (2.106) $\Rightarrow$ $Ae^{-v}=2B\cos u=De^{-v}$, $A/D=1$. (2.107) $\Rightarrow$ $-vAe^{-v}=-2uB\sin u$. We divide this by $Ae^{-v}=2B\cos u$ and obtain $v=u\tan u$, $B/A=e^{-v}/(2\cos u)=\left(ve^{-v}\right)/(2u\sin u)$.

Case $B=-C$: $Ae^{-v}=2jB\sin u=-De^{-v}$, $A/D=-1$, $-vAe^{-v}=2juB\cos u$, $v=-u\cot u$, $B/A=e^{-v}/(2j\sin u)=\left(jve^{-v}\right)/(2u\cos u)$.

These dependencies can be combined into the

$$u\tan(u-m\pi/2)=v=\sqrt{V^2-u^2}\quad \textit{eigenvalue equation for H modes}, \tag{2.109}$$

using $A/D=B/C=(-1)^m$ with mode index $m=0,1,2,3,\ldots$, $\tan(u+\pi)=\tan u$, $\cot u=-\tan(u-\pi/2)$ and $v=\sqrt{V^2-u^2}$ from (2.102).

The second family of solutions contains H_y, E_x, E_z. For these TM or E modes the magnetic field is purely transversal, and a longitudinal component exists for the electric field. We insert

$$E_x=\frac{\beta}{\omega\varepsilon}H_y,\qquad E_z=\frac{-j}{\omega\varepsilon}\frac{\partial H_y}{\partial x} \tag{2.110}$$

from (2.92), (2.94) into (2.96) and find (2.96)

$$-\beta^2H_y+\frac{\partial^2H_y}{\partial x^2}=-\omega^2\varepsilon\mu_0H_y. \tag{2.111}$$

Using the same calculus as before, the two tangential fields become

$$H_y=e^{-j\beta z}\begin{cases}Ae^{-\alpha_{2x}x} & x>d/2\\ Be^{j\beta_{1x}x}+Ce^{-j\beta_{1x}x} & |x|\le d/2\\ De^{\alpha_{2x}x} & x<-d/2\end{cases}, \tag{2.112}$$

$$E_z=\frac{-j}{\omega\varepsilon}\frac{\partial H_y}{\partial x}=\frac{-je^{-j\beta z}}{\omega\varepsilon_0}\begin{cases}-n_2^{-2}\alpha_{2x}Ae^{-\alpha_{2x}x} & x>d/2\\ jn_1^{-2}\beta_{1x}\left(Be^{j\beta_{1x}x}-Ce^{-j\beta_{1x}x}\right) & |x|\le d/2\\ n_2^{-2}\alpha_{2x}De^{\alpha_{2x}x} & x<-d/2\end{cases}. \tag{2.113}$$

Both must be continuous at the boundaries $x=\pm d/2$. This means

$$Ae^{-v}=Be^{ju}+Ce^{-ju}, \tag{2.114}$$

$$De^{-v}=Be^{-ju}+Ce^{ju}, \tag{2.115}$$

$$-n_2^{-2}vAe^{-v}=jn_1^{-2}u\left(Be^{ju}-Ce^{-ju}\right), \tag{2.116}$$

$$n_2^{-2}vDe^{-v}=jn_1^{-2}u\left(Be^{-ju}-Ce^{ju}\right). \tag{2.117}$$

Problem: Solve this for A/D , B/C , $v = f(u)$, $B/A = g(u,v)$.

Solution: (2.114) times (2.117) yields $n_2^{-2} vADe^{-2v} = jn_1^{-2} u\left(B^2 + BCe^{-j2u} - BCe^{j2u} - C^2\right)$. (2.115) times the negative of (2.116) yields $n_2^{-2} vADe^{-2v} = jn_1^{-2} u\left(-B^2 + BCe^{-j2u} - BCe^{j2u} + C^2\right)$. The two right hand sides must be equal, which results in $B/C = \pm 1$.

Case $B = C$: (2.114), (2.115) $\Rightarrow$ $Ae^{-v} = 2B\cos u = De^{-v}$, $A/D = 1$. (2.116) $\Rightarrow$ $-\left(n_1^2/n_2^2\right) vAe^{-v} = -2uB\sin u$. We divide this by $Ae^{-v} = 2B\cos u$ and obtain $v = \left(n_2^2/n_1^2\right) u \tan u$, $B/A = e^{-v}/(2\cos u) = \left(n_1^2/n_2^2\right)\left(ve^{-v}\right)/(2u\sin u)$.

Case $B = -C$: $Ae^{-v} = 2jB\sin u = -De^{-v}$, $A/D = -1$, $-\left(n_1^2/n_2^2\right) vAe^{-v} = 2juB\cos u$, $v = -\left(n_2^2/n_1^2\right) u\cot u$, $B/A = e^{-v}/(2j\sin u) = \left(n_1^2/n_2^2\right)\left(jve^{-v}\right)/(2u\cos u)$.

These dependencies can be combined (2.102)into the

$$u\tan(u - m\pi/2) = n_{12}^{-2} v = n_{12}^{-2}\sqrt{V^2 - u^2} \quad \textit{eigenvalue equation for E modes,} \tag{2.118}$$

where $n_{12} = n_2/n_1$ is the relative refractive index change from medium 1 to medium 2.

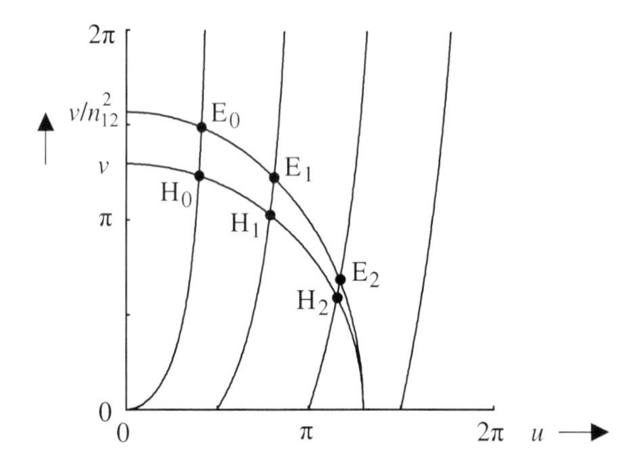

Fig. 2.3 Solution of the eigenvalue equations of the symmetric dielectric slab waveguide with $\lambda = 1{,}55\mu m$, $n_2 = 1{,}46$, $n_1 = 1{,}1 \cdot 1{,}46$, $d = 3\mu m$. Three (transversal electrical) H modes and three (transversal magnetic) E modes can propagate.

The eigenvalue equations are solved numerically, for example by searching the zeros of left expression minus right expression by the Newton method. The propagation constant β and v are obtained from (2.100), (2.101). Alternatively, Fig. 2.3 shows right and left expression separately. The quarter-circle, valid for H modes, represents v in (2.109). The quarter-ellipse, valid for E modes, represents

$n_{12}^{-2}v$ in (2.118)). The steep lines show the left expression for various m. Each intersection point stands for a guided mode. For weakly guiding waveguides ($n_{12} \to 1$) the obtained u, v, β values of H_m and E_m modes become equal. The number of guided modes grows linearly with optical frequency. Waveguides which support only the two fundamental modes (H_0 und E_0) are usually called monomode waveguides.

Eqn. (2.119) gives the complete field amplitudes of the H modes, for $E = B$ and without the multiplicative phasor $e^{j(\omega t - \beta z)}$, using (2.98), $2ux/d = \beta_{1x}x$, $\frac{\beta}{\omega\mu_0} = \frac{\cos\vartheta}{Z_{F1}} = \frac{1}{Z_{F,eff}}$, $\frac{\beta_{1x}}{\omega\mu_0} = \frac{\sin\vartheta}{Z_{F1}}$. E is the electric field amplitude of the two homogeneous plane waves propagating at angles $\pm\vartheta$ in the waveguide center.

H_m modes	medium 1 (center), $\lvert x\rvert \le d/2$	medium 2 (cladding), $\pm x > d/2$
m = 0, 2, 4, ...	$H_z = -j\frac{2E}{Z_{F1}}\sin\vartheta\sin(2ux/d)$ $H_x = -\frac{2E}{Z_{F1}}\cos\vartheta\cos(2ux/d)$ $E_y = 2E\cos(2ux/d)$	$H_z = \mp j\frac{2E}{Z_{F1}}\sin\vartheta\sin u\, e^{v(1\mp 2x/d)}$ $H_x = -\frac{2E}{Z_{F1}}\cos\vartheta\cos u\, e^{v(1\mp 2x/d)}$ $E_y = 2E\cos u\, e^{v(1\mp 2x/d)}$
m = 1, 3, 5, ...	$H_z = -\frac{2E}{Z_{F1}}\sin\vartheta\cos(2ux/d)$ $H_x = -j\frac{2E}{Z_{F1}}\cos\vartheta\sin(2ux/d)$ $E_y = j2E\sin(2ux/d)$	$H_z = -\frac{2E}{Z_{F1}}\sin\vartheta\cos u\, e^{v(1\mp 2x/d)}$ $H_x = \mp j\frac{2E}{Z_{F1}}\cos\vartheta\sin u\, e^{v(1\mp 2x/d)}$ $E_y = \pm j2E\sin u\, e^{v(1\mp 2x/d)}$

(2.119)

Eqn. (2.120) gives the complete field amplitudes of the E modes, for $E/Z_{F1} = B$, using (2.110), $\frac{\beta}{\omega\varepsilon_1} = Z_{F1}\cos\vartheta$, $\frac{\beta_{1x}}{\omega\varepsilon_1} = Z_{F1}\sin\vartheta$.

E_m modes	medium 1 (center), $\lvert x\rvert \le d/2$	medium 2 (cladding), $\pm x > d/2$
m = 0, 2, 4, ...	$E_z = j2E\sin\vartheta\sin(2ux/d)$ $E_x = 2E\cos\vartheta\cos(2ux/d)$ $H_y = \frac{2E}{Z_{F1}}\cos(2ux/d)$	$E_z = \pm j2E\sin\vartheta\sin u\, e^{v(1\mp 2x/d)}$ $E_x = \frac{2E}{n_{12}^2}\cos\vartheta\cos u\, e^{v(1\mp 2x/d)}$ $H_y = \frac{2E}{Z_{F1}}\cos u\, e^{v(1\mp 2x/d)}$

m = 1, 3, 5, ...	$E_z = 2E\sin\vartheta\cos(2ux/d)$ $E_x = j2E\cos\vartheta\sin(2ux/d)$ $H_y = j\dfrac{2E}{Z_{F1}}\sin(2ux/d)$	$E_z = 2E\sin\vartheta\cos u\, e^{v(1\mp 2x/d)}$ $E_x = \pm j\dfrac{2E}{n_{12}^2}\cos\vartheta\sin u\, e^{v(1\mp 2x/d)}$ $H_y = \pm j\dfrac{2E}{Z_{F1}}\sin u\, e^{v(1\mp 2x/d)}$

(2.120)

The longitudinal fields are 90° out of phase with respect to the transversal ones. The resulting complex Pointing vector component points in $\pm x$ direction but is imaginary, which means that no real power is radiated and lost.

The higher the mode index m is for a given multimode waveguide at a fixed frequency, the higher are $u = (d/2)\beta_{1x}$ and $\vartheta = \arcsin(\beta_{1x}/\beta_1)$. In the waveguide center with thickness d there are more than m, but less than $m+1$ standing halfwaves,

$$m\pi \leq 2u = d\beta_{1x} \leq (m+1)\pi\,, \tag{2.121}$$

as can be seen from the eigenvalue equations. The transversal fields have $m+1$ extrema and m zeros in the waveguide center (medium 1). The longitudinal components exhibit m extrema and $m+1$ zeros.

Figs. 2.4 und 2.5 show exemplary normalized field amplitudes as a function of x.

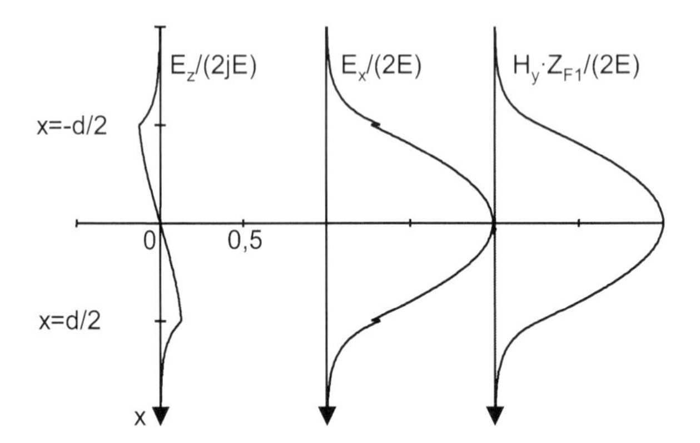

Fig. 2.4 Normalized field amplitudes of the E_0 mode; $\lambda = 1{,}55\mu m$, $n_2 = 1{,}46$, $n_1 = 1{,}1\cdot 1{,}46$, $d = 3\mu m$. The discontinuity of E_x at $x = \pm d/2$ occurs because the normal displacements at both sides of a boundary are identical.

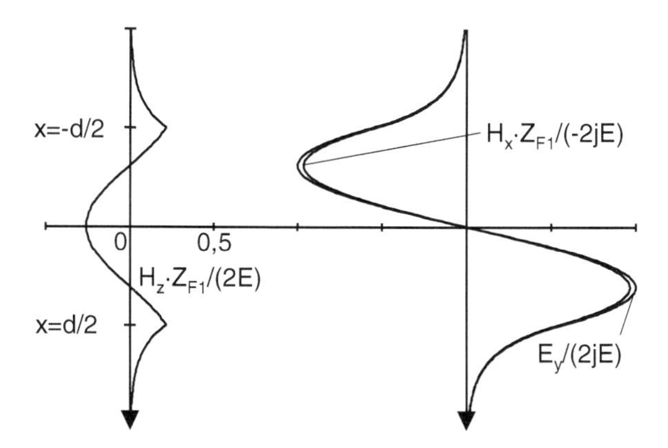

Fig. 2.5 Normalized field amplitudes of the H_1 mode; $\lambda = 1{,}55\mu m$, $n_2 = 1{,}46$, $n_1 = 1{,}1 \cdot 1{,}46$, $d = 3\mu m$.

Under which maximum angle ϑ_{in} with respect to the z axis may a ray impinge onto the cross section $z = const.$ of a waveguide in order to be able to excite a guided mode? Or to be more precise, beyond which angle will guided modes definitely not be excited? For a ray angle ϑ in the medium with refractive index n_1 it holds $n_0 \sin \vartheta_{in} = n_1 \sin \vartheta$ (Fig. 2.6). In most cases the region outside the waveguide contains air, $n_0 \approx 1$. For waveguiding total reflection is required at the boundary $x = \pm d/2$, which means $n_1 \cos \vartheta > n_2$. For $n_0 = 1$ one obtains the

$$A_N := \sin \vartheta_{in} = \sqrt{n_1^2 - n_2^2} \qquad \textit{numerical aperture}. \tag{2.122}$$

This expression is also used for the cylindrical dielectric waveguide.

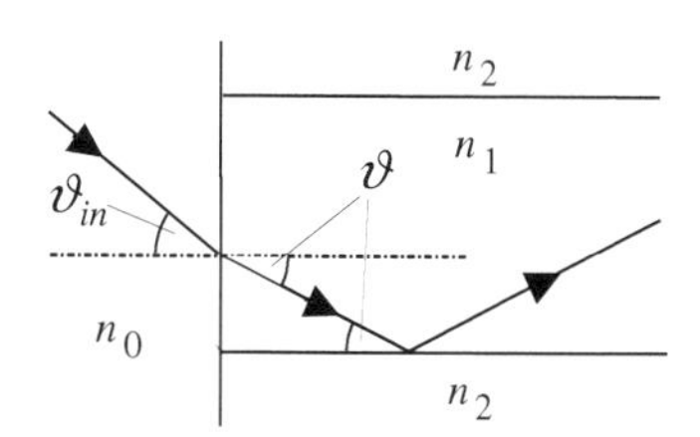

Fig. 2.6 Explanation of the numerical aperture

Under the assumption of weak guiding ($n_1 \approx n_2$) the power fraction of the H_0 mode in the center of a slab waveguide is given by the *confinement factor*

$$\Gamma = \left(\int_{-d/2}^{d/2} \left| E_y \right|^2 dx \right) \Big/ \left(\int_{-\infty}^{\infty} \left| E_y \right|^2 dx \right) = \frac{1 + v/V^2}{1 + 1/v} \approx \frac{V^2}{V^2 + 1/2}. \tag{2.123}$$

2.2.2 Cylindrical Dielectric Waveguide

Most dielectric waveguides feature 2-dimensional confinement, in glass fibers, lasers and optical components. The cylindric dielectric waveguide serves as an example that can be treated analytically, with Bessel functions.

Because of the rotational symmetry one uses cylindrical coordinates r, φ, z for the electromagnetic field calculation. We search all guided modes. The refractive index profile in cross-sectional view is step-shaped. We know already that waveguiding is possible only where the refractive index is higher than elsewhere. In Fig. 2.7 we therefore choose $n_1 > n_2$. The waveguide *core* features a refractive index n_1 and is surrounded by the *cladding*. In our mathematical model the cladding extends until $r = \infty$. Practically the cladding diameter needs not be very large because the fields decay near-exponentially in the cladding.

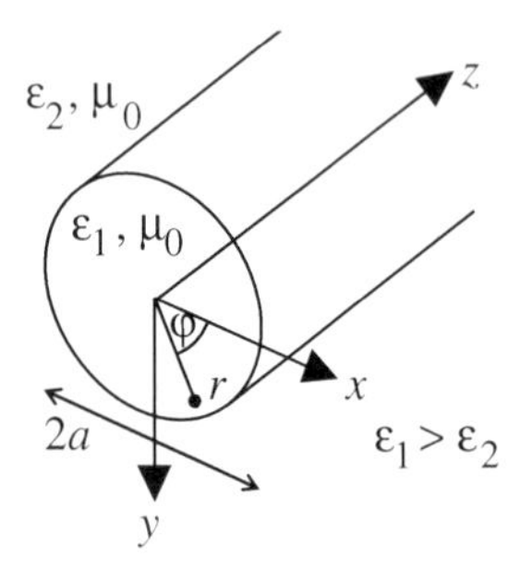

Fig. 2.7 Cylindrical dielectric waveguide, optical step index fiber

We search fields propagating in z direction,

$$\begin{aligned}\mathbf{E}(\mathbf{r},t) &= \mathbf{E}(r,\varphi)e^{j(\omega t-\beta z)} \\ \mathbf{H}(\mathbf{r},t) &= \mathbf{H}(r,\varphi)e^{j(\omega t-\beta z)}\end{aligned}. \tag{2.124}$$

For simplicity we drop the dependence (r,φ) in the following, and assume $\mu = \mu_0$. For (2.124) Maxwell's equations in cylindrical coordinates are

$$j\omega\varepsilon E_r = j\beta H_\varphi + \frac{1}{r}\frac{\partial}{\partial\varphi}H_z, \tag{2.125}$$

$$j\omega\varepsilon E_\varphi = -j\beta H_r - \frac{\partial}{\partial r}H_z, \tag{2.126}$$

$$j\omega\varepsilon E_z = -\frac{1}{r}\frac{\partial}{\partial\varphi}H_r + \frac{1}{r}\frac{\partial}{\partial r}(rH_\varphi), \tag{2.127}$$

$$-j\omega\mu_0 H_r = j\beta E_\varphi + \frac{1}{r}\frac{\partial}{\partial\varphi}E_z, \tag{2.128}$$

$$-j\omega\mu_0 H_\varphi = -j\beta E_r - \frac{\partial}{\partial r} E_z, \tag{2.129}$$

$$-j\omega\mu_0 H_z = -\frac{1}{r}\frac{\partial}{\partial\varphi} E_r + \frac{1}{r}\frac{\partial}{\partial r}\left(rE_\varphi\right). \tag{2.130}$$

It is possible to express the radial and azimuthal field components as a function of the longitudinal ones. If one multiplies both sides of (2.129) by $-\beta$ and inserts it into the eqn. (2.125) that has been multiplied by $\omega\mu_0$, then H_φ is eliminated and one obtains

$$E_r = \frac{-j\beta}{\omega^2\mu_0\varepsilon - \beta^2}\left(\frac{\partial}{\partial r}E_z + \frac{\omega\mu_0}{\beta}\frac{\partial}{r\partial\varphi}H_z\right). \tag{2.131}$$

Analogous eliminations lead to

$$E_\varphi = \frac{-j\beta}{\omega^2\mu_0\varepsilon - \beta^2}\left(\frac{\partial}{r\partial\varphi}E_z - \frac{\omega\mu_0}{\beta}\frac{\partial}{\partial r}H_z\right), \tag{2.132}$$

$$H_r = \frac{-j\beta}{\omega^2\mu_0\varepsilon - \beta^2}\left(-\frac{\omega\varepsilon}{\beta}\frac{\partial}{r\partial\varphi}E_z + \frac{\partial}{\partial r}H_z\right), \tag{2.133}$$

$$H_\varphi = \frac{-j\beta}{\omega^2\mu_0\varepsilon - \beta^2}\left(\frac{\omega\varepsilon}{\beta}\frac{\partial}{\partial r}E_z + \frac{\partial}{r\partial\varphi}H_z\right). \tag{2.134}$$

So, it is sufficient to solve for E_z and H_z because the other field components may be derived from these. We do this with the wave equation, separately for core and cladding regions due to the different refractive indexes. The solutions are adapted to each other with help of the tangential boundary conditions. In the vectorial wave equation $\Delta\mathbf{E} = -\omega^2\frac{n^2(\mathbf{r})}{c^2}\mathbf{E} = -\beta^2(\mathbf{r})\mathbf{E}$ we look into E_z, using the z component of the vector Laplace operator in cylindric coordinates

$$\left(\frac{1}{r}\frac{\partial}{\partial r} + \frac{\partial^2}{\partial r^2} + \frac{1}{r^2}\frac{\partial^2}{\partial\varphi^2} + \left(\beta(r)^2 - \beta^2\right)\right)E_z = 0, \tag{2.135}$$

where $\beta(r) = \beta_0 n(r)$ is the radius-dependent phase constant. The separation ansatz

$$E_z(r,\varphi) = \Psi(r)e^{\pm jl\varphi} \qquad l = 0,1,2,\ldots \tag{2.136}$$

inserted into (2.135) yields Bessel's differential equation

$$\frac{\partial^2\Psi}{\partial r^2} + \frac{1}{r}\frac{\partial\Psi}{\partial r} + \left(\beta(r)^2 - \beta^2 - \frac{l^2}{r^2}\right)\Psi = 0. \tag{2.137}$$

Parameter l must be integer because the solutions must repeat if φ is changed by 2π. Due to the $\pm$ sign we may restrict to non-negative l. The solutions of (2.137) are

$$\Psi(r) = c_1 J_l(hr) + c_2 Y_l(hr) \quad \text{for} \quad \beta(r)^2 - \beta^2 = h^2 > 0\,, \tag{2.138}$$

$$\Psi(r) = c_1 I_l(qr) + c_2 K_l(qr) \quad \text{for} \quad \beta^2 - \beta(r)^2 = q^2 > 0\,. \tag{2.139}$$

$J_l(x)$ and $Y_l(x)$ are the ordinary Bessel function of 1st and 2nd kind, respectively, each of order l. $I_l(x)$ and $K_l(x)$ are the modified Bessel functions of 1st and 2nd kind. From the slab waveguide we know that a guided wave must have a phase constant β that lies in between those of the core region (β_1) and the cladding (β_2),

$$\beta_0 n_2 = \beta_2 < \beta < \beta_1 = \beta_0 n_1\,. \tag{2.140}$$

Condition $\beta_2 < \beta$ assures a decay in the cladding, while $\beta < \beta_1$ prevents this in the core. Since the refractive index depends on r, a common β for core and cladding is possible only if (2.138) holds for the core and (2.139) for the cladding.

The Bessel functions can be approximated by

$$\begin{aligned}
&J_l(x) \to \frac{1}{l!}\left(\frac{x}{2}\right)^l \quad \text{all for } x \ll 1 && I_l(x) \to \frac{1}{l!}\left(\frac{x}{2}\right)^l \\
&Y_0(x) \to \frac{2}{\pi}\left(\ln\frac{x}{2} + 0.5772....\right) && K_0(x) \to -\left(\ln\frac{x}{2} + 0.5772....\right) \\
&Y_l(x) \to -\frac{(l-1)!}{\pi}\left(\frac{2}{x}\right)^l \quad (l = 1,2,3,...) && K_l(x) \to \frac{(l-1)!}{2}\left(\frac{2}{x}\right)^l \quad (l = 1,2,3,...)
\end{aligned} \tag{2.141}$$

for small and by

$$\begin{aligned}
&J_l(x) \to \left(\frac{2}{\pi x}\right)^{1/2} \cos\left(x - \frac{l\pi}{2} - \frac{\pi}{4}\right) \quad \text{all for } x \gg 1, l \\
&Y_l(x) \to \left(\frac{2}{\pi x}\right)^{1/2} \sin\left(x - \frac{l\pi}{2} - \frac{\pi}{4}\right) \\
&I_l(x) \to \left(\frac{1}{2\pi x}\right)^{1/2} e^x \qquad K_l(x) \to \left(\frac{\pi}{2x}\right)^{1/2} e^{-x}
\end{aligned} \tag{2.142}$$

for large arguments. Due to $\lim_{x \to 0+0} Y_l(x) = -\infty$ a $Y_l(x)$ dependence in the core would be unphysical, while $J_l(x)$ is finite. Likewise, $\lim_{x \to \infty} I_l(x) = \infty$ makes $I_l(x)$ unsuitable for the cladding region whereas $K_l(x)$ decays and assures waveguiding.

The wave equation can be solved the same way for H_z. Summarizing, it holds

$$\begin{aligned} E_z(\mathbf{r}) &= AJ_l(hr)e^{j(\pm l\varphi-\beta z)} \\ H_z(\mathbf{r}) &= BJ_l(hr)e^{j(\pm l\varphi-\beta z)} \end{aligned} \quad \text{for } r<a \text{ (core) with } \beta_1^2-\beta^2=h^2, \tag{2.143}$$

$$\begin{aligned} E_z(\mathbf{r}) &= CK_l(qr)e^{j(\pm l\varphi-\beta z)} \\ H_z(\mathbf{r}) &= DK_l(qr)e^{j(\pm l\varphi-\beta z)} \end{aligned} \quad \text{for } r>a \text{ (cladding) with } \beta^2-\beta_2^2=q^2. \tag{2.144}$$

Here $A...D$ are complex constants yet to be determined. In order to fulfill the tangential boundary conditions, E_z, E_φ, H_z, H_φ must be continuous across the core-cladding boundary. Since all l must be considered anyway we may choose the same l for electric and magnetic fields. If this should turn out to be unnecessary then the boundary conditions would admit independent amplitudes of E_z, H_z for the same l. We shall soon see that this case is less frequent than the case where precise ratios of E_z, H_z are needed, leaving only one of $A...D$ to be chosen freely.

Inserting (2.143), (2.144) into (2.131) ... (2.134) yields the other field components in core

$$\begin{aligned}
E_r &= \frac{-j\beta}{h^2}\left(AhJ_l'(hr) \pm \frac{j\omega\mu_0 l}{\beta r}BJ_l(hr)\right)e^{j(\pm l\varphi-\beta z)} \\
E_\varphi &= \frac{-j\beta}{h^2}\left(\pm\frac{jl}{r}AJ_l(hr) - \frac{\omega\mu_0}{\beta}BhJ_l'(hr)\right)e^{j(\pm l\varphi-\beta z)} \\
H_r &= \frac{-j\beta}{h^2}\left(\mp\frac{j\omega\varepsilon_1 l}{\beta r}AJ_l(hr) + BhJ_l'(hr)\right)e^{j(\pm l\varphi-\beta z)} \\
H_\varphi &= \frac{-j\beta}{h^2}\left(\frac{\omega\varepsilon_1}{\beta}AhJ_l'(hr) \pm \frac{jl}{r}BJ_l(hr)\right)e^{j(\pm l\varphi-\beta z)}
\end{aligned} \tag{2.145}$$

with $J_l'(hr) = dJ_l(hr)/d(hr)$, $\varepsilon_1 = \varepsilon_0 n_1^2$, $r<a$ and cladding

$$\begin{aligned}
E_r &= \frac{j\beta}{q^2}\left(CqK_l'(qr) \pm \frac{j\omega\mu_0 l}{\beta r}DK_l(qr)\right)e^{j(\pm l\varphi-\beta z)} \\
E_\varphi &= \frac{j\beta}{q^2}\left(\pm\frac{jl}{r}CK_l(qr) - \frac{\omega\mu_0}{\beta}DqK_l'(qr)\right)e^{j(\pm l\varphi-\beta z)} \\
H_r &= \frac{j\beta}{q^2}\left(\mp\frac{j\omega\varepsilon_2 l}{\beta r}CK_l(qr) + DqK_l'(qr)\right)e^{j(\pm l\varphi-\beta z)} \\
H_\varphi &= \frac{j\beta}{q^2}\left(\frac{\omega\varepsilon_2}{\beta}CqK_l'(qr) \pm \frac{jl}{r}DK_l(qr)\right)e^{j(\pm l\varphi-\beta z)}
\end{aligned} \tag{2.146}$$

with $K_l'(qr) = dK_l(qr)/d(qr)$, $\varepsilon_2 = \varepsilon_0 n_2^2$, $r>a$.

The tangential boundary conditions at $r = a$, applied to E_z, E_φ, H_z, H_φ in (2.143) bis (2.146), yield a linear system of equations (2.149). We define normalized parameters

$$u = ha = a\sqrt{\beta_1^2 - \beta^2} \qquad v = qa = a\sqrt{\beta^2 - \beta_2^2} \,. \tag{2.147}$$

Their geometrical sum, the *normalized frequency*

$$V = \sqrt{u^2 + v^2} = a\sqrt{\beta_1^2 - \beta_2^2} = a\beta_0\sqrt{n_1^2 - n_2^2} = a\frac{\omega}{c}\sqrt{n_1^2 - n_2^2} \,, \tag{2.148}$$

depends only on optical frequency and material parameters. The boundary conditions are fulfilled by

$$\begin{bmatrix} J_l(u) & 0 & -K_l(v) & 0 \\ \pm\frac{jla}{u^2}J_l(u) & -\frac{\omega\mu_0 a}{u\beta}J_l'(u) & \pm\frac{jla}{v^2}K_l(v) & -\frac{\omega\mu_0 a}{v\beta}K_l'(v) \\ 0 & J_l(u) & 0 & -K_l(v) \\ \frac{\omega\varepsilon_1 a}{u\beta}J_l'(u) & \pm\frac{jla}{u^2}J_l(u) & \frac{\omega\varepsilon_2 a}{v\beta}K_l'(v) & \pm\frac{jla}{v^2}K_l(v) \end{bmatrix} \begin{bmatrix} A \\ B \\ C \\ D \end{bmatrix} = \begin{bmatrix} 0 \\ 0 \\ 0 \\ 0 \end{bmatrix}. \tag{2.149}$$

1st and 3rd matrix line yield

$$\frac{C}{A} = \frac{J_l(u)}{K_l(v)} = \frac{D}{B}, \tag{2.150}$$

which is inserted into 2nd and 4th line. This results in a reduced system of equations

$$\begin{bmatrix} \pm jl\left(\frac{1}{u^2} + \frac{1}{v^2}\right) & -\frac{\omega\mu_0}{\beta}\left(\frac{J_l'(u)}{uJ_l(u)} + \frac{K_l'(v)}{vK_l(v)}\right) \\ \frac{\omega\varepsilon_0}{\beta}\left(\frac{n_1^2 J_l'(u)}{uJ_l(u)} + \frac{n_2^2 K_l'(v)}{vK_l(v)}\right) & \pm jl\left(\frac{1}{u^2} + \frac{1}{v^2}\right) \end{bmatrix} \cdot \begin{bmatrix} A \\ B \end{bmatrix} = \begin{bmatrix} 0 \\ 0 \end{bmatrix}. \tag{2.151}$$

A non-trivial solution of (2.151) requires the matrix determinant to be zero,

$$\left(\frac{J_l'(u)}{uJ_l(u)} + \frac{K_l'(v)}{vK_l(v)}\right)\left(\frac{n_1^2 J_l'(u)}{uJ_l(u)} + \frac{n_2^2 K_l'(v)}{vK_l(v)}\right) - l^2\left(\frac{1}{u^2} + \frac{1}{v^2}\right)^2\left(\frac{\beta}{\beta_0}\right)^2 = 0\,. \tag{2.152}$$

One of the three unknowns u, v, β can be expressed by the others. We insert

$$\left(\frac{1}{u^2}+\frac{1}{v^2}\right)\beta^2=\frac{\beta^2-\frac{u^2}{a^2}}{u^2}+\frac{\beta^2+\frac{v^2}{a^2}}{v^2}=\frac{\beta_1^2}{u^2}+\frac{\beta_2^2}{v^2}=\left(\frac{n_1^2}{u^2}+\frac{n_2^2}{v^2}\right)\beta_0^2 . \quad (2.153)$$

and find

$$\left(\frac{J_l'(u)}{uJ_l(u)}+\frac{K_l'(v)}{vK_l(v)}\right)\left(\frac{n_1^2 J_l'(u)}{uJ_l(u)}+\frac{n_2^2 K_l'(v)}{vK_l(v)}\right)=l^2\left(\frac{1}{u^2}+\frac{1}{v^2}\right)\left(\frac{n_1^2}{u^2}+\frac{n_2^2}{v^2}\right). \quad (2.154)$$

Waveguiding is possible if the eigenvalue equation (2.154) and the auxiliary condition $V^2=u^2+v^2$ are fulfilled simultaneously. For example, $v=\sqrt{V^2-u^2}$ is inserted into (2.154), which leaves only one unknown, u. A numerical solution yields u and then v. The phase constant β is found from (2.148). One line of (2.151) and (2.150) determines $A...D$ with exception of one common factor. The fields are then obtained from (2.143) to (2.146).

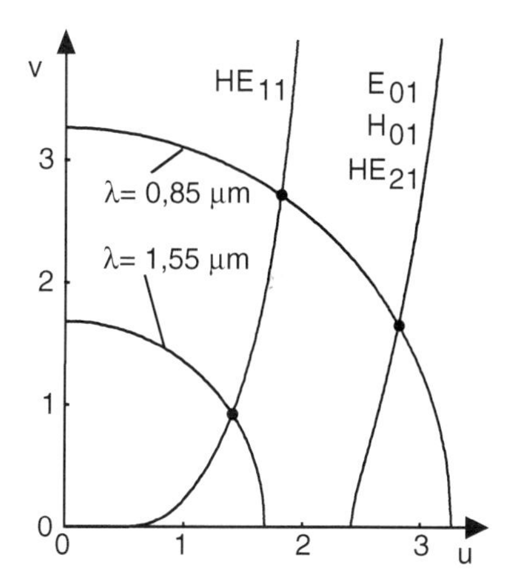

Fig. 2.8 Graphical solution of eigenvalue equation of cylindrical dielectric waveguide with core diameter $2a=9\mu m$ and relative refractive index difference $(n_1-n_2)/n_1=0.002$. For wavelengths above ~ 1.1 µm it is "singlemode" or "monomode", supporting only the two fundamental modes HE_{11}.

Fig. 2.8 presents a graphical solution, i.e., (2.148) (quarter circles) and (2.154) (other curves) are solved separately. Intersections indicated guided modes for a given optical wavelength. Note that for $l\neq 0$ a pair of modes exists for each intersections, corresponding to positive and negative sign of l. These two modes have the same phase constant and differ only in field patterns, notably in polarization. This is seen from the following: Let $E_z(r,\varphi,\pm l)$ be the ansatz (2.136). The linear combinations $E_z(r,\varphi,l)+E_z(r,\varphi,-l)\propto\cos(l\varphi)$ and $-j(E_z(r,\varphi,l)-E_z(r,\varphi,-l))\propto\sin(l\varphi)$ are identical with exception of a rotation about the z axis with an angle $\pi/(2l)$. The same holds for the other field components. So, for any $l \neq 0$ and m there are exactly two independent polarizations or transversal field distributions. For $l = 1$ these have orthogonal directions in space, i.e. these are orthogonally polarized modes.

The term $J_l'(u)/(uJ_l(u))$ oscillates as a function of u between $\pm\infty$. Generally there are several guided modes for one l. In the simplest case $l = 0$ the right side of (2.154) equals zero. For a solution one of the two terms in parentheses must be equal to zero. For

$$\frac{J_0'(u)}{uJ_0(u)} + \frac{K_0'(v)}{vK_0(v)} = 0 \qquad \text{H}_{0m} \text{ modes (TE}_{0m} \text{ modes)} \tag{2.155}$$

the upper line of (2.151) is undetermined while the lower yields $A = 0$, whereupon follows $C = 0$. The only fields are H_z, H_r, E_φ. These modes are called H_{0m} (due to the longitudinal H_z field) or transversal electric (TE_{0m}) modes (since the electric field is transversal). Index m = 1,2,3,... means the m-th solution, starting with small u. Functions $K_l(v)$ are positive and strictly monotous decaying for $v > 0$. So, the term $K_l'(v)/(vK_l(v))$ is negative for all v. According to l'Hospital's rule, (2.141) yields $-\infty$ for this term at $v \to 0$. The term $J_0'(u)/(uJ_0(u))$ has the limit $-1/2$ for $u \to 0$. For increasing u it becomes more and more negative, until it jumps at $J_0(u_0) = 0$, $u_0 \approx 2.405$ from $-\infty$ to $+\infty$. Waveguiding according to (2.148) is therefore possible only for wavelengths shorter than a certain cutoff value, $2.405 \le \frac{2\pi a}{\lambda}\sqrt{n_1^2 - n_2^2}$. This H_{01} mode is shown in Fig. 2.8.

The second solution of (2.154) for $l = 0$ yields

$$\frac{n_1^2 J_l'(u)}{uJ_l(u)} + \frac{n_2^2 K_l'(v)}{vK_l(v)} = 0 \qquad \text{E}_{0m} \text{ modes (TM}_{0m} \text{ modes).} \tag{2.156}$$

Due to the above, the cutoff wavelength of the E_{01} mode is the same as for the H_{01} mode. In usual glass fibers with small refractive index differences the phase constants of E_{0m} and H_{0m} modes are therefore almost identical. Indeed E_{01} and H_{01} can not be distinguished in Fig. 2.8. Similar to the above, the non-vanishing fields are E_z, E_r, H_φ, for which reason these are the E_{0m} (or TM_{0m}) modes.

The state-of-polarization of a wave is given by the direction of the transversal field components. In the special case $l = 0$, i.e., for H_{0m} and E_{0m} modes, these fields depend only on r. E.g., for H_{0m} modes it holds $E_x = -E_\varphi \sin\varphi$,

$E_y = E_\varphi \cos\varphi$. Averaging over the cross section $z = const.$ includes averaging over φ and yields no preferred direction. When the whole mode is considered, these waves are therefore unpolarized even though at each particular point a state-of-polarization of a fully polarized wave exists. This is the reason why there is only one H or E mode of each kind, no degenerate pair as for the hybrid modes.

The eigenvalue equation (2.154) is quadratic in $J_l'(u)/(uJ_l(u))$. Solving for this quantity yields

$$\frac{J_l'(u)}{uJ_l(u)} = \frac{1}{n_1^2}\left(-\frac{n_1^2+n_2^2}{2}\frac{K_l'(v)}{vK_l(v)} \pm \left(\left(\frac{n_1^2-n_2^2}{2}\frac{K_l'(v)}{vK_l(v)}\right)^2 + n_1^2 l^2 \left(\frac{1}{u^2}+\frac{1}{v^2}\right)\left(\frac{n_1^2}{u^2}+\frac{n_2^2}{v^2}\right)\right)^{1/2}\right). \tag{2.157}$$

For the negative sign, the insertion into (2.151) results in smaller values of $|A/B|$. These are called HE_{lm} modes. For the positive sign, EH_{lm} modes result, with larger values of $|A/B|$. HE_{lm} and EH_{lm} modes are *hybrid modes*, where the boundary conditions can be fulfilled only by the presence of both longitudinal field components. All six field components exist. For comparison: Cylindrical hollow waveguides have H_{lm} and E_{lm} modes where one of E_z, H_z vanishes.

The radial dependence of the field amplitudes of the HE_{11} mode is well approximated by a fundamental Gaussian beam. An approximate solution of (2.154) for the HE_{11} mode is [2]

$$v = 1.1428V - 0.996, \quad u = \sqrt{V^2 - v^2} \quad (1 \le V \le 3)\,. \tag{2.158}$$

Note that HE_{11} modes exist down to $u = v = V = 0$, as can be seen from Fig. 2.8. For sufficiently small V only the HE_{11} modes are guided. In this region the waveguide is dubbed as *singlemode* or *monomode*, even though there exists a degenerate pair of two orthogonally polarized HE_{11} modes. For 1.24 µm wavelength, 4.1 µm core radius and a relative refractive index difference of $(n_1 - n_2)/n_1 = 0{,}0036$, the E_{01}, H_{01} and the two HE_{21} modes have set on, yielding a total of six modes (or four if degeneracy is not counted). These modes, including degenerate ones, are illustrated in Figs. 2.9 to 2.12 .

For all modes (H, E, HE, EH), the number of intensity maxima in radial direction, starting at r = 0, yields the second mode index m. The first intensity maximum occurs at $r > 0$, except for the HE_{1m} modes where it is found at $r = 0$.

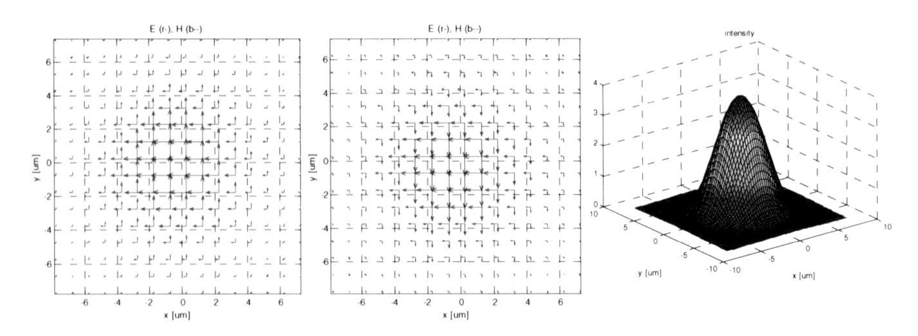

Fig. 2.9 HE_{11} mode fields (left, center) and intensity (right). Vertical polarization (left) is obtained by a superposition of right and left circular polarizations with equal amplitudes and phases ($A = 1$ for $l = \pm 1$). Horizontal polarization requires their superposition with opposite polarities ($A = \pm 1$ for $l = \pm 1$). Intensity profile (right) is the same for all polarizations.

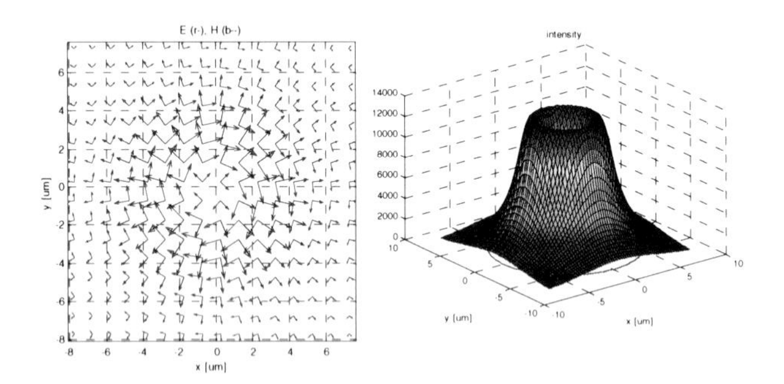

Fig. 2.10 H_{01} mode fields (left) and intensity (right)

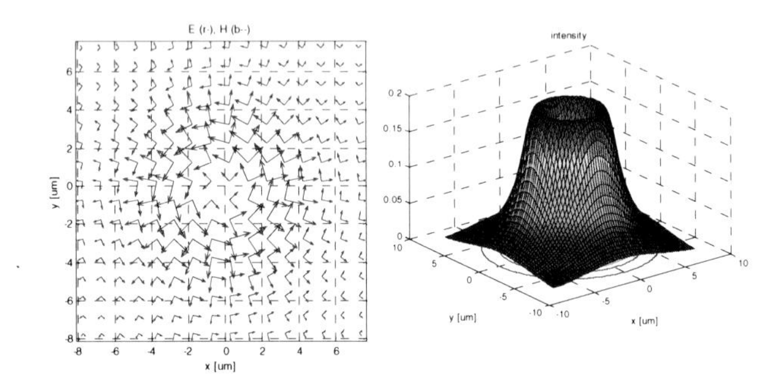

Fig. 2.11 E_{01} mode fields (left) and intensity (right)

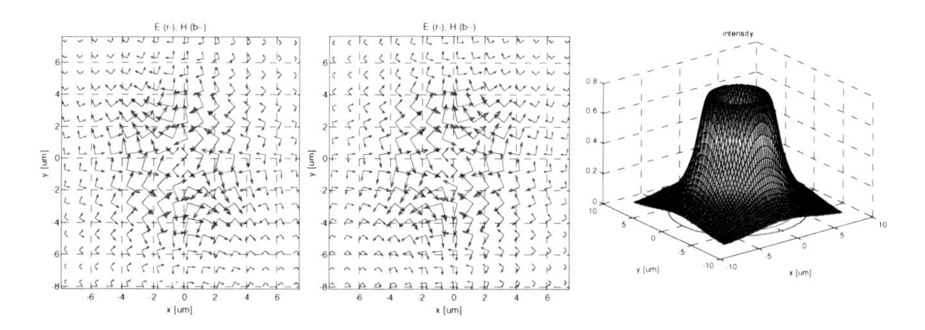

Fig. 2.12 HE_{21} mode fields (left: $A = 1$ for $l = \pm 2$; center: $A = \pm 1$ for $l = \pm 2$) and intensity (right)

The number of guided modes grows roughly proportional to $(Va)^2$, the square of normalized frequency times core radius. This is because of the two-dimensional confinement. In the slab waveguide with one-dimensional confinement the number of modes grows only proportional to the product Vd of frequency times thickness.

Except for hybrid mode pairs, different modes have different group velocities. If several modes are excited their respective intensity contributions arrive at different times at the end of the waveguide. Digital pulses are thereby broadened. This normally undesirable property is called *mode dispersion* or *intermode dispersion*. Intermode dispersion in multimode waveguides causes length-dependent group delay differences of about 0.5 .. 50 ns/km. The permissible product of bitrate times distance is thereby strongly limited.

With exception of short-reach links (< 1 km long or so) with low bit rates, singlemode waveguides are therefore preferred. In principle the two HE_{11} polarization modes travel with the same group velocity. In real, deployed glass fibers this is not exactly the case, due to fabrication-induced deviations from the circular core shape, bending and lateral pressure. So, the fiber is slightly birefringent. As a consequence, two principal states-of-polarization propagate with slightly different group velocities. This *polarization mode dispersion* (PMD) can be detrimental at data rates ≥ 10 Gbit/s.

In singlemode (and all other) fibers, there is additional *material dispersion* and *waveguide dispersion*. The former is caused by the frequency-dependence of the material refractive index, and the latter is due to the fact that even for constant refractive indexes the phase constant in the waveguide depends not only linearly but also quadratically on the optical frequency. Both influences added together constitute the *chromatic dispersion*. If the transfer function of a fiber with length L is $e^{-(\alpha(\omega)+j\beta(\omega))L}$, then β/ω is the phase delay per unit length, $\partial\beta/\partial\omega = \tau_g/L$

is the group delay per unit length, and $\partial^2\beta/\partial\omega^2 \equiv \beta''$ causes chromatic dispersion. More frequently, the chromatic dispersion parameter

$$D = \frac{\partial(\tau_g/L)}{\partial\lambda} = \frac{\partial^2\beta}{\partial\lambda\partial\omega} = \frac{\partial\omega}{\partial\lambda}\frac{\partial^2\beta}{\partial\omega^2} = -\frac{2\pi c}{\lambda^2}\frac{\partial^2\beta}{\partial\omega^2} \tag{2.159}$$

is used. For a standard singlemode fiber used at λ = 1.55 μm, with parameters similar to those of Fig. 2.8, D is about 17 ps/(nm·km), and attenuation is lowest, less than 0.2 dB/km. So, if the wavelength is increased by 1 nm, then the group delay of a 1 km long fiber increases by 17 ps. Light pulses are generally broadened by chromatic dispersion. Optical fiber dispersion is a tiny fraction of hollow waveguide dispersion but since fiber bandwidth is so high and attenuation so low it impairs transmission at data rates of 10 Gb/s and beyond. Chromatic dispersion is by far the largest transmission impairment in most high-bitrate, medium- and long-haul optical fiber communication links!

Around λ = 1.31 μm, there is a dispersion zero but the attenuation is much higher, about 0.35 dB/km.

Different refractive index profiles, for example with several refractive index steps in the core region, allow to taylor dispersion properties.

For optical fiber fabrication initially a cylindrical preform of amorphous SiO_2 (silica, quartz glass) is prepared, with a diameter of several cm. Proportionally it must have the same refractive index profile as the end product, the optical fiber. The small necessary refractive index differences with respect to that of pure silica ($n \approx 1.47$) are usually achieved by doping the later core region with GeO_2 or Al_2O_3, which increase n. Most power is transported in the core and doping always increases transmission loss. From this standpoint it is preferable to dope instead the cladding by F or B_2O_3, which reduce n. The vertically mounted preform is heated at its bottom until it melts and a thin glass fiber can be drawn due to the glass viscosity. Its diameter is usually 125 μm. The core diameter of standard singlemode fibers is a = 8.2 μm. According to (2.148), V and hence the number of modes grows with a. In order to keep this large a for easier light coupling a very small refractive index difference is needed, $(n_1 - n_2)/n_1 = 0{,}0036$ in standard singlemode fiber. Yet the required alignment tolerance is smaller than 1 μm.

Due to the large cladding diameter the fields are well attenuated at the fiber periphery. For large wavelengths, 1.65 μm and beyond, this is not sufficiently the case. Bending losses rise sharply, and in order to prevent this the permissible bending radius must be increased. The fiber is covered by a primary acrylate coating with a diameter of usually 250 μm. In an optical cable usually several or many fibers are bundled, and a stress member allows to pull and bend it without destruction.

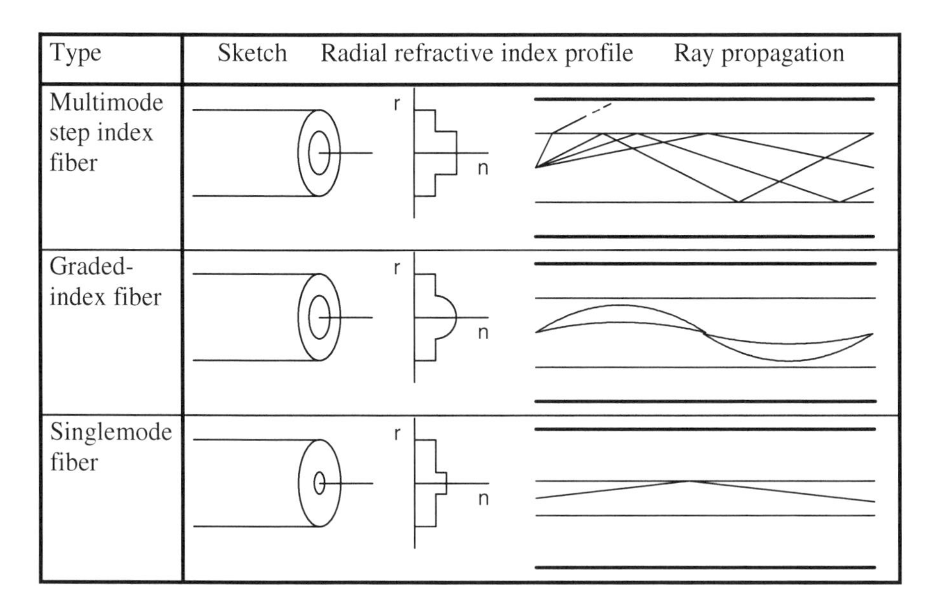

Fig. 2.13 Optical fiber types

In short-reach links with small bit rates even intermode dispersion can not cause significant signal degradation. This means that multimode waveguides can be used, with drastically relaxed alignment tolerances. Multimode step index waveguides are often fabricated as polymer fibers. Fig. 2.13 schematically shows the function, including a non-guided ray with under an angle where reflection is not total. The other rays are totally reflected at the core-cladding boundaries. Of interest are also *graded-index fibers*, the refractive index of which decays with the square (or another power) of the radius. This means that off-axis rays are bent back to the center. The propagation delay of center rays is essentially the same as that of off-axis rays because the latter travel significant distances in regions with low refractive index. The intermodal dispersion of graded index fibers is much smaller than that of multimode step index fibers, but much larger than the chromatic dispersion of singlemode fibers.

At 1.55 μm wavelength binary optical signals can be transmitted over more than 200 km without the need for intermediate amplification or regeneration. In long wavelength-division multiplex (WDM) links fiber lengths between optical amplifiers are 80...100 km (terrestrial) or less (transoceanic). For comparison: In coaxial cables the standard distance between regenerators was 1.8 km.

Important fiber types are summarized or explained in the following and in Table 2.2.

Table 2.2 Properties of important singlemode fibers

	SSMF	DSF[1]	DCF	NZDSF	NZDSF	RDF
Attenuation [dB/km] at 1.31 μm	0.3 ... 0.4	0.5				
Attenuation [dB/km] at 1.55 μm	0.15 ... 0.2	0.25	0.5	0.21[2]		<1.0[3]
Dispersion [ps/nm/km] at 1.31 μm	0	−15[4]		−12[5]	−18	
Dispersion [ps/nm/km] at 1.55 μm	17	0	−100 ... −200	5	−5	−17

- *Standard single-mode fiber, SSMF:* Optimized for lowest attenuation. Pure silica (SiO_2) core fiber is doped in the cladding, for lowest attenuation. Chromatic dispersion at 1.55 μm is high. Very common. In Europe and USA most links are composed of SSMF. Fibers which were deployed before 1995 can exhibit significant amounts of polarization-mode dispersion.
- *Dispersion-shifted fiber, DSF:* Dispersion zero at 1.55 μm but attenuation is slightly higher than in SSMF. Widely deployed in Japan. Neighbor channels of WDM links have almost the same group velocities. Therefore four-wave mixing (FWM) is strong. In order to still transmit high aggregate data rates, single channel data rates must be high or longer wavelengths must be used.
- *Dispersion-compensating fiber, DCF:* Optimized for very strong negative dispersion at still acceptable attenuation. Is inserted in optical amplifier modules between preamplifier and main amplifier. Makes SSMF links usable for high aggregate data rates and WDM. Very common.
- *Non-zero dispersion-shifted fiber, NZDSF, NDSF:* Types with positive and negative dispersion are cascaded, in order to keep total dispersion small while suppressing FWM.
- *Reverse-dispersion fiber, RDF:* Modern NZDSF, can be cascaded with SSMF in approximately equal lengths. Very good suppression of FWM.

A large core area increases the tolerance against nonlinear effects. A small core area allows for a large Raman amplification.

At 1.4 μm OH groups have strong resonances and increase fiber attenuation. However, newest fabrication and cabling technologies allow to prevent the intrusion of OH groups by moisture very well. In the far infrared silica itself has molecule resonances, and these dominate the excess loss from 1.6 μm on. At its

[1] http://www.pofc.com.tw/en/products/optical-fiber/singlemode-fiber/dispersion-shifted-fiber
[2] http://www.ofsoptics.com/resources/HowLongisLongHaul.pdf
[3] http://v3.espacenet.com/publicationDetails/biblio?CC=EP&NR=0554714&KC=&FT=E
[4] http://www.ciscopress.com/articles/article.asp?p=170740&seqNum=7 (from graphic)
[5] http://ptgmedia.pearsoncmg.com/images/ch03_1587051052/elementLinks/fig14.gif (estimation)

solidification, silica becomes amorphous, and as a consequence there are microscopically small regions with density and hence refractive index variations. These scatter the light into all possible directions (*Rayleigh scattering*). The smaller the wavelength, the less can the size of the scattering centers be neglected against it. The strength of the Rayleigh scattering scales with λ^{-4}. A small portion P_r of the reflected power excites a HE_{11} mode in backward direction. The reflection of long fibers ($L \to \infty$) with an attenuation of 0.2 dB/km at 1.55 μm is about −32 dB. The temporal average of the relative backreflection, even of short fibers, is constant. The time-dependent backreflected power is not constant because a quasi infinite number of scattering centers is distributed along the fiber, and due to laser phase noise the backreflected wave contributions vary in phase all the time. The scattering events are (almost) statistically independent. In the following polarization effects are not included, or, to be more precise, it is assumed that the fiber does not transform the polarization, which is linear. In a narrowband attenuation the backreflected wave can be written as $x_p \cos(\omega t) - x_q \sin(\omega t)$. Under these assumptions x_p, x_q are zero-mean Gaussian statistically independent random variables. Their envelope $r = \sqrt{x_p^2 + x_q^2}$ is therefore Rayleigh-distributed and the phase is equidistributed.

Problem: At position $z = 0$ an optical power P_0 is fed into a fiber. How large is the backreflected power P_R at $z = 0$ as a function of fiber length L, if for $L \to \infty$ it holds $P_R = 0.63 \cdot 10^{-3} \cdot P_0$? Multiple reflections (forward-backward-forward) may be neglected.

Solution: Power that is reflected at position $z = z_0$ and transmitted back toward $z = 0$ has traveled a distance $2z_0$ and been attenuated by the factor $e^{-2\alpha \cdot 2z_0}$. If the fiber length is increased by dz, in which piece Rayleigh scattering occurs, P_R is increased by dP_R :

$$\frac{dP_R}{dz} = \zeta P_0 e^{-4\alpha z} \quad \Rightarrow \quad P_R = \int_0^L \frac{dP_R}{dz} dz = P_0 \frac{\zeta}{4\alpha}\left(1 - e^{-4\alpha L}\right) \tag{2.160}$$

Additionally it holds $P_R(\infty) \approx 0.63 \cdot 10^{-3} P_0 \Rightarrow \frac{\zeta}{4\alpha} \approx 0.63 \cdot 10^{-3}$. The relation between fiber attenuation and α is $10^{-\left(\frac{0.2\,\text{dB/km}}{10\,\text{dB}} z\right)} = e^{-2\alpha z} \Rightarrow 2\alpha = 0.046/\text{km}$. Now all unknowns for the calculation of P_R have been determined.

2.3 Polarization

The term polarization is used to denominate various things. The electrical polarization **P** of a material is the electrical dipole moment per unit volume, and according to $\mathbf{D} = \varepsilon_0 \mathbf{E} + \mathbf{P}$ it is the difference vector between the dielectric displacements in the material and in vacuum. As will be shown later, **E**, **P**, **D** need not be parallel. Similarly, using $\mathbf{B} = \mu_0 \mathbf{H} + \mathbf{M}$ the vector **M** is sometimes called the magnetic polarization or magnetic dipole moment per unit volume.

In this chapter we deal with another, though similar meaning of the word. The state-of-polarization (SOP) or polarization of a plane monochromatic electromagnetical wave denotes the spatial evolution of the dielectric displacement **D** at a certain time or its temporal evolution at a certain place. If the wave is not homogeneous or not plane it is useful to consider just the transversal component of the dielectric displacement. In isotropic media **E** and **D** are parallel which means the behavior of **E** may be investigated instead of that of **D**. But even if there is an anisotropic medium in the propagation path of an electromagnetic wave one usually chooses **E** in polarization algebra. Sufficient conditions for this to be valid are given in subchapter 2.3.3.

2.3.1 Representing States-of-Polarization

A plane monochromatic transversal electromagnetic wave in an isotropic medium which travels in the positive z direction can be respresented as

$$\mathbf{E}(z,t) = \mathbf{E} e^{j(\omega t - kz)} \tag{2.161}$$

with real $k \equiv \beta$ or complex $\underline{k} = \beta - j\alpha$ and the complex *cartesian Jones vector* or simply *Jones vector*

$$\mathbf{E} = \begin{bmatrix} E_x \\ E_y \end{bmatrix}. \tag{2.162}$$

Other basis vectors may be used instead of the cartesian unit vectors. If the basis vectors are not explicitly given one usually means the cartesian system.

The periodic dependence of the wave on $\omega t - kz$ is often known and does not influence the state-of-polarization. So it can often be dropped and one uses just the Jones vector. Yet the Jones vector may be time and space-dependent.

With real and imaginary parts of two components a Jones vector has four degrees-of-freedom. Suitably expressed, two degrees-of-freedom denote the state-of-polarization and two more the wave amplitude and phase, respectively.

For an observer at a position z = const. the real part **E** of the electric field vector generally describes an elliptical trajectory, the polarization ellipse. For $\mathrm{Im}\left(E_x E_y^*\right) = 0$, i.e. one vector component equals the other times a real number, the vibration occurs in one plane. The vibration plane is defined by the propagation direction z and an azimuth angle ϑ with respect to the x axis,

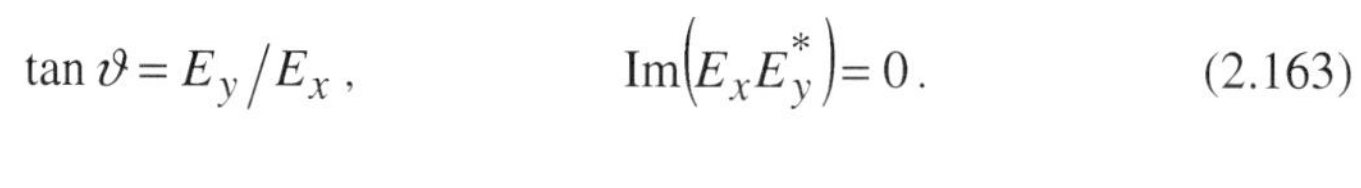

$$\tan \vartheta = E_y / E_x , \qquad \mathrm{Im}\left(E_x E_y^*\right) = 0 . \tag{2.163}$$

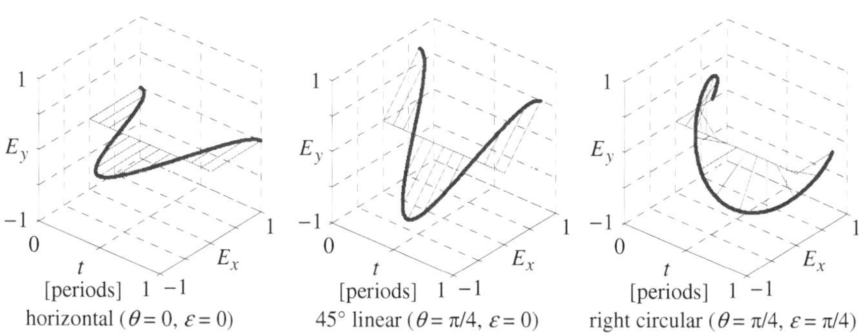

Fig. 2.14 Temporal field evolution for characteristic states-of-polarization

For $E_x = \pm jE_y$ the locus of the electric field vector $\mathbf{E}$ is a circle, corresponding to circular polarization. The rotation direction is defined from the temporal impression of an observer at a position z = const. For this observer x, y and t axes are a right-handed coordinate system. For the positive sign[6] ($E_x = jE_y$) he sees a clockwise rotating pointer, i.e., right circular polarization. The negative sign holds for left circular polarization. Characteristic states-of-polarization are illustrated in Fig. 2.14. Note that for frozen time (t = const.) the electric field vector rotates in the right-handed system x, y, z opposed to the rotation direction in the right-handed system x, y, t because the terms ωt and kz influence the wave phase in opposite directions.

When (2.162) is expressed as

$$\mathbf{E} = |\mathbf{E}| e^{j\varphi} \begin{bmatrix} \cos\vartheta \cos\varepsilon + j \sin\vartheta \sin\varepsilon \\ \sin\vartheta \cos\varepsilon - j \cos\vartheta \sin\varepsilon \end{bmatrix} \tag{2.164}$$

then ϑ is the azimuth angle of the large polarization ellipse axis with respect to the x axis and ε the ellipticity angle. It holds $\tan\varepsilon = \pm a/c$ where a and c denote small and large half axis of the polarization ellipse, respectively (Fig. 2.15a). The posititive and negative sign holds for right and left elliptical polarization, respectively.

Instead of the Jones vector the complex variable

$$\chi = E_y / E_x \tag{2.165}$$

can as well represent the state-of-polarization. The two unused degrees-of-freedom of the Jones vector have been eliminated and there remain only the two degrees-of-freedom for the state-of-polarization.

---

[6] The opposite definition is also being used, see for example [3].

A powerful representation of states-of-polarization is given by the *Stokes vector* $\mathbf{S} = [S_0, S_1, S_2, S_3]^T$ or its components, the *Stokes parameters*

$$\begin{aligned} S_0 &= \overline{|E_x|^2 + |E_y|^2} = \overline{|\mathbf{E}|^2} && \text{(power)} \\ S_1 &= \overline{|E_x|^2 - |E_y|^2} = \overline{|\mathbf{E}|^2 \cdot \cos 2\varepsilon \cdot \cos 2\vartheta} && \text{(horizontal/vertical)} \\ S_2 &= \overline{2\,\mathrm{Re}\left(E_x E_y^*\right)} = \overline{|\mathbf{E}|^2 \cdot \cos 2\varepsilon \cdot \sin 2\vartheta} && (45°/-45° \text{ linear}) \\ S_3 &= \overline{2\,\mathrm{Im}\left(E_x E_y^*\right)} = \overline{|\mathbf{E}|^2 \cdot \sin 2\varepsilon} && \text{(right/left circular)} \end{aligned} \quad . \qquad (2.166)$$

If the sense of ellipticity is defined in the x, y, z rather than x, y, t system then the sign of S_3 is inverted, which means the term $E_x E_y^*$ in (2.166) is replaced against $E_x^* E_y$ [3].

The Stokes parameters can also be cast with help of the *Pauli spin matrices*

$$\boldsymbol{\sigma}_0 = \begin{bmatrix} 1 & 0 \\ 0 & 1 \end{bmatrix} \quad \boldsymbol{\sigma}_1 = \begin{bmatrix} 1 & 0 \\ 0 & -1 \end{bmatrix} \quad \boldsymbol{\sigma}_2 = \begin{bmatrix} 0 & 1 \\ 1 & 0 \end{bmatrix} \quad \boldsymbol{\sigma}_3 = \begin{bmatrix} 0 & -j \\ j & 0 \end{bmatrix} \qquad (2.167)$$

into the compact form

$$S_i = \overline{\mathbf{E}^+ \boldsymbol{\sigma}_i^* \mathbf{E}} \quad (i = 0 \ldots 3). \qquad (2.168)$$

For the oppositely defined sign of S_3 we use instead $S_i = \overline{\mathbf{E}^+ \boldsymbol{\sigma}_i \mathbf{E}}$.

The overbar means temporal or spectral, or possibly spatial averaging over a range so large that the Stokes parameter change no longer when the range is widened. The terms behind the first equality sign are the Stokes parameter definitions. The terms behind the second equality sign are obtained by inserting (2.164) into the definition. Preponderance of horizontal (vertical) linear polarization is expressed by $S_1 > 0$ ($S_1 < 0$), of linear polarization with 45° (–45°) azimuth angle by $S_2 > 0$ ($S_2 < 0$), and of right (left) circular polarization by $S_3 > 0$ ($S_3 < 0$), respectively.

Proofs: If the main axes of the polarization ellipse coincide with the cartesisian coordinate system the Jones Vector is proportional to $[\cos\varepsilon \quad -j\sin\varepsilon]^T$. A rotation of the coordinate system by the angle ϑ results in

$$\mathbf{E} = |\mathbf{E}| e^{j\varphi} \begin{bmatrix} \cos\vartheta & -\sin\vartheta \\ \sin\vartheta & \cos\vartheta \end{bmatrix} \begin{bmatrix} \cos\varepsilon \\ -j\sin\varepsilon \end{bmatrix}, \qquad (2.169)$$

which is identical with (2.164). If the rotation matrix is brought to the left side,

$$\begin{bmatrix} E_x \cos\vartheta + E_y \sin\vartheta \\ -E_x \sin\vartheta + E_y \cos\vartheta \end{bmatrix} = |\mathbf{E}| \begin{bmatrix} \cos\varepsilon \\ -j\sin\varepsilon \end{bmatrix} e^{j\varphi}, \qquad (2.170)$$

and if the upper is multiplied by the conjugated lower line one obtains

$$\left(|E_y|^2 - |E_x|^2\right)\sin 2\vartheta + 2\,\mathrm{Re}\left(E_x^* E_y\right)\cos 2\vartheta - j2\,\mathrm{Im}\left(E_x^* E_y\right) = j|\mathbf{E}|^2 \sin 2\varepsilon \,. \tag{2.171}$$

Splitting into real and imaginary parts results in

$$\begin{aligned} 2\,\mathrm{Im}\left(E_x E_y^*\right) &= |\mathbf{E}|^2 \sin 2\varepsilon \\ 2\,\mathrm{Re}\left(E_x E_y^*\right)\cos 2\vartheta &= \left(|E_x|^2 - |E_y|^2\right)\sin 2\vartheta \end{aligned} . \tag{2.172}$$

On the other hand it may be verified

$$\left(|E_x|^2 - |E_y|^2\right)^2 + \left(2\,\mathrm{Re}\left(E_x E_y^*\right)\right)^2 + \left(2\,\mathrm{Im}\left(E_x E_y^*\right)\right)^2 = |\mathbf{E}|^{2^2} . \tag{2.173}$$

It follows

$$\sqrt{\left(|E_x|^2 - |E_y|^2\right)^2 + \left(2\,\mathrm{Re}\left(E_x E_y^*\right)\right)^2} = |\mathbf{E}|^2 \cos 2\varepsilon \,. \tag{2.174}$$

The right hand sides of (2.166) can now be verified.

Without averaging $S_0^2 = S_1^2 + S_2^2 + S_3^2$ would always hold. In general,

$$p = \sqrt{S_1^2 + S_2^2 + S_3^2} \Big/ S_0 \tag{2.175}$$

is the *degree-of-polarization* of the wave. It ranges between 0 for unpolarized and 1 for fully polarized light. In a sufficiently small spectral and spatial range any electromagnetic wave is fully polarized (p = 1) because averaging has no effect. Yet we call unpolarized (p = 0) those waves or wave mixtures where a predominant polarization can not be found in the chosen observation range, be it temporal, spectral or spatial. This happens while averaging in calculating the Stokes parameters. Such waves may have different polarizations in their various frequency components (cxamples: sun, light bulb). In the time domain, the beating between the frequency components corresponds to fast, chaotic changes of the instantaneous polarization state of the whole wave. Even a monochromatic signal can be unpolarized. Though a polarization state is defined anywhere in the cross section of a multimode optical waveguide the integration over this cross section may result in an almost vanishing degree-of-polarization. The four Stokes parameters contain information about power (1 degree-of-freedom) and polarization (2 degrees-of-freedom). The fourth degree-of-freedom is the degree-of-polarization.

In many cases the power of the wave is known and can be dropped. It is therefore useful to normalize the Stokes parameters with respect to S_0 (to divide them by S_0). One writes only the three then-normalized parameters S_1, S_2, S_3 and drops the "0-th" parameter $S_0 = 1$. When averaging has no effect in (2.166) then the doubled azimuth and ellipticity angles form spherical coordinates on the so-called *Poincaré sphere* (Fig. 2.15b). If its radius is halved it becomes a Riemann sphere, onto the surface of which the complex variable $\chi = E_y / E_x$ is stereographically projected.

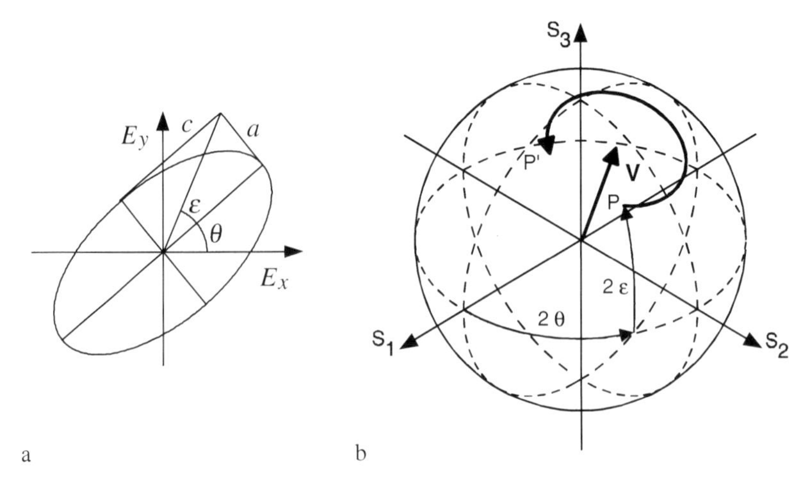

Fig. 2.15 Polarization ellipse a and Poincaré sphere b (For P', **V** see section 2.3.3)

Fully polarized light is represented by a point P on the surface of the sphere having a unit radius. Partly or unpolarized light is represented by a point inside or in the center of the sphere, respectively. p is the length of the normalized Stokes vector. An azimuth angle ϑ of the polarization ellipse corresponds to a spherical coordinate 2ϑ because an ellipse which is physically rotated by 180° falls onto itself, only its phase is changed by 180°. The other spherical coordinate is 2ε, the doubled ellipticity angle.

As an alternative to the Stokes parameters there exist *coherency or correlation matrices* $\overline{\mathbf{EE}^+}$. The elements of these Hermitian matrices are simply linear combinations of the Stokes parameters,

$$\overline{\mathbf{EE}^+} = \begin{bmatrix} \overline{E_x E_x^*} & \overline{E_x E_y^*} \\ \overline{E_y E_x^*} & \overline{E_y E_y^*} \end{bmatrix} = \frac{1}{2} \begin{bmatrix} S_0 + S_1 & S_2 + jS_3 \\ S_2 - jS_3 & S_0 - S_1 \end{bmatrix} = \overline{\mathbf{EE}^+}^+ . \tag{2.176}$$

Two polarization states are *orthogonal* if their Jones vectors $\mathbf{E}_a$, $\mathbf{E}_b$ or complex variables χ_a, χ_b fulfill one of the conditions

$$\mathbf{E}_a^+ \mathbf{E}_b = 0 \quad \Leftrightarrow \quad E_{ax}^* E_{bx} + E_{ay}^* E_{by} = 0 \quad \Leftrightarrow \quad \frac{E_{by}}{E_{bx}} = -\left(\frac{E_{ax}}{E_{ay}} \right)^* . \tag{2.177}$$

$$\Leftrightarrow \quad E_{bx} = -E_{ay}^* \cdot h, \quad E_{by} = E_{ax}^* \cdot h, \quad (h = const.)$$

$$\chi_a \chi_b^* = -1 . \tag{2.178}$$

The two polarization ellipses are rotated with respect to each other by 90° and have opposite senses of ellipticity. For linearly polarized light the orthogonality is immediately understandable.

In section 2.3.3 it will be shown that the orthogonality of two polarizations corresponds to the orthogonality relation

$$\mathbf{S}_a \cdot \mathbf{S}_b = 0 \qquad \mathbf{S}_i = \left[S_{0i}, S_{1i}, S_{2i}, S_{3i}\right]^T, \quad i = a,b \tag{2.179}$$

of the respective *Stokes vectors* $\mathbf{S}_a$, $\mathbf{S}_b$. For two *normalized Stokes vectors* $\mathbf{S}_a$, $\mathbf{S}_b$ the orthogonality relation can be written as

$$\mathbf{S}_b = -\mathbf{S}_a, \qquad p_b = p_a = 1, \qquad \mathbf{S}_i = \left[S_{1i}, S_{2i}, S_{3i}\right]^T, \quad i = a,b. \tag{2.180}$$

Two orthogonal polarization states are therefore antipodal on the Poincaré sphere. Only fully polarized waves can be orthogonal to each other in the strict sense of the word.

2.3.2 Anisotropy, Index Ellipsoid

In anisotropic media the propagation constant $\underline{k}$ depends on polarization state and propagation direction. The anisotropy can relate to the phase constant β (birefringence, double refraction), to the attenuation constant α (dichroism, lit.: two-coloredness), or to both. This general case is covered by the following calculus. An important case are lossless anisotropic media. For these the general calculus can be visualized by the index ellipsoid [1].

For the optical materials and frequencies $\mu = \mu_0$ is a scalar, i.e., these materials are magnetically isotropic. In electrically anisotropic media a *second rank tensor* is inserted into the material equation instead of a dielectricity constant. Any second rank tensor is a quadratic matrix. For our 3-dimensional space the quadratic matrices have the size 3×3. In complex notation,

$$\mathbf{P} = \varepsilon_0 \boldsymbol{\chi}^{(1)} \mathbf{E} \qquad \mathbf{D} = \varepsilon_0 \mathbf{E} + \mathbf{P} = \boldsymbol{\varepsilon} \mathbf{E} = \varepsilon_0 \boldsymbol{\varepsilon}_r \mathbf{E} \qquad \boldsymbol{\varepsilon}_r = \mathbf{1} + \boldsymbol{\chi}^{(1)} \tag{2.181}$$

holds. The *relative dielectricity tensor* $\boldsymbol{\varepsilon}_r$ is the sum of unity matrix $\mathbf{1}$ and the *susceptibilitity tensor* $\boldsymbol{\chi}^{(1)}$. In non-magnetically but electrically anisotropic media the following differential form of Poynting's theorem (2.79) holds,

$$-\operatorname{div} \mathbf{T} = \frac{j\omega}{2}\left(\mathbf{H}^+ \mathbf{B} - \mathbf{E}^T \mathbf{D}^*\right) = \frac{j\omega}{2}\left(\mu_0 |\mathbf{H}|^2 - \left(\mathbf{E}^+ \boldsymbol{\varepsilon} \mathbf{E}\right)^*\right). \tag{2.182}$$

If in addition the medium shall be lossless then the effective power flow into a differential volume element, i.e., $-\operatorname{Re}(\operatorname{div} \mathbf{T})$ must disappear. To fulfill this condition $\mathbf{E}^+ \boldsymbol{\varepsilon} \mathbf{E}$ must be a *Hermitian form*, i.e., the dielectric material tensors are *Hermitian*,

$$\boldsymbol{\varepsilon}^{+} = \boldsymbol{\varepsilon} \qquad \boldsymbol{\varepsilon}_r^{+} = \boldsymbol{\varepsilon}_r \qquad \boldsymbol{\chi}^{(1)+} = \boldsymbol{\chi}^{(1)} . \tag{2.183}$$

Their real parts are *symmetric* while the imaginary parts are *skew symmetric*. In the case of real matrices (2.183) implies matrix symmetry. Any Hermitian (real symmetric) matrix can be transformed by a unitary (orthogonal) transformation into a real diagonal matrix. Orthogonal transformations are coordinate transformations. This means that a real, symmetric dielectricity tensor can always be diagonalized by choosing a suitable coordinate system for the field quantities. It then assumes the form

$$\boldsymbol{\varepsilon} = \begin{bmatrix} \varepsilon_x & 0 & 0 \\ 0 & \varepsilon_y & 0 \\ 0 & 0 & \varepsilon_z \end{bmatrix} . \tag{2.184}$$

The desired general solution of the wave equation for arbitrary dielectricity tensors is best conducted with respect to the dielectric displacement **D**. We define the complex impermeability tensor

$$\boldsymbol{\eta} = \left(\boldsymbol{\varepsilon}/\varepsilon_0\right)^{-1} = \boldsymbol{\varepsilon}_r^{-1}, \qquad \mathbf{E} = \frac{1}{\varepsilon_0}\boldsymbol{\eta}\mathbf{D} \tag{2.185}$$

as the inverse of the relative dielectricity tensor and, using $\mu = \mu_0$, write the wave equation (2.37) as

$$\operatorname{curl}\left(\operatorname{curl}\boldsymbol{\eta}\mathbf{D}\right) = \omega^2 \mu_0 \varepsilon_0 \mathbf{D} = \frac{\omega^2}{c^2}\mathbf{D} . \tag{2.186}$$

Since wave propagation in the anisotropic medium depends on propagation direction and polarization state we postulate a solution

$$\mathbf{D}(t, \mathbf{r}) = \mathbf{D} e^{j(\omega t - \mathbf{k}\cdot\mathbf{r})}, \tag{2.187}$$

where **D** is a position- and time-independent constant. This will be seen to be valid in general only for two solutions in each propagation direction. Assuming a homogeneous plane wave the amplitude vector is parallel to the phase vector, see chapter 2.1.4, and the wave vector is therefore

$$\mathbf{k} = \left[k_x \quad k_y \quad k_z\right]^T = \frac{\omega}{c} n\mathbf{s} . \tag{2.188}$$

The constant dielectric displacement **D** and the wave vector are linked in each solution by an eigenvalue equation. The effective refractive index n is initially unknown. Vector $\mathbf{r}$ is the position. Insertion of (2.187) into $\operatorname{curl}(F\mathbf{A}) = F \operatorname{curl}\mathbf{A} - \mathbf{A}\times\operatorname{grad} F$ results in $\operatorname{curl}(\boldsymbol{\eta}\mathbf{D}) = -j(\mathbf{k}\times(\boldsymbol{\eta}\mathbf{D}))$. Eqn. (2.186) is thereby transformed into

$$-\mathbf{s}\times\left(\mathbf{s}\times\left(\boldsymbol{\eta}\mathbf{D}\right)\right) = \frac{1}{n^2}\mathbf{D} . \tag{2.189}$$

We supplement the unit vector in propagation direction $\mathbf{s}$ by two other vectors $\boldsymbol{\tau}_1$, $\boldsymbol{\tau}_2$ to an orthonormal coordinate system ($\mathbf{s}\times\boldsymbol{\tau}_1=\boldsymbol{\tau}_2$). We transform (2.189) into a new coordinate system, in which the vectors $\boldsymbol{\tau}_1$, $\boldsymbol{\tau}_2$, $\mathbf{s}$ become basis vectors $\boldsymbol{\tau}_1'=\begin{bmatrix}1 & 0 & 0\end{bmatrix}^T$, $\boldsymbol{\tau}_2'=\begin{bmatrix}0 & 1 & 0\end{bmatrix}^T$, $\mathbf{s}'=\begin{bmatrix}0 & 0 & 1\end{bmatrix}^T$. An orthogonal transformation matrix $\mathbf{A}$ allows to express the old vectors in the new coordinate system and vice versa, e.g.,

$$\begin{aligned} &\boldsymbol{\tau}_1=\mathbf{A}\boldsymbol{\tau}_1', \qquad \boldsymbol{\tau}_2=\mathbf{A}\boldsymbol{\tau}_2', \qquad \mathbf{s}=\mathbf{A}\mathbf{s}', \qquad \underline{\mathbf{D}}=\mathbf{A}\underline{\mathbf{D}}', \qquad \underline{\mathbf{E}}=\mathbf{A}\underline{\mathbf{E}}', \\ &\mathbf{A}=\begin{bmatrix}\boldsymbol{\tau}_1 & \boldsymbol{\tau}_2 & \mathbf{s}\end{bmatrix}, \qquad \mathbf{A}^{-1}=\mathbf{A}^T. \end{aligned} \tag{2.190}$$

Since $\varepsilon_0\mathbf{E}=\boldsymbol{\eta}\mathbf{D}$ must hold both in the old and the new coordinate system one further obtains

$$\boldsymbol{\eta}=\mathbf{A}\boldsymbol{\eta}'\mathbf{A}^{-1}=\mathbf{A}\boldsymbol{\eta}'\mathbf{A}^T. \tag{2.191}$$

The wave equation (2.189) reads now as

$$-\begin{bmatrix}0\\0\\1\end{bmatrix}\times\left(\begin{bmatrix}0\\0\\1\end{bmatrix}\times\left(\boldsymbol{\eta}'\mathbf{D}'\right)\right)=\frac{1}{n^2}\mathbf{D}'. \tag{2.192}$$

Using $-\mathbf{u}\times(\mathbf{v}\times\mathbf{w})=-\mathbf{v}(\mathbf{u}\cdot\mathbf{w})+\mathbf{w}(\mathbf{u}\cdot\mathbf{v})$ one obtains

$$\begin{bmatrix}\underline{\eta}'_{11} & \underline{\eta}'_{12} & \underline{\eta}'_{13}\\ \underline{\eta}'_{21} & \underline{\eta}'_{22} & \underline{\eta}'_{23}\\ 0 & 0 & 0\end{bmatrix}\mathbf{D}'=\frac{1}{n^2}\mathbf{D}'. \tag{2.193}$$

Since both sides are equal the third component of $\mathbf{D}'$ must be zero, which amounts to saying that the dielectric displacement is always perpendicular to the propagation vector,

$$\mathbf{s}'\cdot\mathbf{D}'=\mathbf{s}\cdot\mathbf{D}=0. \tag{2.194}$$

This statement is also obtained directly if one assumes a harmonic magnetic field $\mathbf{H}=\mathbf{H}_0e^{j(\omega t-\mathbf{k}\cdot\mathbf{r})}$ and inserts it into $\operatorname{curl}\mathbf{H}=j\omega\mathbf{D}$ (2.21), using $\operatorname{curl}\mathbf{H}=-j(\mathbf{k}\times\mathbf{H})$ and (2.188). The dielectric displacement is fully determined by its transversal components, and since the wave normal is known (2.193) can be simplified into

$$\boldsymbol{\eta}'_t\mathbf{D}'_t=\frac{1}{n^2}\mathbf{D}'_t \qquad \boldsymbol{\eta}'_t=\begin{bmatrix}\eta'_{11} & \eta'_{12}\\ \eta'_{21} & \eta'_{22}\end{bmatrix}. \tag{2.195}$$

The transversal component of the dielectric displacement, $\mathbf{D}'_t$, is a vector having two components, and $\boldsymbol{\eta}'_t$ is the transversal impermeability tensor (both in the new coordinate system). The simplified wave equation (2.195) is a simple two-dimensional eigenvalue problem with the eigenvalues $\left(1/n^2\right)_{1,2}$ and the eigenvectors $\mathbf{D}'_{t1,2} = \mathbf{D}'_{1,2}$.

The two eigenvectors of the dielectric displacement in the original coordinate system are obtained by backtransformation by $\mathbf{A}$ according to (2.190). The eigenvalues and eigenvectors are inserted into (2.187) to obtain the two superimposable solutions of the wave equation. For isotropic media the simplification $n_1^2 = n_2^2 = \varepsilon_r$ results from the definition of the impermeability tensor (2.185).

For lossless media the following holds: $\boldsymbol{\varepsilon}$ is Hermitian, so $\boldsymbol{\eta}$, $\boldsymbol{\eta}'$ and $\boldsymbol{\eta}'_t$ must also be Hermitian. Therefore the eigenvalues $\left(1/n^2\right)_{1,2}$ are real, the eigenvectors (eigenpolarizations, eigenmodes) are complex orthogonal, i.e., orthogonally polarized, and the matrix of (normalized) eigenvectors, the transformation matrix, is unitary. In the case of real, symmetric tensors $\boldsymbol{\varepsilon}$, $\boldsymbol{\eta}$, $\boldsymbol{\eta}'$, $\boldsymbol{\eta}'_t$ the eigenvectors are real orthogonal, also linearly (and mutually orthogonally) polarized, and the matrix of (normalized) eigenvectors is orthogonal.

In general media the eigenvectors need not be orthogonal and the eigenvalues need not be real. Usually the eigenvectors are normalized.

Beside the dielectric displacement we are of course interested in the electric field. The $\mathbf{s}'$ component of the dielectric displacement equals zero. If we insert in the coordinate system τ'_1, τ'_2, $\mathbf{s}'$ a dielectric displacement eigenvector into $\mathbf{E}' = \boldsymbol{\eta}'\mathbf{D}'/\varepsilon_0$ then we obtain, with (2.195)

$$\mathbf{E}'_i = \frac{1}{\varepsilon_0}\left(\frac{1}{n_i^2}\mathbf{D}'_{ti} + \mathbf{s}'\begin{bmatrix}\eta'_{31} & \eta'_{32}\end{bmatrix}\mathbf{D}'_{ti}\right) \quad (i = 1,2)\,. \tag{2.196}$$

The first term is the transversal part of $\underline{\mathbf{E}}'_i$. It is oriented in the same direction as the dielectric displacement. The second term is a component along the propagation direction $\mathbf{s}'$, the magnitude of which will be estimated in the following:

Any real symmetric dielectricity tensor can be diagonalized. For that propagation direction the $\mathbf{s}'$ electric field componente is zero. We write the dielectricity tensor in a new coordinate system,

$$\boldsymbol{\varepsilon}'_r = \mathbf{A}^T\boldsymbol{\varepsilon}_r\mathbf{A} = \varepsilon_{ii}\mathbf{A}^T(\mathbf{1}+\mathbf{K})\mathbf{A} = \varepsilon_{ii}\left(\mathbf{1}+\mathbf{K}'\right) \quad \text{with} \quad \mathbf{K}' = \mathbf{A}^T\mathbf{K}\mathbf{A}\,. \tag{2.197}$$

Here $\underline{\varepsilon}_{ii}$ is one of the diagonal elements. If all eigenvalues of the dielectricity tensor differ only little then the matrix $\mathbf{K}$ contains only element with small magnitudes, $|k_{ij}| << 1$. This holds for almost all optical media. The transformed matrix $\mathbf{K}'$ contains then also only small elements. This allows to use an approximation formula for matrix inversion,

$$\mathbf{\eta}' = \left(\mathbf{\varepsilon}_r'\right)^{-1} = \left(\varepsilon_{ii}\left(\mathbf{1}+\mathbf{K}'\right)\right)^{-1} = \frac{1}{\varepsilon_{ii}}\left(\mathbf{1}+\mathbf{K}'\right)^{-1} \approx \frac{1}{\varepsilon_{ii}}\left(\mathbf{1}-\mathbf{K}'\right). \qquad (2.198)$$

Since the diagonal elements dominate also the impermeability tensor the longitudinal electric field is small in comparison to the transversal electric field in any weakly anisotropic medium.

The wave propagation calculus in anisotropic media having real dielectricity tensors can be geometrically explained using the *index ellipsoid* (Fig. 2.16). This tool is legitimated solely by the formal identity of mathematical steps with the afore-given.

The temporal average of the electric energy density in an anisotropic medium is in general

$$\overline{w_\mathrm{e}} = \frac{1}{4}\mathrm{Re}\left(\mathbf{E}^{+}\mathbf{D}\right) = \frac{1}{4\varepsilon_0}\mathrm{Re}\left(\mathbf{D}^{+}\mathbf{\eta}^{+}\mathbf{D}\right). \qquad (2.199)$$

In the case of losslessness the tensors are Hermitian and it is not necessary to take the real part. In order to allow a geometrical interpretation let all quantities in (2.200)–(2.204) be real. We substitute $\mathbf{D}/\sqrt{\varepsilon_0 4\overline{w_e}}$ by the position vector $\mathbf{r}$ and obtain the index ellipsoid equation

$$\mathbf{r}^T\mathbf{\eta}\mathbf{r} = 1. \qquad (2.200)$$

In the coordinate system in which the impermeability tensor is also diagonal the simple expression

$$\frac{x^2}{n_x^2} + \frac{y^2}{n_y^2} + \frac{z^2}{n_z^2} = 1 \text{ with } n_i^2 = \varepsilon_i/\varepsilon_0 \;\; (i = x, y, z) \qquad (2.201)$$

holds. The half axes in x, y and z directions represent the *main refractive indices* n_x, n_y, n_z. In an isotropic medium the index ellipsoid is a sphere.

Now that intersection ellipse between the index ellipsoid and the plane $\mathbf{r}\cdot\mathbf{s} = 0$ must be determined, with respect to which the vector $\mathbf{s}$ is perpendicular. For this purpose (2.200) is orthogonally transformed, to obtain

$$\mathbf{r}'^T\mathbf{\eta}'\mathbf{r}' = 1 \quad \text{with } \mathbf{\eta}' = \mathbf{A}^T\mathbf{\eta}\mathbf{A},\ \mathbf{r} = \mathbf{A}\mathbf{r}',\ \mathbf{A} = \left[\mathbf{\tau}_1 \quad \mathbf{\tau}_2 \quad \mathbf{s}\right]. \qquad (2.202)$$

Due to $\mathbf{r}\cdot\mathbf{s} = 0$ the third component of $\mathbf{r}'$ vanishes. One may replace $\mathbf{r}'$ by its transversal part $\mathbf{r}_t'$ having two components and finds

$$\mathbf{r}_t'^T\mathbf{\eta}_t'\mathbf{r}_t' = 1. \qquad (2.203)$$

In order to obtain the normal form of the ellipse equation the coordinates must again be orthogonally transformed,

$$\mathbf{r}_t''^T \boldsymbol{\eta}''_t \mathbf{r}_t'' = 1$$

$$\text{with } \boldsymbol{\eta}''_t = \Delta^T \boldsymbol{\eta}'_t \Delta = \begin{bmatrix} n_1^{-2} & 0 \\ 0 & n_2^{-2} \end{bmatrix}, \ \mathbf{r}_t' = \Delta \mathbf{r}_t'', \ \Delta^T \Delta = \mathbf{1}, \ \Delta = \begin{bmatrix} \mathbf{D}_{1t} & \mathbf{D}_{2t} \end{bmatrix}. \quad (2.204)$$

The half axis lengths n_1, n_2 of the intersection ellipse correspond to the eigenvalues of (2.195), and their directions are those of the dielectric displacement eigenvectors $\mathbf{D}_{1,2t}$. Amplitudes and signs of the eigenvectors may be chosen at will. If in an anisotropic medium the intersection ellipse degenerates into a circle the medium is isotropic in that particular propagation direction, and there exists a twofold eigenvalue while any orthogonal pair of eigenvectors may be chosen in the intersection ellipse. We summarize the steps of a geometrical wave equation solution for real, symmetric dielectricity tensors:

- Determine the index ellipsoid using the dielectricity tensor. This is trivial since the main refractive indices are found in tables and the coordinate system may be chosen identical to the main ellipsoid axes.
- Determine the intersection ellipse between the ellipsoid and the plane with respect to which the propagation direction is perpendicular.
- The half axis directions of the intersection ellipse denote the two orthogonal eigenpolarizations.
- The half axis lengths denote the corresponding refractive indices.

Fig. 2.16 Index ellipsoid

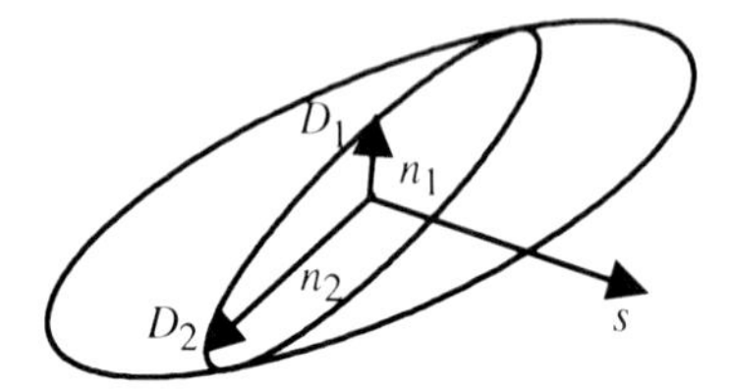

Snell's law has to be modified in anisotropic media. In particular, the refractive index in the anisotropic medium depends on the incident polarization state. The two eigenpolarizations are refracted under different angles (*birefringence*). Any anisotropic medium acts usually as a birefringent device.

Problem: Lithium Niobate ($LiNbO_3$) is a *uniaxial* crystal, i.e., two main refractive indices are identical. $n_x = n_y = n_o \approx 2.21283$ is refractive index of the ordinary, $n_z = n_e \approx 2.13739$ that of the extraordinary ray in this essentially lossless crystal. Which are the eigenpolarizations and associated refractive indices when the wave propagates under an angle ζ with respect to the *optical axis*, i.e., the z axis?

Solution: The index ellipsoid degenerates into an ellipsoid of revolution. In the main (principal) coordinate system the ellipsoid equation reads

$$\frac{x^2+y^2}{n_o^2}+\frac{z^2}{n_e^2}=1\,. \tag{2.205}$$

The two real eigenvectors $\mathbf{D}_{1,2}$ and the wave normal $\mathbf{s}$ form an orthonormal system. The plane $\mathbf{s}\cdot\mathbf{r}=0$ intersects with the ellipsoid in an ellipse. One half axis, equal to one of the eigenvectors, has a length $n_1=n_o$ since it is a uniaxial crystal. Geometrical considerations allow to calculate the other half axis length n_2,

$$x^2+y^2=n_2^2\cos^2\zeta \qquad z^2=n_2^2\sin^2\zeta\,. \tag{2.206}$$

Insertion into (2.205) results in

$$n_2=\frac{n_o n_e}{\sqrt{n_o^2\sin^2\zeta+n_e^2\cos^2\zeta}} \quad \text{with } n_o\ge n_2\ge n_e\,. \tag{2.207}$$

Let $\mathbf{s}$ lie in the x-z plane, i.e., $\mathbf{s}=\mathbf{e}_x\sin\zeta+\mathbf{e}_z\cos\zeta$. The normalized eigenvectors are therefore

$$\mathbf{D}_1=\mathbf{e}_y\,,\quad \mathbf{D}_2=-\mathbf{e}_x\cos\zeta+\mathbf{e}_z\sin\zeta\,. \tag{2.208}$$

The signs have been chosen to make the coordinate system $\mathbf{D}_1$, $\mathbf{D}_2$, $\mathbf{s}$ right-handed.

Problem: Faraday media are in the optical domain non-magnetic while they are magnetic at low frequencies. A static magnetic field in z direction generates imaginary off-diagonal elements in the dielectricity tensor of the otherwise isotropic material,

$$\boldsymbol{\varepsilon}=\varepsilon_0\begin{bmatrix}\varepsilon_r & -j\chi & 0\\ j\chi & \varepsilon_r & 0\\ 0 & 0 & \varepsilon_r\end{bmatrix}. \tag{2.209}$$

Which are the eigenvectors and refractive indices if the signal propagates in the z direction?

Solution:

$$\boldsymbol{\eta}=\left(\frac{\boldsymbol{\varepsilon}}{\varepsilon_0}\right)^{-1}=\frac{1}{\varepsilon_r\left(\varepsilon_r^2-\chi^2\right)}\begin{bmatrix}\varepsilon_r^2 & j\chi\varepsilon_r & 0\\ -j\chi\varepsilon_r & \varepsilon_r^2 & 0\\ 0 & 0 & \varepsilon_r^2-\chi^2\end{bmatrix} \tag{2.210}$$

$$\boldsymbol{\eta}_t=\frac{1}{\varepsilon_r^2-\chi^2}\begin{bmatrix}\varepsilon_r & j\chi\\ -j\chi & \varepsilon_r\end{bmatrix} \tag{2.211}$$

A coordinate system transformation is unnecessary. Solving the eigenvalue equation

$$\det\left(\mathbf{\eta}_t - n_{1,2}^{-2}\mathbf{1}\right) = 0 \tag{2.212}$$

results in

$$n_{1,2} = \sqrt{\varepsilon_r \pm \chi}\ , \quad \mathbf{D}_{1,2} = \frac{1}{\sqrt{2}}\begin{bmatrix} 1 \\ \pm j \end{bmatrix}. \tag{2.213}$$

The two eigenmodes are therefore left and right circular polarization states. The material is lossless ($\mathbf{\varepsilon}^+ = \mathbf{\varepsilon}$) and non-reciprocal ($\mathbf{\varepsilon}^T \neq \mathbf{\varepsilon}$).

2.3.3 Jones Matrices, Müller Matrices

In general media which are anisotropic in different manners with respect to phase constant β and attenuation constant α the eigenvectors need not be orthogonal. Since such cases are less frequent we have so far mostly assumed orthonormal eigenvectors and will stick to this convention. Taking the inverse of the transformation matrix (matrix of eigenvectors) is thereby simplified to taking its Hermitian conjugate.

We have written the wave equation as an eigenvalue problem (2.189), (2.195). Let the z axis of the cartesian coordinate system be the propagation direction. One has to consider the dielectric displacement or the transversal component of the electric field. Both are parallel and are perpendicular to the propagation direction. Vectors with two components are sufficient for the description. The two solutions of the wave equation are in this coordinate system, which is not specified now by an apostroph ' but by (x,y),

$$\mathbf{D}_i^{(x,y)} = \mathbf{D}_{ti}^{(x,y)} e^{j\omega(t - n_i z/c)} \quad \text{with} \quad n_i\, \omega/c = k_i, \quad i = 1,2\,. \tag{2.214}$$

Any wave $\mathbf{D}$ can be expressed as a linear combination of the two orthogonal eigenmodes. At a fixed place, e.g., $z = 0$, it may hence be described by a Jones vector

$$\mathbf{D}_{z=0}^{(1,2)} = \begin{bmatrix} A_1 \\ A_2 \end{bmatrix} e^{j\omega t}, \tag{2.215}$$

the basis of which are not x and y axes of the cartesian coordinate system (x,y) but the two eigenvectors (1,2). The complex scalars A_1, A_2 are the amplitudes of the eigenvectors of the dielectric displacement. When switching between the vector bases (1,2) and (x,y)

$$\mathbf{D}^{(x,y)} = \left[\mathbf{D}_1^{(x,y)} \quad \mathbf{D}_2^{(x,y)}\right]\mathbf{D}^{(1,2)}, \quad \mathbf{D}^{(1,2)} = \left[\mathbf{D}_1^{(x,y)} \quad \mathbf{D}_2^{(x,y)}\right]^+ \mathbf{D}^{(x,y)} \tag{2.216}$$

holds. The second equation follows from the first because the eigenvectors form an orthonormal basis[7]. When the wave propagates to the place $z = l$ only its phasor changes. In a matrix equation,

$$\mathbf{D}_{z=l}^{(1,2)} = \mathbf{J}^{(1,2)} \mathbf{D}_{z=0}^{(1,2)} \qquad \mathbf{J}^{(1,2)} = \begin{bmatrix} \lambda_1 & 0 \\ 0 & \lambda_2 \end{bmatrix} = \begin{bmatrix} e^{-jk_1 l} & 0 \\ 0 & e^{-jk_2 l} \end{bmatrix} \tag{2.217}$$

holds. The matrix $\mathbf{J}^{(1,2)}$ is a *Jones matrix* with respect to the eigenvectors (1,2). Due to the diagonal form of $\mathbf{J}^{(1,2)}$ the eigenvectors in the chosen coordinate system are identical with the basis: $\mathbf{D}_1^{(1,2)} = [1 \quad 0]^T$, $\mathbf{D}_2^{(1,2)} = [0 \quad 1]^T$. By insertion of (2.216) into (2.217) we transform into the cartesian coordinate system,

$$\begin{aligned} &\mathbf{D}_{z=l}^{(x,y)} = \mathbf{J}^{(x,y)} \mathbf{D}_{z=0}^{(x,y)} \\ &\mathbf{J}^{(x,y)} = \left[\mathbf{D}_1^{(x,y)} \quad \mathbf{D}_2^{(x,y)}\right] \begin{bmatrix} \lambda_1 & 0 \\ 0 & \lambda_2 \end{bmatrix} \left[\mathbf{D}_1^{(x,y)} \quad \mathbf{D}_2^{(x,y)}\right]^+ . \end{aligned} \tag{2.218}$$

The eigenvectors $\underline{\mathbf{D}}_{1,2}^{(x,y)}$ of the Jones matrix $\mathbf{J}^{(x,y)}$ are the cartesian basis vectors.

On the other hand, a given Jones matrix $\mathbf{J}^{(x,y)}$ can be diagonalized: The solutions of $\det\left(\mathbf{J}^{(x,y)} - \lambda \mathbf{1}\right) = 0$ are the eigenvalues $\lambda_i = e^{-jk_i l}$, $i = 1,2$. The eigenvectors $\mathbf{D}_i^{(x,y)}$ are obtained after inserting the eigenvalues into $\left(\mathbf{J}^{(x,y)} - \lambda_i \mathbf{1}\right)\mathbf{D}_i^{(x,y)} = \mathbf{0}$. The same holds in other coordinate systems.

If we don't specify the coordinate system, as in $\mathbf{J}$, we mean the cartesian. For $\mathrm{Im}(k_1) = \mathrm{Im}(k_2)$ the eigenvalue magnitudes are identical ($|\lambda_1| = |\lambda_2|$) and the (power) transmission is polarization-independent. Such a component, the anisotropy of which relates only to the phase constant β, is a *retarder* because the phase of the *slow eigenmode* is delayed with respect to the phase of the *fast eigenmode*. The *retardation* is the angle $\delta = \mathrm{Re}(k_2 - k_1) l = \arg(\lambda_1 / \lambda_2)$. For positive retardations $\mathbf{D}_1$ is the fast eigenmode, having the eigenvalue $\lambda_1 = e^{-jk_1 l}$.

For lossless retarders[8] (2.240) $\mathrm{Im}(k_1) = \mathrm{Im}(k_2) = 0$ holds and the eigenvalues are on the unit circle, $|\lambda_1| = |\lambda_2| = 1$. When diagonalized, (2.217), such a Jones matrix is apparently unitary. Since unitary matrices stay unitary under unitary

---

[7] From $\mathbf{D}_1^{(x,y)*} \cdot \mathbf{D}_2^{(x,y)} = 0 \Leftrightarrow \mathbf{D}_1^{(x,y)+} \mathbf{D}_2^{(x,y)} = 0$ and $\left|\mathbf{D}_i^{(x,y)}\right|^2 = 1$, $(i = 1,2)$ it follows $\left[\mathbf{D}_1^{(x,y)} \quad \mathbf{D}_2^{(x,y)}\right]^+ \left[\mathbf{D}_1^{(x,y)} \quad \mathbf{D}_2^{(x,y)}\right] = \mathbf{1}$; the transformation matrix is also unitary.

[8] In many cases a polarization-independent loss can be neglected.

transformations the Jones matrix of a lossless retarder is always unitary $\mathbf{JJ}^+ = \mathbf{1}$, no matter in which orthonormal coordinate system.

Anisotropy with respect to the attenuation constant α, i.e. $\mathrm{Im}(k_1) \neq \mathrm{Im}(k_2)$, $|\lambda_1| \neq |\lambda_2|$, results in polarization-dependent transmission loss. Such media act as polarizers. The polarizer extinction is the ratio of the power transfer factors of the strongly and of the weakly transmitted eigenmodes. It is usually given in dB,

$$\text{Extinction} = 10\left|\log|\lambda_1/\lambda_2|^2\right|\text{dB} = 20\log(e)|\alpha_2 - \alpha_1|l\text{ dB}. \quad (2.219)$$

The polarization state of a wave is generally altered by propagation. The output polarization state is identical to the input polarization state only if

- the incident wave is an eigenmode (eigenpolarization) or
- the eigenvalues are identical, $e^{-jk_1 l} = e^{-jk_2 l}$. The latter is not only the case for $k_1 = k_2$ – the medium is isotropic in the chosen propagation direction and the eigenvectors are any orthonormal basis –, but also for $(k_1 - k_2)l = m \cdot 2\pi$ in a retarder – the phase retardation between the two eigenmodes is an integer multiple (m) of 2π.

Jones vectors and matrices allow to calculate the propagation of electromagnetic waves with known mode profile. Amplitude and phase informations are correctly given.

It is common to use transversal electrical field vectors $\mathbf{E}$ instead of dielectric displacement vectors $\mathbf{D}$. This is allowed in many cases, for example always when the media at input and output are identical. In those cases the same proportionalities hold between $\mathbf{D}$ and the transversal $\mathbf{E}$ at input and output. We adopt this convention from here on but one should be aware of possible errors.

Later we will show that the polarization transformation specified by a retarder is explained quite simply on the Poincaré sphere, see Fig. 2.15b:

- Find the point P which represents the retarder input polarization.
- Determine the gyration vector $\mathbf{V}$ which points from the coordinate origin to that point Ω which represents the fast retarder eigenmode.
- Rotate P clockwise about the vector $\mathbf{V}$ by the absolute value $|\delta|$ of the retardation. ($\mathbf{V}$ and the rotation direction form a right hand screw. The sign of δ is already taken into account by the choice of the fast eigenmode for $\mathbf{V}$.) The resulting point P' represents the retarder output polarization.

The rotation symmetry of the Poincaré sphere suggests that there are no states-of-polarization with special meaning in linear optics. Horizontal and vertical polarizations are not more important than any other pair of orthogonal basis vectors.

Next the power transfer characteristic of an ideal polarizer is determined. We want it to be passive, choose $\alpha_1 = 0$, $|\lambda_1| = 1$, and enforce complete extinction by $\alpha_2 \to \infty$, $\lambda_2 = 0$. The Jones matrix in the coordinate system of the eigenvectors is

$$\mathbf{J}^{(1,2)} = \begin{bmatrix} e^{-j\beta_1 l} & 0 \\ 0 & 0 \end{bmatrix}. \tag{2.220}$$

Let the two orthonormal electric field eigenvectors of the polarizer be $\mathbf{E}_{1,2}$ in the (x,y) coordinate system. The media before and behind the polarizer be identical, and $\mathbf{E}_{ein}$ be a Jones vector of the signal before the polarizer. We find

$$\begin{aligned} \mathbf{J}_{pol} &= e^{-j\beta_1 l}[\mathbf{E}_1 \quad \mathbf{E}_2]\begin{bmatrix} 1 & 0 \\ 0 & 0 \end{bmatrix}[\mathbf{E}_1 \quad \mathbf{E}_2]^+ = e^{-j\beta_1 l}\mathbf{E}_1\mathbf{E}_1^+ \\ \mathbf{E}_{out} &= \mathbf{J}_{pol}\mathbf{E}_{in} = e^{-j\beta_1 l}\mathbf{E}_1\mathbf{E}_1^+\mathbf{E}_{in} = e^{-j\beta_1 l}\mathbf{E}_1\left(\mathbf{E}_1^* \cdot \mathbf{E}_{in}\right) \end{aligned}. \tag{2.221}$$

The index 1 of the transmitted polarizer eigenmode is in the following replaced by *pol*. The output signal has the Jones vector $\mathbf{E}_{pol}$ with the amplitude multiplicator $e^{-j\beta_{pol} l}\left(\mathbf{E}_{pol}^* \cdot \mathbf{E}_{in}\right)$. A comparison with the definitions (2.166) shows that the Stokes vector $\mathbf{E}_{out}$ of the output signal is equal to that of the transmitted eigenmode $\mathbf{E}_{pol}$, multiplied by the squared magnitude of this multiplicator,

$$\mathbf{S}_{out} = \mathbf{S}_{pol}\left|\mathbf{E}_{pol}^* \cdot \mathbf{E}_{in}\right|^2. \tag{2.222}$$

With

$$\begin{aligned} &\left|\mathbf{E}_{pol}^* \cdot \mathbf{E}_{in}\right|^2 = \left|E_{x,pol}^* E_{x,in} + E_{y,pol}^* E_{y,in}\right|^2 \\ &= \left|E_{x,pol}\right|^2\left|E_{x,in}\right|^2 + \left|E_{y,pol}\right|^2\left|E_{y,in}\right|^2 + 2\,\mathrm{Re}\left(E_{x,pol}^* E_{x,in} E_{y,pol} E_{y,in}^*\right) \\ &= \frac{1}{2}\left(S_{0,pol} + S_{1,pol}\right)\frac{1}{2}\left(S_{0,in} + S_{1,in}\right) + \frac{1}{2}\left(S_{0,pol} - S_{1,pol}\right)\frac{1}{2}\left(S_{0,in} - S_{1,in}\right) \\ &+ \frac{1}{2}\mathrm{Re}\left(\left(S_{2,in} + jS_{3,in}\right)\left(S_{2,pol} - jS_{3,pol}\right)\right) \\ &= \frac{1}{2}S_{0,pol}S_{0,in} + \frac{1}{2}S_{1,pol}S_{1,in} + \frac{1}{2}S_{2,in}S_{2,pol} + \frac{1}{2}S_{3,in}S_{3,pol} \\ &= \frac{1}{2}\mathbf{S}_{in} \cdot \mathbf{S}_{pol} \end{aligned} \tag{2.223}$$

it follows

$$\mathbf{S}_{out} = \mathbf{S}_{pol}\left(\frac{1}{2}\mathbf{S}_{in} \cdot \mathbf{S}_{pol}\right) = \frac{1}{2}\mathbf{S}_{pol}\mathbf{S}_{pol}^T\mathbf{S}_{in}. \tag{2.224}$$

The power behind the polarizer is proportional to $\left|\mathbf{E}_{out}\right|^2$. Maximum power is transmitted if the input polarization is identical to the not attenuated polarizer eigenmode

$$\mathbf{E}_{pol} = h\mathbf{E}_{in} \qquad h = \text{const.}. \tag{2.225}$$

The transmitted power divided by its maximum possible value is called the *intensity*

$$I = \frac{P_{out}}{P_{out,\max}} = \frac{\left|\mathbf{E}_{out}\right|^2}{\left|\mathbf{E}_{out}\right|^2_{\max}} = \frac{\left|\mathbf{E}_{pol}^* \cdot \mathbf{E}_{in}\right|^2}{\left|\mathbf{E}_{pol}\right|^2 \left|\mathbf{E}_{in}\right|^2} = \frac{\frac{1}{2}\mathbf{S}_{in} \cdot \mathbf{S}_{pol}}{S_{0,pol} S_{0,in}}. \tag{2.226}$$

It holds $I = 0$ if the input polarization is orthogonal to the transmitted polarizer eigenmode. With our assumption of normalized eigenvectors there is $\left|\mathbf{E}_{pol}\right|^2 = 1$, $S_{0,pol} = 1$. For normalized Stokes vectors we obtain

$$I = \frac{1}{2}\left(1 + \mathbf{S}_{in} \cdot \mathbf{S}_{pol}\right). \tag{2.227}$$

For orthogonal polarizations, $I = 0$, one obtains the equivalent conditions (2.177) to (2.180).

The angle φ between two points on the Poincaré sphere can be expressed by the scalar product of the *normalized* Stokes vectors,

$$\mathbf{S}_{in} \cdot \mathbf{S}_{pol} = \cos\varphi. \tag{2.228}$$

Another expression for the intensity is therefore

$$I = \frac{1}{2}(1 + \cos\varphi) = \cos^2(\varphi/2). \tag{2.229}$$

When only a polarization transfer is of interest it is disadvantageous that the Jones vector has 4 degrees-of-freedom. The calculation effort of the Jones calculus is unnecessarily high if the phase is not of importance.

Depolarization must be understood as a temporal, or a spectral or spatial averaging process because electromagnetic waves, which are transversal, are in principle always fully polarized ($p = 1$). Example: A linearly polarized wave with an azimuth angle that varies slowly with the angular frequency Ω possess the complex time-domain function $\left(\cos(\Omega t)\mathbf{e}_x + \sin(\Omega t)\mathbf{e}_y\right)e^{j\omega t}$ and the Jones vector $\left[\cos\Omega t \quad \sin\Omega t\right]^T$. If the wave is averaged over one period of the slow vibration one obtains zero as if no wave were there. Obviously Jones vectors and matrices can not take the degree-of-polarization (DOP) into account.

To eliminate phase and introduce the DOP one can calculate the Stokes vector $\mathbf{S}_{out} \equiv \mathbf{S}_{z=l}$ behind a component or medium from the Stokes vector $\mathbf{S}_{in} \equiv \mathbf{S}_{z=0}$ at its input, using the *Müller matrix* **M**, a real 4×4 matrix:

$$\mathbf{S}_{out} = \mathbf{M}\mathbf{S}_{in} \tag{2.230}$$

This is analogous to the Jones matrix equations $\mathbf{D}_{z=l}^{(x,y)} = \mathbf{J}^{(x,y)}\mathbf{D}_{z=0}^{(x,y)}$ or $\mathbf{E}_{z=l}^{(x,y)} = \mathbf{J}^{(x,y)}\mathbf{E}_{z=0}^{(x,y)}$ or $\mathbf{E}_{out} = \mathbf{J}\mathbf{E}_{in}$.

Which is the Müller matrix of an optical component with the Jones matrix $\mathbf{J} = \mathbf{J}^{(x,y)} = \begin{bmatrix} J_{11} & J_{12} \\ J_{21} & J_{22} \end{bmatrix}$? Using the output field quantities

$$E_{x,out} = J_{11}E_{x,in} + J_{12}E_{y,in}, \qquad E_{y,out} = J_{21}E_{x,in} + J_{22}E_{y,in} \tag{2.231}$$

and (2.166) the output Stokes parameters $S_{i,out}$ ($i = 0...3$) may be calculated as a function of $E_{x,in}, E_{y,in}, J_{ij}$. After suitable separation the terms with $E_{x,in}, E_{y,in}$ are replaced by the input Stokes parameters.

One obtains the Müller matrix elements m_{ij}, where the indices $i, j = 0...3$ (instead of the usual range 1...4) correspond to those of the Stokes parameters. In the following, means of products have been replaced by products of means, e.g.,

$$\overline{\mathrm{Re}\left(J_{11}J_{21}^{*} - J_{12}J_{22}^{*}\right)\left(\left|E_{x,ein}\right|^{2} - \left|E_{y,ein}\right|^{2}\right)} = m_{21}S_{1,ein}. \tag{2.232}$$

A sufficient condition is that variations of both factors, if existent, are uncorrelated! This is often the case, for which reason Müller matrices are very useful. The result can be written as

$$\mathbf{M} = \overline{\mathbf{M}_J}$$

$$\mathbf{M}_J = \mathbf{A}^{*}\left(\mathbf{J} \otimes \mathbf{J}^{*}\right)\mathbf{A}^{*-1} = \frac{1}{2}\mathbf{A}^{*}\left(\mathbf{J} \otimes \mathbf{J}^{*}\right)\mathbf{A}^{*+}$$

$$= \frac{1}{2}\begin{bmatrix} \begin{array}{c} |J_{11}|^2 + |J_{12}|^2 \\ + |J_{21}|^2 + |J_{22}|^2 \end{array} & \begin{array}{c} |J_{11}|^2 - |J_{12}|^2 \\ + |J_{21}|^2 - |J_{22}|^2 \end{array} & 2\,\mathrm{Re}\begin{pmatrix} J_{11}J_{12}^{*} \\ + J_{21}J_{22}^{*} \end{pmatrix} & 2\,\mathrm{Im}\begin{pmatrix} - J_{11}J_{12}^{*} \\ - J_{21}J_{22}^{*} \end{pmatrix} \\ \begin{array}{c} |J_{11}|^2 + |J_{12}|^2 \\ - |J_{21}|^2 - |J_{22}|^2 \end{array} & \begin{array}{c} |J_{11}|^2 - |J_{12}|^2 \\ - |J_{21}|^2 + |J_{22}|^2 \end{array} & 2\,\mathrm{Re}\begin{pmatrix} J_{11}J_{12}^{*} \\ - J_{21}J_{22}^{*} \end{pmatrix} & 2\,\mathrm{Im}\begin{pmatrix} - J_{11}J_{12}^{*} \\ + J_{21}J_{22}^{*} \end{pmatrix} \\ 2\,\mathrm{Re}\begin{pmatrix} J_{11}J_{21}^{*} \\ + J_{12}J_{22}^{*} \end{pmatrix} & 2\,\mathrm{Re}\begin{pmatrix} J_{11}J_{21}^{*} \\ - J_{12}J_{22}^{*} \end{pmatrix} & 2\,\mathrm{Re}\begin{pmatrix} J_{11}J_{22}^{*} \\ + J_{12}J_{21}^{*} \end{pmatrix} & 2\,\mathrm{Im}\begin{pmatrix} - J_{11}J_{22}^{*} \\ + J_{12}J_{21}^{*} \end{pmatrix} \\ 2\,\mathrm{Im}\begin{pmatrix} J_{11}J_{21}^{*} \\ + J_{12}J_{22}^{*} \end{pmatrix} & 2\,\mathrm{Im}\begin{pmatrix} J_{11}J_{21}^{*} \\ - J_{12}J_{22}^{*} \end{pmatrix} & 2\,\mathrm{Im}\begin{pmatrix} J_{11}J_{22}^{*} \\ + J_{12}J_{21}^{*} \end{pmatrix} & 2\,\mathrm{Re}\begin{pmatrix} J_{11}J_{22}^{*} \\ - J_{12}J_{21}^{*} \end{pmatrix} \end{bmatrix} \tag{2.233}$$

with

$$\mathbf{A}=\begin{bmatrix}1&0&0&1\\1&0&0&-1\\0&1&1&0\\0&j&-j&0\end{bmatrix},\quad \mathbf{A}^{-1}=\frac{1}{2}\mathbf{A}^{+}=\frac{1}{2}\begin{bmatrix}1&1&0&0\\0&0&1&-j\\0&0&1&j\\1&-1&0&0\end{bmatrix}. \tag{2.234}$$

The Kronecker product $\otimes$ is defined as the multiplication of each element of the first matrix by the complete second matrix, thereby multiplying matrix sizes,

$$\begin{aligned}&\mathbf{Z}=\mathbf{X}\otimes\mathbf{Y}\\&\mathbf{X}=\left[x_{i,j}\right]\quad i=1...m,\quad j=1...n\\&\mathbf{Y}=\left[y_{k,l}\right]\quad k=1...p,\quad l=1...q\\&\mathbf{Z}=\begin{bmatrix}x_{1,1}\mathbf{Y}&x_{1,2}\mathbf{Y}&...&x_{1,n}\mathbf{Y}\\x_{2,1}\mathbf{Y}&x_{2,2}\mathbf{Y}&...&x_{2,n}\mathbf{Y}\\...&...&...&...\\x_{m,1}\mathbf{Y}&x_{m,2}\mathbf{Y}&...&x_{m,n}\mathbf{Y}\end{bmatrix}=\left[z_{a,b}\right]\quad a=1...mp,\quad b=1...nq\\&z_{a,b}=x_{i,j}y_{k,l}\quad a=(i-1)p+k,\quad b=(j-1)q+l\end{aligned}\;. \tag{2.235}$$

Müller-Jones matrices $\mathbf{M}_J$ are Müller matrices $\mathbf{M}=\mathbf{M}_J$ where the averaging operator in (2.233) has no effect. Only for Müller-Jones matrices a constant Jones matrix $\mathbf{J}$ exists. $\mathbf{J}$ has 8 degrees-of-freedom. $\mathbf{M}_J$ has only 7 because the absolute phase is lost.

Problem: Calculate $\mathbf{J}$ from $\mathbf{M}_J$ with exception of this phase.

Solution: Linear combinations of all m_{ik} with $i,k\in\{0,1\}$, or elements of matrix

$$2\mathbf{A}^{*+}\mathbf{M}_J\mathbf{A}^{*}=\left(\mathbf{J}\otimes\mathbf{J}^{*}\right)=\left[\begin{array}{cc|cc}J_{11}J_{11}^{*}&J_{11}J_{12}^{*}&J_{12}J_{11}^{*}&J_{12}J_{12}^{*}\\J_{11}J_{21}^{*}&J_{11}J_{22}^{*}&J_{12}J_{21}^{*}&J_{12}J_{22}^{*}\\\hline J_{21}J_{11}^{*}&J_{21}J_{12}^{*}&J_{22}J_{11}^{*}&J_{22}J_{12}^{*}\\J_{21}J_{21}^{*}&J_{21}J_{22}^{*}&J_{22}J_{21}^{*}&J_{22}J_{22}^{*}\end{array}\right], \tag{2.236}$$

yield all $|J_{lm}|^2$ and hence $|J_{lm}|$ with $l,m\in\{1,2\}$. At least one of them is non-zero, for instance $|J_{11}|^2=\frac{1}{2}(m_{00}+m_{01}+m_{10}+m_{11})$, which we use to define $J_{11}=|J_{11}|e^{j\varphi}$ with arbitrary phase φ. Subsequently the other elements $J_{12}=\frac{\left(J_{12}J_{11}^{*}\right)J_{11}}{|J_{11}|^2}$ etc. are found. The needed quantity $J_{12}J_{11}^{*}=\frac{1}{2}(m_{02}+m_{12}+jm_{03}+jm_{13})$ is also contained in $\mathbf{J}\otimes\mathbf{J}^{*}$.

Note that if the sign of S_3 is defined oppositely (see p. 43), $\mathbf{A}^*$ is replaced by $\mathbf{A}$ in (2.233), (2.236). As a result the last line and the last column of $\mathbf{M}$ change signs, like in (2.252).

All depolarizing elements (i.e, which decrease the DOP) possess no constant Jones Matrix, but a variable one, for example time-variable. A particular example is a

Lossless ideal depolarizer

$$\text{A constant } \mathbf{J} \text{ does not exist, } \mathbf{M} = \begin{bmatrix} 1 & 0 & 0 & 0 \\ 0 & 0 & 0 & 0 \\ 0 & 0 & 0 & 0 \\ 0 & 0 & 0 & 0 \end{bmatrix} \tag{2.237}$$

Its output DOP equals zero, no matter what the input polarization is. In the following we characterize more optical components.

Lossless retarder

$$\mathbf{J}\mathbf{J}^+ = \mathbf{1}, \ \mathbf{M} = \left[\begin{array}{c|ccc} 1 & 0 & 0 & 0 \\ \hline 0 & & & \\ 0 & & \mathbf{G} & \\ 0 & & & \end{array}\right], \qquad \mathbf{G}\mathbf{G}^T = \mathbf{1}, \tag{2.238}$$

holds, where the 3×3 submatrix $\mathbf{G}$ contains the elements m_{ij} $(i, j = 1...3)$ from (2.233). Let the retardation be δ, and the eigenmodes be transmitted with the eigenvalues $e^{\pm j\delta/2}$. The upper exponent sign corresponds to an eigenmode (for $0 < \delta < \pi$: the fast eigenmode) with the normalized Jones vector $\mathbf{E}_1 = \frac{1}{\sqrt{2(1+V_1)}} \begin{bmatrix} 1+V_1 \\ V_2 - jV_3 \end{bmatrix}$ and the normalized Stokes vector $\mathbf{V} = \begin{bmatrix} V_1 & V_2 & V_3 \end{bmatrix}^T$ ($V_1^2 + V_2^2 + V_3^2 = 1$). The Jones matrix

$$\mathbf{J} = \begin{bmatrix} \cos\delta/2 + jV_1 \sin\delta/2 & j(V_2 + jV_3)\sin\delta/2 \\ j(V_2 - jV_3)\sin\delta/2 & \cos\delta/2 - jV_1 \sin\delta/2 \end{bmatrix} \tag{2.239}$$

is obtained from (2.218), the Müller matrix $\mathbf{M}$ is given by (2.238) with

$$\mathbf{G} = \left[\begin{array}{c|c|c} V_1^2 + \left(V_2^2 + V_3^2\right)\cos\delta & \begin{array}{c} V_1V_2(1-\cos\delta) \\ -V_3\sin\delta \end{array} & \begin{array}{c} V_1V_3(1-\cos\delta) \\ +V_2\sin\delta \end{array} \\ \hline \begin{array}{c} V_1V_2(1-\cos\delta) \\ +V_3\sin\delta \end{array} & V_2^2 + \left(V_1^2 + V_3^2\right)\cos\delta & \begin{array}{c} V_2V_3(1-\cos\delta) \\ -V_1\sin\delta \end{array} \\ \hline \begin{array}{c} V_1V_3(1-\cos\delta) \\ -V_2\sin\delta \end{array} & \begin{array}{c} V_2V_3(1-\cos\delta) \\ +V_1\sin\delta \end{array} & V_3^2 + \left(V_1^2 + V_2^2\right)\cos\delta \end{array}\right],$$

$$\det \mathbf{G} = 1 \,. \tag{2.240}$$

This is a rotation matrix in the three-dimensional space of the normalized Stokes parameters, which rotates a point about the rotation axis **V** by an angle δ, as has been described on p. 55. Particular retarders are:

Linear retarder or waveplate

$\mathbf{E}_1 = [\cos\alpha \quad \sin\alpha]^T$, $\mathbf{V} = [\cos 2\alpha \quad \sin 2\alpha \quad 0]^T$ (linear polarization having an azimuth angle α with respect to the x axis), e.g., *quarterwave plate* ($\delta = \pi/2$) or *halfwave plate* ($\delta = \pi$),

$$\mathbf{J} = \begin{bmatrix} \cos\delta/2 + j\cos 2\alpha \sin\delta/2 & j\sin 2\alpha \sin\delta/2 \\ j\sin 2\alpha \sin\delta/2 & \cos\delta/2 - j\cos 2\alpha \sin\delta/2 \end{bmatrix}, \tag{2.241}$$

$$\mathbf{G} = \begin{bmatrix} \left((1+\cos 4\alpha) + (1-\cos 4\alpha)\cos\delta\right)/2 & \sin 4\alpha (1-\cos\delta)/2 & \sin 2\alpha \sin\delta \\ \sin 4\alpha (1-\cos\delta)/2 & \left((1-\cos 4\alpha) + (1+\cos 4\alpha)\cos\delta\right)/2 & -\cos 2\alpha \sin\delta \\ -\sin 2\alpha \sin\delta & \cos 2\alpha \sin\delta & \cos\delta \end{bmatrix}. \tag{2.242}$$

(TE-TM) phase shifter: $\alpha = 0$,

$$\mathbf{J} = \begin{bmatrix} e^{j\delta/2} & 0 \\ 0 & e^{-j\delta/2} \end{bmatrix}, \qquad \mathbf{G} = \begin{bmatrix} 1 & 0 & 0 \\ 0 & \cos\delta & -\sin\delta \\ 0 & \sin\delta & \cos\delta \end{bmatrix}. \tag{2.243}$$

(TE-TM) mode converter: $\alpha = \pi/4$,

$$\mathbf{J} = \begin{bmatrix} \cos\delta/2 & j\sin\delta/2 \\ j\sin\delta/2 & \cos\delta/2 \end{bmatrix}, \quad \mathbf{G} = \begin{bmatrix} \cos\delta & 0 & \sin\delta \\ 0 & 1 & 0 \\ -\sin\delta & 0 & \cos\delta \end{bmatrix}. \tag{2.244}$$

Circular retarder: $\mathbf{E}_1 = \left[j/\sqrt{2} \quad 1/\sqrt{2}\right]^T$, $\mathbf{V} = [0 \quad 0 \quad 1]^T$ (right circular),

$$\mathbf{J} = \begin{bmatrix} \cos\delta/2 & -\sin\delta/2 \\ \sin\delta/2 & \cos\delta/2 \end{bmatrix}, \quad \mathbf{G} = \begin{bmatrix} \cos\delta & -\sin\delta & 0 \\ \sin\delta & \cos\delta & 0 \\ 0 & 0 & 1 \end{bmatrix}. \tag{2.245}$$

Ideal polarizer

The eigenmodes are transmitted with the eigenvalues 1 and 0. The not attenuated eigenmode have the normalized Jones vector $\mathbf{E}_{pol}$ and the Stokes vector $\mathbf{S}_{pol} = [1 \quad V_1 \quad V_2 \quad V_3]^T$ ($V_1^2 + V_2^2 + V_3^2 = 1$). From (2.221), (2.224) it follows

$$\mathbf{J} = \mathbf{E}_{pol}\mathbf{E}_{pol}^+, \qquad \mathbf{M} = \frac{1}{2}\mathbf{S}_{pol}\mathbf{S}_{pol}^T. \tag{2.246}$$

Ideal linear polarizer

$\mathbf{E}_{pol} = [\cos\alpha \quad \sin\alpha]^T$, $\mathbf{S}_{pol} = [1 \quad \cos 2\alpha \quad \sin 2\alpha \quad 0]^T$ (linear polarization having an azimuth angle α with respect to the x axis),

$$\mathbf{J} = \begin{bmatrix} \cos^2\alpha & \sin\alpha\cos\alpha \\ \sin\alpha\cos\alpha & \sin^2\alpha \end{bmatrix}, \tag{2.247}$$

$$\mathbf{M} = \frac{1}{2}\begin{bmatrix} 1 & \cos 2\alpha & \sin 2\alpha & 0 \\ \cos 2\alpha & \cos^2 2\alpha & \cos 2\alpha \sin 2\alpha & 0 \\ \sin 2\alpha & \cos 2\alpha \sin 2\alpha & \sin^2 2\alpha & 0 \\ 0 & 0 & 0 & 0 \end{bmatrix}. \tag{2.248}$$

Partial 0°/90° (TE/TM) polarizer with retardation

$$\mathbf{J} = \begin{bmatrix} e^{(\gamma + j\delta)/2} & 0 \\ 0 & e^{-(\gamma + j\delta)/2} \end{bmatrix}, \tag{2.249}$$

$$\mathbf{M} = \begin{bmatrix} \cosh\gamma & \sinh\gamma & 0 & 0 \\ \sinh\gamma & \cosh\gamma & 0 & 0 \\ 0 & 0 & \cos\delta & -\sin\delta \\ 0 & 0 & \sin\delta & \cos\delta \end{bmatrix}. \tag{2.250}$$

Note that in this definition the better-transmitted eigenmode (horizontal for $\gamma > 0$) experiences a gain. Even though anisotropy extends to both amplitude and phase the eigenmodes of this particular component are orthogonal. If the eigenmodes are not horizontal/vertical then given matrices **J** and **M** must be replaced by $\mathbf{KJK}^+$ and $\mathbf{NMN}^+$. Here **K** and **N** are Jones and Müller matrices of retarders as defined in (2.239) and (2.238) with (2.240), respectively, which transform horizontal/vertical polarizations into the desired eigenmodes. Therefore the first column vectors of **K** and the rotation part **G** of **N** contain the Jones and normalized Stokes vectors, respectively, of that eigenmode into which horizontal polarization is mapped.

Eqns. (2.233), (2.239), (2.240), (2.246) allow to verify the equivalence of substitutions,

$$\mathbf{J} \to \mathbf{J}^+ \quad \Leftrightarrow \quad \mathbf{M} \to \mathbf{M}^T \quad \Leftrightarrow \quad \delta \to -\delta \text{ or } \mathbf{E}_{pol} \to \mathbf{E}_{pol}, \tag{2.251}$$

$$\mathbf{J} \to \mathbf{J}^T \quad \Leftrightarrow \quad \mathbf{M} \to \begin{bmatrix} m_{00} & m_{10} & m_{20} & -m_{30} \\ m_{01} & m_{11} & m_{21} & -m_{31} \\ m_{02} & m_{12} & m_{22} & -m_{32} \\ -m_{03} & -m_{13} & -m_{23} & m_{33} \end{bmatrix}, \tag{2.252}$$

$$\Leftrightarrow \quad V_3 \to -V_3 \text{ or } \mathbf{E}_{pol} \to \mathbf{E}_{pol}^*$$

where the last substitution before the "or" applies for retarders and the one behind for ideal polarizers. In (2.252), **M** is transposed and the sign is inverted for all elements where the index 3 occurs an odd number of times. If **M** were only transposed instead then the signs of S_3 at input and output would have to be inverted.

The Müller matrix can be measured as follows: At the input of an optical element we apply sequentially $n \geq 4$ polarization states which span a non-zero volume in the Poincaré sphere. For best accuracy it is recommended to maximize this volume. The 4 polarization states at the corners of a tetrahedron fulfill this condition. We form a matrix

$$\mathbf{S}_{in} = \lfloor \mathbf{S}_{in,1} \quad \mathbf{S}_{in,2} \quad \ldots \quad \mathbf{S}_{in,n} \rfloor \qquad (k = 1 \ldots n) \tag{2.253}$$

with 4 lines and n columns. The columns $\mathbf{S}_{in,k}$ are the full input Stokes vectors with 4 components.

The output Stokes vectors measured for the various input Stokes vectors can be arranged in a similar $4 \times n$ matrix

$$\mathbf{S}_{out} = \mathbf{M}\mathbf{S}_{in} \, . \tag{2.254}$$

We multiply from the right side with the transpose (or Hermitian conjugate) of $\mathbf{S}_{in}$, then with the inverse of $\mathbf{S}_{in}\mathbf{S}_{in}^{+}$ to obtain the Müller matrix

$$\mathbf{M} = \mathbf{S}_{out}\mathbf{S}_{in}^{+}\left(\mathbf{S}_{in}\mathbf{S}_{in}^{+}\right)^{-1} . \tag{2.255}$$

Sometimes it is known that an element is non-depolarizing, while its measured Müller matrix **M** is to some degree depolarizing, due to measurement errors. It is possible to purge this Müller matrix, i.e., to obtain from it a non-depolarizing Müller-Jones matrix $\mathbf{M}_J$, using the following calculus [4, 5]. We define a Hermitian operator

$$\mathbf{K} = \mathbf{H}(\mathbf{F}) = \frac{1}{2}\sum_{i=0}^{3}\sum_{j=0}^{3}\left(f_{ij}\left(\boldsymbol{\sigma}_i^{*} \otimes \boldsymbol{\sigma}_j\right)\right) \tag{2.256}$$

on the elements f_{ij} of a matrix **F**, see (2.167), (2.235). The result is a Hermitian matrix **K**. The matrix elements f_{ij} and hence the matrix **F** can be obtained from **K** by the inverse operation

$$f_{ij} = h_{ij}^{-1}(\mathbf{K}) = \frac{1}{2}\sum\sum\left(\mathbf{K} \circ \left(\boldsymbol{\sigma}_i \otimes \boldsymbol{\sigma}_j^{*}\right)\right). \tag{2.257}$$

Here $\circ$ defines an element-wise matrix multiplication: $\mathbf{Z} = \mathbf{X} \circ \mathbf{Y}$ means $z_{kl} = x_{kl} y_{kl}$. The double summation $\sum\sum\mathbf{Z} = \sum_{k=0}^{3}\sum_{l=0}^{3} z_{kl}$ adds the elements of a matrix $\mathbf{Z} = \left[z_{kl}\right]$ over all rows and columns.

For the opposite sign of S_3 (see also pp. 43, 59) we can exchange the terms $\left(\boldsymbol{\sigma}_i^* \otimes \boldsymbol{\sigma}_j\right)$, $\left(\boldsymbol{\sigma}_i \otimes \boldsymbol{\sigma}_j^*\right)$ against each other in (2.256), (2.257). As long as we don't care what **K** actually is this exchange is not even necessary.

Our operator is now applied to the elements of the Müller matrix **M**,

$$\mathbf{T} = \mathbf{H}(\mathbf{M}). \tag{2.258}$$

Being Hermitian, matrix **T** can be diagonalized

$$\mathbf{T} = \mathbf{B}\boldsymbol{\Lambda}\mathbf{B}^+ \tag{2.259}$$

with a real diagonal eigenvalue matrix Λ and a unitary eigenvector matrix **B**. A non-depolarizing **M** corresponds to a matrix **T** with only one, positive non-zero eigenvalue λ_{max}, which is $\lambda_{max} = 2$ if there is no average gain or loss. So, for the experimentally obtained matrix we can define a new eigenvalue matrix Λ_J where the largest positive eigenvalue λ_{max} is kept (or set to $\lambda_{max} = 2$) while the three other eigenvalues are set equal to zero. The resulting new matrix can be written as

$$\mathbf{T}_J = \mathbf{B}\boldsymbol{\Lambda}_J\mathbf{B}^+ = \lambda_{max}\mathbf{B}_{max}\mathbf{B}_{max}^+ \tag{2.260}$$

where in the rightmost, simpler expression $\mathbf{B}_{max}$ is the eigenvector corresponding to λ_{max}. Finally, with (2.257) we calculate the elements $(m_J)_{ij}$ of the sought non-depolarizing Müller-Jones matrix $\mathbf{M}_J$

$$(m_J)_{ij} = h_{ij}^{-1}(\mathbf{T}_J). \tag{2.261}$$

The 3×3 Müller submatrix $\mathbf{G}_m$ of a retarder may be determined very similar to the Müller matrix: At the input of the retarder we apply sequentially $n \geq 3$ polarization states. The 3 cartesian unit vectors are a good choice. We form a matrix $\mathbf{S}_{in}$ of type (2.253) but with 3 lines and n columns because now its column vectors $\mathbf{S}_{in,k}$ are normalized, 3-component Stokes vectors. The output Stokes vectors measured for the various $\mathbf{S}_{in,k}$ are arranged in a 3×n matrix

$$\mathbf{S}_{out} = \mathbf{G}_m\mathbf{S}_{in}. \tag{2.262}$$

The retarder matrix itself is

$$\mathbf{G}_m = \mathbf{S}_{out}\mathbf{S}_{in}^+\left(\mathbf{S}_{in}\mathbf{S}_{in}^+\right)^{-1}. \tag{2.263}$$

The measured matrix $\mathbf{G}_m$ may contain errors and therefore does not describe a pure retarder rotation. But it can be purged. An orthogonal matrix has singular values equal to 1. So, we decompose the experimental $\mathbf{G}_m$ for its singular values,

$$\mathbf{G}_m = \mathbf{U}\mathbf{S}\mathbf{V}^+. \tag{2.264}$$

The unitary matrices **U**, **V** are even real orthogonal here because $\mathbf{G}_m$ is real. **S** is a diagonal matrix containing the nonnegative singular values. The purged, truly orthogonal rotation matrix **G** is obtained as

$$\mathbf{G} = \mathbf{U}\mathbf{V}^{+}. \tag{2.265}$$

Next we now consider cascaded optical components. Let the light traverse component 1 and then component 2. Overall Jones and Müller matrices (written without index) are the products of the individual matrices in reverse order of light propagation:

$$\begin{aligned} \mathbf{E}_{2,out} &= \mathbf{J}_2\mathbf{E}_{1,out} = \mathbf{J}_2\mathbf{J}_1\mathbf{E}_{1,in} \qquad \mathbf{J} = \mathbf{J}_2\mathbf{J}_1 \\ \mathbf{S}_{2,out} &= \mathbf{M}_2\mathbf{S}_{1,out} = \mathbf{M}_2\mathbf{M}_1\mathbf{S}_{1,in} \qquad \mathbf{M} = \mathbf{M}_2\mathbf{M}_1 \end{aligned} \tag{2.266}$$

These equations can easily be generalized for more components.

How does the signal polarization change if several signals are added? We distinguish 2 cases:

- *Signals with equal frequencies: Addition of Jones vectors;* the common phasor with time and position dependence may be omitted.
- *Signals with different frequencies: Addition of Jones vector while taking space and position dependence into account.* When this sum is inserted into (2.166) the signals with different frequencies are uncorrelated if the observation time is large against the inverse of the frequency difference. The cross terms vanish and it remains, as an alternative, the *sum of the (not normalized) Stokes vectors.*

The Müller calculus can be replaced by the manipulation of coherency matrices $\overline{\mathbf{E}\mathbf{E}^{+}}$. For $\mathbf{E}_{out} = \mathbf{J}\mathbf{E}_{in}$ it holds

$$\overline{\mathbf{E}_{out}\mathbf{E}_{out}^{+}} = \mathbf{J}\overline{\mathbf{E}_{in}\mathbf{E}_{in}^{+}}\mathbf{J}^{+}. \tag{2.267}$$

In spite of the extra effort needed to calculate Müller matrices we prefer Stokes vectors over coherency matrices, because normalized Stokes vectors allow geometric interpretations.

2.3.4 Monochromatic Polarization Transmission

We investigate polarization transmission issues at a given optical frequency, namely preservation of polarization orthogonality, polarization-dependent loss and depolarization. Frequency-dependent retardation and eigenmodes, called polarization mode dispersion, will be covered from p. 72 on.

Let's look into more properties of Jones matrices. The Jones matrix can that of a single optical component as well as that of a complete transmission link.

Two orthogonally polarized optical signals with Jones vectors $\mathbf{E}_{1,in}$, $\mathbf{E}_{2,in}$ are fed into an optical component with Jones matrix **J**. Under which condition are the two output signals also orthogonally polarized?

Orthogonality at the input means $\mathbf{E}_{1,in}^{+}\mathbf{E}_{2,in} = 0$. To answer the question we check the orthogonality at the output,

$$\mathbf{E}_{1,out}^{+}\mathbf{E}_{2,out} = \left(\mathbf{J}\mathbf{E}_{1,in}\right)^{+}\left(\mathbf{J}\mathbf{E}_{2,in}\right) = \mathbf{E}_{1,in}^{+}\left(\mathbf{J}^{+}\mathbf{J}\right)\mathbf{E}_{2,in}. \tag{2.268}$$

If the Jones matrix can be written as $\mathbf{J} = h\mathbf{J}'$ ($h = const.$) where $\mathbf{J}'$ is a unitary matrix then the output signals are orthogonal,

$$\mathbf{E}_{1,out}^{+}\mathbf{E}_{2,out} = \mathbf{E}_{1,in}^{+}\left(|h|^{2}\mathbf{1}\right)\mathbf{E}_{2,in} = 0. \tag{2.269}$$

In lossless multiports the scattering matrix is unitary. If the ports are non-reflecting in groups, so that the submatrices $\mathbf{S}_{11}$, $\mathbf{S}_{22}$ vanish, the unitarity condition reduces to the unitarty of the submatrices $\mathbf{S}_{12}$, $\mathbf{S}_{21}$. If the component is non-reflecting the Jones matrix is a scattering submatrix: $\mathbf{S}_{21} = \mathbf{J}$. The factor h is a polarization-independent loss (or gain). It follows:

The polarization orthogonality of two codirectional waves is preserved during the transmission through a non-reflecting medium having no polarization-dependent loss,

$$\mathbf{E}_{1,in}^{+}\mathbf{E}_{2,in} = 0 \quad \Leftrightarrow \quad \mathbf{E}_{1,out}^{+}\mathbf{E}_{2,out} = 0 \quad \text{for } \mathbf{J}^{+}\mathbf{J} \sim \mathbf{1}. \tag{2.270}$$

Fortunately the polarization-dependent loss (PDL) of optical fibers is very small. Two signals can be combined at the input of an optical fiber with orthogonal polarizations, using a normal coupler or simply a polarization beam splitter. At the fiber end the signals will in general have random polarizations but will still be orthogonal to each other and can be separated without loss. This *polarization division multiplex* allows to double the information capacity of fibers.

In special cases it may be expedient to transmit two signals in different directions over a single optical fiber. In order to combine and separate the modes without los we want again to use polarization beam splitters. Here the following holds:

Let the polarization of a wave which propagates forward in z direction be defined in some coordinate system. To represent the backward-transmitted polarization state the coordinate system may be mirrored at an axis which can be freely chosen in the x-y plane. The right-handedness is thereby preserved. This procedure describes also a mirror reflection. When the coordinate system is mirrored, e.g., at the y axis, i.e., direction inversion of z and x axes, the longitudinal field dependence continues to be a phasor e^{-jkz}, and the electrical field sign in x direction must be inverted. If we mirror at the x or at the y axis then he signs of the Stokes parameters S_2, S_3 are inverted. If one mirrors at the 45° or at the −45° axis the signs of S_1, S_3 are inverted. No matter at which of the x and y axes of the x-y plane the coordinate system is mirrored: In the new propagation direction the reflection will change the sign of S_3 and hence the sense of the polarization ellipse.

Other than just described we will now use the same coordinate system for both directions. Now the sign of the propagation constant k is changed by the reflection. The Jones matrix of an ideal metallic mirror is here $\mathbf{J} = -\mathbf{1}$. The minus sign results from the tangential boundary conditions for the electric field, which vanishes in an ideal conductor.

A signal with Jones vector $\mathbf{E}_{f,in}$ is fed into an optical component which has a Jones matrix $\mathbf{J}_f$ in forward direction. The output signal is $\mathbf{E}_{f,out} = \mathbf{J}_f \mathbf{E}_{f,in}$. At the output a second signal $\mathbf{E}_{b,in}$ is injected in backward direction. It exits as $\mathbf{E}_{b,out}$ at the input. Assume $\mathbf{E}_{b,in}^T \mathbf{E}_{f,out} = 0$. Under which condition holds $\mathbf{E}_{b,out}^T \mathbf{E}_{f,in} = 0$? We require

$$0 = \mathbf{E}_{b,in}^T \mathbf{E}_{f,out} = \mathbf{E}_{b,in}^T \mathbf{J}_f \mathbf{E}_{f,in} = \left(\mathbf{J}_f^T \mathbf{E}_{b,in}\right)^T \mathbf{E}_{f,in} . \tag{2.271}$$

If the refractive indices are identical at all ports of the optical four-port under consideration then the scattering submatrix $\mathbf{S}_{21} = \mathbf{J}_f$ is responsible for transmission in forward direction. In *reciprocal* media the scattering matrix is symmetric, so that $\mathbf{S}_{12} = \mathbf{S}_{21}^T$ holds. The transmission matrix in backward direction $\mathbf{J}_b$ is then $\mathbf{J}_b = \mathbf{S}_{12} = \mathbf{J}_f^T$ and the term of (2.271) in parentheses is $\mathbf{E}_{b,out}^T$. It follows:

The orthogonality of linear polarizations of two counterdirectional waves is preserved in reciprocal media,

$$\mathbf{E}_{b,in}^T \mathbf{E}_{f,out} = 0 \quad \Leftrightarrow \quad \mathbf{E}_{b,out}^T \mathbf{E}_{f,in} = 0 \quad \text{for } \underline{\mathbf{J}}_b = \underline{\mathbf{J}}_f^T . \tag{2.272}$$

This is intuitively clear for linearly polarized waves where the orthogonality relation may be written as $\mathbf{E}_1^T \mathbf{E}_2 = 0$. The theorem even holds for the orthogonality of all polarizations if the orthogonality of counterpropagating waves is defined by $\mathbf{E}_b^T \mathbf{E}_f = 0$.

Problem: Optical fibers are reciprocal. Due to unavoidable core asymmetries the polarization along the propagation direction z evolves with an equidistribution on the Poincaré sphere. Calculate the DOP of the Rayleigh-scattered waves! In doing so, neglect attenuation because the equidistribution occurs already after short distances. Without loss of generality horizontal input polarization may be assumed.

Solution: Equidistribution on the Poincaré sphere can be represented by an equidistributed double azimuth angle 2ϑ and a cos-distributed double ellipticity angle 2ε. In order to generate it by a very simple Jones matrix it is helpful to define instead a longitudinal angle $\chi = \arg(-S_3 + jS_2)$ on the S_2 - S_3 great circle and an azimuthal angle δ from the positive S_1 axis of the Poincaré sphere, with $S_1 = \cos\delta$, $p_\delta(\delta) = (1/2)\sin\delta$ $(0 \le \delta \le \pi)$, $p_\chi(\chi) = 1/(2\pi)$ $(|\chi| \le \pi)$. Here $S_{1...3}$ define the polarization state at the point of reflection, not an eigenmode of the retarder. The z-dependent Jones vector is $\mathbf{E}_f(z) = \begin{bmatrix} \cos\delta/2 \\ je^{-j\chi}\sin\delta/2 \end{bmatrix}$, as can be verified from the eigenmodes of (2.239) or by insertion into (2.166). It may be generated from the horizontal polarization $\mathbf{E}_f(0) = \begin{bmatrix} 1 \\ 0 \end{bmatrix}$ by a retarder with the Jones matrix $\mathbf{J} = \begin{bmatrix} \cos\delta/2 & je^{j\chi}\sin\delta/2 \\ je^{-j\chi}\sin\delta/2 & \cos\delta/2 \end{bmatrix}$ (2.239), where the retardation is δ and the fast eigenmode is given by the normalized Stokes parameters $V_1 = 0$, $V_2 = \cos\chi$, $V_3 = \sin\chi$. The Rayleigh reflection causes a simple phase shift and an attenuation (which is not of concern here). In the reciprocal medium the Jones matrix in backward direction is $\mathbf{J}^T$. The wave which is reflected from a position z back to the fiber input therefore exhibits the Jones vector $\mathbf{E}_r(0) = \mathbf{J}^T\mathbf{J}\mathbf{E}_v(0)$ $= \begin{bmatrix} (1/2)(1+\cos\delta) - e^{-j2\chi}(1/2)(1-\cos\delta) \\ j\cos\chi\sin\delta \end{bmatrix}$. The associated normalized Stokes vector is $\hat{\hat{\mathbf{S}}} = \begin{bmatrix} (1/2)\left(1+\cos^2\delta - \cos 2\chi\left(1-\cos^2\delta\right)\right) - \cos^2\chi\sin^2\delta \\ -\sin 2\chi(1-\cos\delta)\cos\chi\sin\delta \\ -((1+\cos\delta) - \cos 2\chi(1-\cos\delta))\cos\chi\sin\delta \end{bmatrix}$. Averaging over all χ results in $\hat{\mathbf{S}} = \int \hat{\hat{\mathbf{S}}} p_\chi(\chi) d\chi = \begin{bmatrix} (1/2)(1+\cos 2\delta) \\ 0 \\ 0 \end{bmatrix}$. Additional averaging over δ results in the overall averaged normalized Stokes vector $\mathbf{S} = \int \hat{\mathbf{S}} p_\delta(\delta) d\delta = \begin{bmatrix} 1/3 & 0 & 0 \end{bmatrix}^T$. In agreement with (2.175) the DOP is calculated from the normalized Stokes vector as $p = |\mathbf{S}| = 1/3$. This can be intuitively understood: For linear polarization at the place of reflection ($\mathbf{E}_f(z)$), $\mathbf{J}$ is real and the backreflected polarization ($\mathbf{E}_b(0)$) is identical to the transmitted horizontal polarization ($\mathbf{E}_f(0)$) due to $\mathbf{J}^T = \mathbf{J}^+ = \mathbf{J}^{-1}$. However, for circular polarization at the place of reflection the backreflected polarization is vertical, and it holds $\mathbf{J}^T \ne \mathbf{J}^{-1}$. The three normalized Stokes parameters averaged over all places of reflection have zero means, are uncorrelated and possess equal variances of 1/3 each. In the polarization which is backreflected to the input there are two positive and one negative contribution of 1/3 each to S_1 and hence to p.

Polarization division multiplex is impaired by a possible *polarization-dependent loss (PDL)* or *polarization-dependent gain (PDG)*, for the quantification of which the Stokes parameter quotient $S_{0,out,\max}/S_{0,out,\min}$ at the output is of use. A system with PDL can be understood as a partial polarizer with a finite extinction $10\log\left(S_{0,aus,\max}/S_{0,aus,\min}\right)$dB with leading and/or trailing retarders.

The necessary temporal averaging is left out for simplicity in the following. If the system Jones matrix is constant,

$$S_{0,out} = \left|\mathbf{E}_{out}\right|^2 = \left|\mathbf{J}\mathbf{E}_{in}\right|^2 = \mathbf{E}_{in}^+\left(\mathbf{J}^+\mathbf{J}\right)\mathbf{E}_{in}, \qquad \left(\mathbf{J}^+\mathbf{J}\right)^+ = \left(\mathbf{J}^+\mathbf{J}\right). \tag{2.273}$$

holds. Since $\left(\mathbf{J}^+\mathbf{J}\right)$ is Hermitian, $S_{0,out}$ assumes only real values. In addition $S_{0,out}$ is non-negative. $\left(\mathbf{J}^+\mathbf{J}\right)$ is therefore positive semidefinite. This is also seen if the Jones matrix is represented by column vectors,

$$\mathbf{J} = \left[\mathbf{e}_1 \quad \mathbf{e}_2\right] \;\Rightarrow\; \left(\mathbf{J}^+\mathbf{J}\right) = \left[\left(\mathbf{e}_i^+\mathbf{e}_j\right)\right]. \tag{2.274}$$

The principal minors of $\left(\mathbf{J}^+\mathbf{J}\right)$ are non-negative,

$$\begin{aligned} &\left(\mathbf{e}_1^+\mathbf{e}_1\right) = \left|\mathbf{e}_1\right|^2 \ge 0 \qquad \left(\mathbf{e}_2^+\mathbf{e}_2\right) = \left|\mathbf{e}_2\right|^2 \ge 0 \\ &\left(\mathbf{e}_1^+\mathbf{e}_1\right)\left(\mathbf{e}_2^+\mathbf{e}_2\right) - \left(\mathbf{e}_1^+\mathbf{e}_2\right)\left(\mathbf{e}_2^+\mathbf{e}_1\right) = \left|\mathbf{e}_1\right|^2\left|\mathbf{e}_2\right|^2 - \left|\mathbf{e}_1^* \cdot \mathbf{e}_2\right|^2 \ge 0 \end{aligned}. \tag{2.275}$$

The eigenvalues are non-negative, $\lambda_i \ge 0$. The eigenvectors $\mathbf{a}_i$ of the Hermitian matrix $\left(\mathbf{J}^+\mathbf{J}\right)$ are complex orthogonal; if they are also normalized the transformation matrix $\left[\mathbf{a}_1 \quad \mathbf{a}_2\right]$ is unitary. One thereby obtains

$$\begin{aligned} &S_{0,out} = \mathbf{E}_{in}^+\left[\mathbf{a}_1 \quad \mathbf{a}_2\right]\Lambda\left[\mathbf{a}_1 \quad \mathbf{a}_2\right]^+\mathbf{E}_{in} = \mathbf{X}^+\Lambda\mathbf{X} \\ &\text{with} \qquad \mathbf{X} = \left[\mathbf{a}_1 \quad \mathbf{a}_2\right]^+\mathbf{E}_{in} \qquad \Lambda = \begin{bmatrix} \lambda_1 & 0 \\ 0 & \lambda_2 \end{bmatrix} \end{aligned}. \tag{2.276}$$

Maximum and minimum are assumed for

$$\begin{aligned} &\mathbf{X}_1 = h\begin{bmatrix} 1 \\ 0 \end{bmatrix} \;\Leftrightarrow\; \mathbf{E}_{1,in} = h\mathbf{a}_1 \\ &\mathbf{X}_2 = h\begin{bmatrix} 0 \\ 1 \end{bmatrix} \;\Leftrightarrow\; \mathbf{E}_{2,in} = h\mathbf{a}_2 \end{aligned} \qquad (h = const.). \tag{2.277}$$

The required input polarizations are orthogonal. The associated values of $S_{0,out}$ are

$$S_{0,out} = \lambda_i\left|\left[\mathbf{a}_1 \quad \mathbf{a}_2\right]^+\mathbf{E}_{i,in}\right|^2 = \lambda_i\left|\mathbf{E}_{i,in}\right|^2 = \lambda_i S_{0,in}\,. \tag{2.278}$$

The desired quotient is

$$S_{0,out,\max}/S_{0,out,\min} = \lambda_{\max}/\lambda_{\min} \quad (\lambda_i\text{: eigenvalues of } \left(\mathbf{J}^+\mathbf{J}\right)). \tag{2.279}$$

The two polarization states which are transmitted by a component having the Jones Matrix $\mathbf{J}$ *with maximum and minimum power to the output are the two orthognal eigenvectors of the matrix* $(\mathbf{J}^+\mathbf{J})$. *The powers are multiplied by the respective eigenvalues.*

Practical sources of PDL in some transmission links are optical couplers, amplifiers and other optical components.

We search boundaries for the loss of polarization orthogonality in a medium with PDL [6]. Let the input signals be orthogonally polarized, $\mathbf{E}_{1,in}^+\mathbf{E}_{2,in}=0$, and have equal powers, $|\mathbf{E}_{1,in}|=|\mathbf{E}_{2,in}|$. For simplicity we normalize $|\mathbf{E}_{1,in}|=|\mathbf{E}_{2,in}|=1$. Since the transformation matrix $[\mathbf{a}_{\max} \quad \mathbf{a}_{\min}]$ is unitary, the $\mathbf{X}_{1,2}$ are also orthonormal. Let an ideal polarizer be provided at the output for the polarization state $\mathbf{E}_{2,out}$. We check the transmitted output intensity (2.226)

$$I=\frac{\left|\mathbf{E}_{2,out}{}^+\mathbf{E}_{1,out}\right|^2}{\left|\mathbf{E}_{2,out}\right|^2\left|\mathbf{E}_{1,out}\right|^2}=\frac{\left|\mathbf{X}_2^+\mathbf{\Lambda}\mathbf{X}_1\right|^2}{\left|\mathbf{X}_2^+\mathbf{\Lambda}\mathbf{X}_2\right|\left|\mathbf{X}_1^+\mathbf{\Lambda}\mathbf{X}_1\right|} \quad \text{with } \mathbf{\Lambda}=\begin{bmatrix}\lambda_{\max} & 0\\ 0 & \lambda_{\min}\end{bmatrix}, \tag{2.280}$$

of the signal $\mathbf{E}_{1,out}$. These two polarization states result from orthogonal input polarizations $\mathbf{E}_{2,in}$, $\mathbf{E}_{1,in}$, and the definition in (2.276) has been applied. The PDL can also be expressed by the variable

$$\gamma=\frac{\lambda_{\max}-\lambda_{\min}}{\lambda_{\max}+\lambda_{\min}}. \tag{2.281}$$

The orthonormality of the $\mathbf{X}_{1,2}$ allows to write

$$\mathbf{X}_1=\begin{bmatrix}\sqrt{a}\\ e^{j\varphi}\sqrt{1-a}\end{bmatrix},\ \mathbf{X}_2=\begin{bmatrix}-e^{-j\varphi}\sqrt{1-a}\\ \sqrt{a}\end{bmatrix}, \tag{2.282}$$

where an arbitrary leading phasor has been neglected. It follows

$$I=\frac{4a(1-a)}{\gamma^{-2}-(2a-1)^2}\le\gamma^2. \tag{2.283}$$

The maximum value γ^2 holds for $a=1/2$. Maximum intensity, i.e., "smallest orthogonality", is obtained if each input signal power is split equally into the maximally and minimally transmitted polarization state.

If a polarization division multiplex receiver selects $\underline{E}_{2,out}$ then signal 2 generated the wanted signal $I_2=1$ and signal 1 the incoherent crosstalk $I_1\le\gamma^2$. In addition there is a strong beat term (coherent crosstalk). If the orthogonal polarization state is chosen, I is replaced by $1-I$ and the signals 1, 2 exchange functions. The wanted signal is then $I_1 \geq 1-\gamma^2$, the incoherent crosstalk $I_2 = 0$ vanishes (and also the coherent crosstalk). *In the presence of PDL the absence of loss and crosstalk can not be guaranteed simultaneously.*

Even a completely monochromatic signal can be *depolarized* by a time-variable system. We consider in the following a monochromatic signal with constant Stokes vector $\mathbf{S}_{in}$, which is passed through a lossless, time-variable retarder with a 3×3 rotation matrix $\mathbf{G}(t)$. Let the time-variability be slow compared to the optical oscillation. For the normalized Stokes vectors it holds

$$\mathbf{S}_{out}(t) = \mathbf{G}(t)\mathbf{S}_{in} . \tag{2.284}$$

If one is interested in the DOP which can be observed over larger averaging times then the temporally averaged vector $\overline{\mathbf{S}_{out}} \equiv \overline{\mathbf{S}_{out}(t)}$ must be determined. According to (2.175) the DOP is the length of the normalized Stokes vector, $p = \left|\overline{\mathbf{S}_{out}}\right|$. For a constant (time-invariable) $\mathbf{S}_{in}$,

$$p = \left|\overline{\mathbf{S}_{out}}\right| = \left|\overline{\mathbf{G}}\mathbf{S}_{in}\right| . \tag{2.285}$$

An ideal, polarization-independent depolarizer offers an output DOP $p = 0$ for any incident $\mathbf{S}_{in}$. This is only possible if the averaged 3×3 matrix vanishes,

$$\overline{\mathbf{G}} = \mathbf{0} . \tag{2.286}$$

Within which limits does the output DOP vary as function of $\mathbf{S}_{in}$? The expression

$$p^2 = \left|\overline{\mathbf{S}_{out}}\right|^2 = \mathbf{S}_{in}^T \left(\overline{\mathbf{G}}^T \overline{\mathbf{G}} \right) \mathbf{S}_{in} \tag{2.287}$$

is the real case of a Hermitian form. Due to $p^2 \geq 0$, $\overline{\mathbf{G}}^T \overline{\mathbf{G}}$ must be positive semidefinite. In analogy to (2.273)*ff* one obtains

$$\sqrt{\lambda_{\min}} \leq p \leq \sqrt{\lambda_{\max}} , \tag{2.288}$$

where $\lambda_{\min}$, $\lambda_{\max}$ are minimum and maximum eigenvalues of $\overline{\mathbf{G}}^T \overline{\mathbf{G}}$. The input polarizations $\mathbf{S}_{in}$, for which these DOP extrema occur are the corresponding eigenvectors of $\overline{\mathbf{G}}^T \overline{\mathbf{G}}$.

Problem: Which input polarizations are fully depolarized by a rotating quarterwave plate (QWP)? And which by a rotating halfwave plate (HWP)? Use both results to construct an input-polarization independent depolarizer. For a given depolarization time T, how small can the maximum absolute value F_{rot} of required physical rotation frequencies be made?

Solution: From (2.242) we obtain $\mathbf{G}_1 = \begin{bmatrix} (1/2)(1+\cos 4\alpha_1) & (1/2)\sin 4\alpha_1 & \sin 2\alpha_1 \\ (1/2)\sin 4\alpha_1 & (1/2)(1-\cos 4\alpha_1) & -\cos 2\alpha_1 \\ -\sin 2\alpha_1 & \cos 2\alpha_1 & 0 \end{bmatrix}$,

$\mathbf{G}_2 = \begin{bmatrix} \cos 4\alpha_2 & \sin 4\alpha_2 & 0 \\ \sin 4\alpha_2 & -\cos 4\alpha_2 & 0 \\ 0 & 0 & -1 \end{bmatrix}$, where "1" means QWP and "2" means HWP. With time-variable (t) Poincaré sphere eigenmode rotation angles $2\alpha_1 = 2\pi f_1 t$, $2\alpha_2 = 2\pi f_2 t$ it follows

$$\overline{\mathbf{G}_1} = \begin{bmatrix} 1/2 & 0 & 0 \\ 0 & 1/2 & 0 \\ 0 & 0 & 0 \end{bmatrix}, \quad \overline{\mathbf{G}_2} = \begin{bmatrix} 0 & 0 & 0 \\ 0 & 0 & 0 \\ 0 & 0 & -1 \end{bmatrix}.$$

The QWP depolarizes circular, but not linear polarizations whereas the HWP depolarizes all linear but not circular polarizations. We cascade a QWP and a HWP. The resulting rotation matrix is $\mathbf{G} = \mathbf{G}_2\mathbf{G}_1$

$$= \begin{bmatrix} (1/2)(\cos 4\alpha_2 + \cos(4\alpha_1 - 4\alpha_2)) & (1/2)(\sin 4\alpha_2 + \sin(4\alpha_1 - 4\alpha_2)) & \sin(2\alpha_1 - 4\alpha_2) \\ (1/2)(\sin 4\alpha_2 - \sin(4\alpha_1 - 4\alpha_2)) & (1/2)(-\cos 4\alpha_2 + \cos(4\alpha_1 - 4\alpha_2)) & \cos(2\alpha_1 - 4\alpha_2) \\ \sin 2\alpha_1 & -\cos 2\alpha_1 & 0 \end{bmatrix}.$$

We need to achieve $\overline{\mathbf{G}} = \mathbf{0}$ for a given depolarization time T. This is the case if we choose $f_1 T$, $2f_2 T$, $(f_1 - 2f_2)T$ and $2(f_1 - f_2)T$ as integers unequal zero. The maximum absolute value of the required physical rotation frequency is at least $F_{rot} = 1/(2T)$. This is achieved for $f_1 = |T|^{-1}$ and $f_2 = f_1/2$ or $f_2 = -f_1$. In the latter case, QWP and HWP rotate at the same speed in opposite directions. In a different but analogous form, this has been realized in [7].

Regarding the application of depolarizers, Erbium-doped fiber amplifiers (EDFAs) are usually operated somewhat in saturation. In that case the amplification of the signal mode decreases slightly more than the amplification in the orthogonally polarized mode which carries less power. This special manifestation of polarization dependent gain (PDG) is called *polarization holeburning (PHB)*. Even though PHB is just 0.1 ... 0.2 dB per EDFA it can significantly accumulate over ultra-long haul transmission links having many EDFAs. A depolarizer which consists of a time-variable retarder can prevent PHB. Depolarization in the frequency domain by a number of WDM channels with unequal polarizations is however quite as good, and is normally good enough.

2.3.5 Polarization Mode Dispersion

Now we consider the *frequency dependence* of Jones matrices in so far as it influences the polarization at the output of the medium. This is the *polarization mode dispersion (PMD)* [8, 9, 3]. We assume that there is no polarization-dependent loss so that the Jones matrix can be written as

$$\mathbf{J}(\omega) = e^{-(\alpha(\omega)+j\beta(\omega))z}\mathbf{U}(\omega), \qquad \mathbf{U}(\omega) = \begin{bmatrix} u_1(\omega) & u_2(\omega) \\ -u_2^*(\omega) & u_1^*(\omega) \end{bmatrix},$$
$$|u_1(\omega)|^2 + |u_2(\omega)|^2 = 1, \qquad \mathbf{U}^+(\omega)\mathbf{U}(\omega) = \mathbf{1}. \tag{2.289}$$

The matrix $\mathbf{U}(\omega)$ is unitary, the propagation constant $k(\omega)$ is frequency-dependent. For constant input polarization $\mathbf{E}_{in}$ the output polarization $\mathbf{E}_{out}(\omega) = \mathbf{J}(\omega)\mathbf{E}_{in}$ generally varies as a function of the input signal frequency. A broadband signal is thereby depolarized.

However, there exist for each frequency two orthogonal input polarizations for which the corresponding orthogonal output polarizations do not vary as a function of frequency in a first-order approximation. They are called *principal states-of-polarization (PSPs).*

In order to drop complex amplitude changes by the term $e^{-(\alpha(\omega)+j\beta(\omega))z}$ we consider only $\mathbf{U}(\omega)$ and write a matrix equation with normalized vectors,

$$\mathbf{e}_{out} = \mathbf{U}(\omega)\mathbf{e}_{in}, \qquad |\mathbf{e}_{in}| = |\mathbf{e}_{out}| = 1. \tag{2.290}$$

The 2 degrees-of-freedom of polarization are contained in the complex variable $\chi = E_y/E_x = e_y/e_x$. It follows

$$\frac{d\mathbf{e}_{out}(\omega)}{d\omega} \equiv \mathbf{e}_{out,total}' = \mathbf{U}'\mathbf{e}_{in}. \tag{2.291}$$

But also $\mathbf{U}(\omega)$ may cause complex amplitude changes which do not influence polarization. Such overall phase changes of $\mathbf{e}_{out}$ of the kind

$$\mathbf{e}_{out,\varphi} = e^{j\varphi(\omega)} \cdot \mathbf{e}_{const.} \quad \Rightarrow \quad \mathbf{e}_{out,\varphi}' = j\varphi'\mathbf{e}_{out} \tag{2.292}$$

let χ_{out} indeed unchanged. The total output Jones vector change is composed of this contribution and that ($\mathbf{e}_{out,\chi}'$) from a change of χ_{out},

$$\mathbf{e}_{out,total}' = \mathbf{e}_{out,\varphi}' + \mathbf{e}_{out,\chi}' \tag{2.293}$$

We solve for $\mathbf{e}_{out,\chi}'$ and set it equal to zero,

$$\mathbf{e}_{out,\chi}' = \mathbf{e}_{out,total}' - \mathbf{e}_{out,\varphi}' = \left(-j\varphi'\mathbf{U} + \mathbf{U}'\right)\mathbf{e}_{in} = \mathbf{0}. \tag{2.294}$$

The frequency-dependent quantity φ' is yet unknown. Solutions exist for

$$\det\left(-j\varphi'\mathbf{U} + \mathbf{U}'\right) = 0. \tag{2.295}$$

This leads to a quadratic equation. With

$$u_1'u_1^* + u_1u_1^{*'} + u_2'u_2^* + u_2u_2^{*'} = \left(|u_1|^2 + |u_2|^2\right)' = 1' = 0 \tag{2.296}$$

the solutions

$$\varphi_\pm' = \pm\sqrt{\left|u_1'\right|^2 + \left|u_2'\right|^2} \tag{2.297}$$

are found. The corresponding solutions of (2.294) are the PSPs $\mathbf{e}_{in\pm}$.

If $e^{j\psi(\omega)}$ is the phasor of a transfer function then $\tau_g = -\psi'$ is the group delay. When the PSPs are transmitted through the component with the Jones matrix $\mathbf{J}(\omega)$ (2.289) the group delay $\tau_g = -\psi' = \beta' z - \varphi'$ is composed of a polarization-independent part $\beta' z$, where β' is the inverse of the mean group velocity, and a polarization-dependent part $\varphi_\pm'$. The term

$$\tau = \tau_{g-} - \tau_{g+} = 2\sqrt{\left|u_1'\right|^2 + \left|u_2'\right|^2} \tag{2.298}$$

is the *differential group delay (DGD)* between the two PSPs. The existence and (not correctly) sometimes also the DGD value is the PMD. The smaller group delay and higher group velocity applies for the PSP $\underline{\mathbf{e}}_{in+}$. The first matrix linc of (2.294) leads to the solution

$$\mathbf{e}_{in\pm} = \begin{bmatrix} u_2' - j\varphi_\pm' u_2 \\ -\left(u_1' - j\varphi_\pm' u_1\right) \end{bmatrix} \frac{1}{D_\pm}, \qquad D_\pm = \sqrt{2\varphi_\pm'\left(\varphi_\pm' - \operatorname{Im}\left(u_1' u_1^* + u_2' u_2^*\right)\right)}. \tag{2.299}$$

The normalization by $D_\pm$ can be omitted, and the PSP Jones vectors may be multiplied by arbitrary phasors. One easily shows $\underline{\mathbf{e}}_{in-}^+ \underline{\mathbf{e}}_{in+} = 0$. The PSPs are hence orthogonal at the input, and due to the assumed polarization-independence of the losses also at the output. For mixed polarizations the group delay lies always between the extrema τ_{g-} and τ_{g+}, and an impulse broadening occurs. PMD depolarizes a signal unless it is strictly monochromatic.

In modern fibers PMD is usually near the lower boarder of the range $\tau \approx \sqrt{l} \cdot (0{,}05 \ldots 2)\,\mathrm{ps}/\sqrt{\mathrm{km}}$. The dependence of the square root of the fiber length l results from the addition of many random perturbations which can add positive or negative contributions. In fibers with an exact core symmetry PMD would vanish.

Fiber attenuation can be compensated by optical amplifiers. The chromatic dispersion of ~17 ps/nm/km in standard fibers, ~(–1...1) ps/nm/km in dispersion-shifted fibers and ±(3...6) ps/nm/km in non-zero dispersion-shifted fibers

(NZDSF) can be compensated by cascading fibers with negative and positive chromatic dispersion. PMD however depends for example on the bending stress in the fiber and hence on temperature. Therefore PMD is difficult to compensate.

The above derivation is the classical entry into the PMD topic [8]. But PMD can more intuitively be defined via the small-signal intensity transfer function in a polarization mode dispersive medium [10]. Let the transmitted signal with the frequency ω be modulated in its power to this purpose with the modulation frequency ω_m and the modulation index $|a| << 1$. The optical field at the input (in) of the fiber is

$$\mathbf{E}_{in} = \mathbf{e}_{in}\left(1 + (a/4)\left(e^{j\omega_m t} + e^{-j\omega_m t}\right)\right)e^{j\omega t} \qquad \text{with} \qquad |\mathbf{e}_{in}| = 1. \tag{2.300}$$

Define the optical intensity furthermore as $I = |\mathbf{E}|^2$,

$$I_{in}(t) = |\mathbf{E}_{in}|^2 = 1 + a\cos(\omega_m t). \tag{2.301}$$

The signal is transmitted through a medium with the unitary Jones matrix $\mathbf{J}(\omega)$ to the output (out). From $\mathbf{E}_{out}(\omega) = \mathbf{J}(\omega)\mathbf{E}_{in}(\omega)$ it follows

$$\mathbf{E}_{out}(t) = \begin{pmatrix} \mathbf{J}(\omega) + e^{j\omega_m t}(a/4)\mathbf{J}(\omega + \omega_m) \\ + e^{-j\omega_m t}(a/4)\mathbf{J}(\omega - \omega_m) \end{pmatrix} e^{j\omega t}\mathbf{e}_{in}, \tag{2.302}$$

$$I_{out}(t) = |\mathbf{E}_{out}(t)|^2 =$$
$$\mathbf{e}_{in}^{+}\begin{pmatrix} \mathbf{1} + \mathbf{J}^{+}(\omega)e^{j\omega_m t}(a/4)\mathbf{J}(\omega + \omega_m) + \mathbf{J}^{+}(\omega)e^{-j\omega_m t}(a/4)\mathbf{J}(\omega - \omega_m) \\ + e^{-j\omega_m t}(a/4)\mathbf{J}^{+}(\omega + \omega_m)\mathbf{J}(\omega) + e^{j\omega_m t}(a/4)\mathbf{J}^{+}(\omega - \omega_m)\mathbf{J}(\omega) \\ + O\left((a/4)^2\right) \end{pmatrix}\mathbf{e}_{in}. \tag{2.303}$$

Due to $a << 1$ the term $O\left((a/4)^2\right)$ can be neglected. One thereby obtains

$$\frac{I_{out}(t) - 1}{a} = (1/4)\mathbf{e}_{in}^{+}\begin{pmatrix} e^{j\omega_m t}\left(\mathbf{J}^{+}(\omega)\mathbf{J}(\omega + \omega_m) + \mathbf{J}^{+}(\omega - \omega_m)\mathbf{J}(\omega)\right) \\ + e^{-j\omega_m t}\left(\mathbf{J}^{+}(\omega)\mathbf{J}(\omega - \omega_m) + \mathbf{J}^{+}(\omega + \omega_m)\mathbf{J}(\omega)\right) \end{pmatrix}\mathbf{e}_{in}. \tag{2.304}$$

This is a Hermitian form, as is easily verified. The Hermitian matrix is composed of the first term plus its Hermitian conjugate. Therefore one may write

$$\frac{I_{out}(t) - 1}{a} = (1/2)\mathrm{Re}\left(e^{j\omega_m t}\mathbf{e}_{in}^{+}\left(\mathbf{J}^{+}(\omega)\mathbf{J}(\omega + \omega_m) + \mathbf{J}^{+}(\omega - \omega_m)\mathbf{J}(\omega)\right)\mathbf{e}_{in}\right). \tag{2.305}$$

The (complex) intensity transfer function is therefore

$$\mathbf{H}_m(\omega_m)=(1/2)\mathbf{e}_{in}^{+}\left(\mathbf{J}^{+}(\omega)\mathbf{J}(\omega+\omega_m)+\mathbf{J}^{+}(\omega-\omega_m)\mathbf{J}(\omega)\right)\mathbf{e}_{in}\,. \tag{2.306}$$

Now we write a Taylor series which is truncated after the first-order term,

$$\mathbf{J}(\omega+\omega_m)=\mathbf{J}(\omega)+\omega_m\mathbf{J}'(\omega). \tag{2.307}$$

Under this condition the transfer function can be approximated by

$$\mathbf{H}_m(\omega_m)=1+\omega_m(1/2)\mathbf{e}_{in}^{+}\left(\mathbf{J}^{+}\mathbf{J}'-\left(\mathbf{J}^{+}\mathbf{J}'\right)^{+}\right)\mathbf{e}_{in}\,. \tag{2.308}$$

In order to investigate only PMD effects we limit ourselves to the anisotropic and unitary part $\underline{\mathbf{U}}(\omega)$ of the Jones matrix (2.289), and obtain after a few manipulations

$$\mathbf{H}_m(\omega_m)=1+j\omega_m\mathbf{e}_{in}^{+}\begin{bmatrix}\mathrm{Im}\left(u_1^{*}u_1'+u_2u_2^{*\prime}\right) & j\left(u_2u_1^{*\prime}-u_1^{*}u_2'\right)\\ -j\left(u_2^{*}u_1'-u_1u_2^{*\prime}\right) & \mathrm{Im}\left(u_1u_1^{*\prime}+u_2^{*}u_2'\right)\end{bmatrix}\mathbf{e}_{in} \tag{2.309}$$

The imaginary part of $\mathbf{H}_m(\omega_m)$ is a Hermitian form, as is easily seen. Its extrema and the corresponding input polarizations are the eigenvalues and corresponding eigenvectors of the Hermitian matrix. The solutions are (2.297), (2.299). At the extrema the transfer function assumes the values $\mathbf{H}_m(\omega_m)=1\pm j\omega_m\tau/2\approx e^{\pm j\omega_m\tau/2}$. The approximation holds for small ω_m. The extrema differ in their group delay by τ.

With (2.296) one can verify

$$\mathrm{Im}\left(u_1^{*}u_1'+u_2u_2^{*\prime}\right)=-j\left(u_1^{*}u_1'+u_2u_2^{*\prime}\right). \tag{2.310}$$

As a next step PMD shall be expressed by normalized Stokes vectors. This is easily possible by insertion of (2.299) into (2.166). Alternatively we rewrite the Hermitian form in such a way that it contains a scalar product of normalized Stokes vectors,

$$\mathbf{e}_{in}^{+}\begin{bmatrix}-j\left(u_1^{*}u_1'+u_2u_2^{*\prime}\right) & j\left(u_2u_1^{*\prime}-u_1^{*}u_2'\right)\\ -j\left(u_2^{*}u_1'-u_1u_2^{*\prime}\right) & -j\left(u_1u_1^{*\prime}+u_2^{*}u_2'\right)\end{bmatrix}\mathbf{e}_{in}=\mathbf{e}_{in}^{+}\begin{bmatrix}A & B\\ B^{*} & -A\end{bmatrix}\mathbf{e}_{in}$$

$$=AS_{1,in}+\mathrm{Re}(B)S_{2,in}+\mathrm{Im}(B)S_{3,in}=(1/2)\mathbf{\Omega}_{in}^{T}\mathbf{S}_{in}\,, \tag{2.311}$$

$$\mathbf{\Omega}_{in} = 2\begin{bmatrix} A \\ \mathrm{Re}(B) \\ \mathrm{Im}(B) \end{bmatrix} = 2\begin{bmatrix} -j\left(\underline{u}_1^*\underline{u}_1' + \underline{u}_2^*\underline{u}_2'\right) \\ \mathrm{Re}\left(j\left(\underline{u}_2\underline{u}_1^{*\prime} - \underline{u}_1^*\underline{u}_2'\right)\right) \\ \mathrm{Im}\left(j\left(\underline{u}_2\underline{u}_1^{*\prime} - \underline{u}_1^*\underline{u}_2'\right)\right) \end{bmatrix}, \qquad |\mathbf{\Omega}_{in}| = \tau . \tag{2.312}$$

The (input-referred) PMD vector $\mathbf{\Omega}_{in}$ *points in the space of normalized Stokes vectors in the direction of the fast PSP. Its length is the DGD.* The squared magnitude of its length can be calculated using (2.296); from this it follows $|\mathbf{\Omega}_{in}| = \tau$. The relative group delay for arbitrary input polarization is given by $-(1/2)\mathbf{\Omega}_{in}^T\mathbf{S}_{in}$.

It is useful to reference PMD vectors to a fixed location. Let us denote PMD vectors referenced to the input of an optical fiber or to the input of a cascade of retarders by a "~". Very important is the addition property of PMD vectors referenced to the same location when several retarders are cascaded. We start by the claim that the input-referred overall PMD vector resulting from a cascade of two retarders 1, 2 is obtained as $\tilde{\mathbf{\Omega}} = \mathbf{\Omega}_1 + \mathbf{G}_1^T\mathbf{\Omega}_2$. Here $\mathbf{G}_1$ is the rotation submatrix of the Müller matrix of retarder 1. The input-referred vector $\tilde{\mathbf{\Omega}}$ is the sum of the two vectors $\mathbf{\Omega}_1, \mathbf{\Omega}_2$, but the latter is multiplied by the inverse of the polarization transformation in the retarder 1 so that it is referenced to the input of retarder 1. In order to show that this is true one defines the Jones matrices $\mathbf{U}_1(\omega)$, $\mathbf{U}_2(\omega)$ of two retarders in analogy to $\mathbf{U}(\omega)$ in (2.289). For the left hand side of the claim the PMD vector of the matrix product $\mathbf{U}(\omega) = \mathbf{U}_2(\omega)\mathbf{U}_1(\omega)$ is calculated according to (2.289) and (2.312). At the right hand side one calculates the individual PMD vectors, and with the help of (2.233), (2.238) the rotation matrix $\mathbf{G}_1$. After a few manipulations it is seen that the equation is fulfilled. Complete induction allows to generalize for an arbitrary number of retarders,

$$\tilde{\mathbf{\Omega}} = \sum_{i=1}^{n}\left(\prod_{j=1}^{i-1}\mathbf{G}_j^T\right)\mathbf{\Omega}_i = \sum_{i=1}^{n}\tilde{\mathbf{\Omega}}_i \qquad \tilde{\mathbf{\Omega}}_i = \left(\prod_{j=1}^{i-1}\mathbf{G}_j^T\right)\mathbf{\Omega}_i . \tag{2.313}$$

For the eigenmodes of a retarder chain no such equation exists! A multiplication form the left side by the product $\prod_{j=n}^{1}\mathbf{G}_j$ of all rotation matrices (note the descending order of the indices!) one obtains the output-referred PMD vector

$$\tilde{\tilde{\mathbf{\Omega}}} = \sum_{i=1}^{n}\left(\prod_{j=n}^{i}\mathbf{G}_j\right)\mathbf{\Omega}_i . \tag{2.314}$$

$\mathbf{\Omega}_i$ is the locally input-referred PMD vector. The term $\mathbf{G}_i\mathbf{\Omega}_i$ is therefore the locally output-referred PMD vector. This vector is therefore multiplied onto the rotation matrices of all subsequent retarders from $j=i+1$ to n. The length $\left|\tilde{\Omega}\right|=\left|\tilde{\tilde{\Omega}}\right|$ of input- and output-referred PMD vectors yields the differential group delay (DGD) of the whole retarder cascade.

The frequency-dependent output polarization is related to a fixed input polarization by the rotation matrix equation $\mathbf{S}_{out}(\omega)=\mathbf{G}(\omega)\mathbf{S}_{in}$. According to one of the definitions given above, an output-referred PSP is characterized by $d\mathbf{S}_{out}(\omega)/d\omega=\mathbf{0}$. This results in

$$\frac{d\mathbf{G}(\omega)}{d\omega}\tilde{\mathbf{\Omega}}_n\overset{!}{=}\mathbf{0}. \tag{2.315}$$

This homogeneous system of 3 equations can be solved and yields the normalized input-referred PSP Stokes vector $\tilde{\mathbf{\Omega}}_n$. In order to calculate the corresponding DGD we excite both PSPs with equal powers: An input Stokes vector $\mathbf{S}_m$ is chosen which is perpendicular in the Stokes space to the two PSPs. The phase difference between the two PSPs changes as a function of frequency. The output Stokes vector travels once around the Poincaré sphere on a great circle if the phase difference $\omega\tau$ changes by 2π. This is seen from the definition of the Stokes parameters S_2, S_3, and the fact that a Jones vector basis other than x and y polarizations will simply rotate the Poincaré sphere. It is therefore possible to write

$$|\tau|=\left|\frac{d\mathbf{G}(\omega)}{d\omega}\mathbf{S}_m\right|. \tag{2.316}$$

Problem: A lossless transmission fiber is modeled as a first and a last frequency-independent polarization transformer with Jones matrices $\mathbf{J}_1$, $\mathbf{J}_3$, separated by one DGD section with Jones matrix $\mathbf{J}_2(\omega)=\begin{bmatrix} e^{j\omega\tau/2} & 0 \\ 0 & e^{-j\omega\tau/2}\end{bmatrix}$. Derive the small-signal intensity transfer function and express its polarization dependence in terms of the input-referred PMD vector. Express the input-referred PMD vector also in terms of elements of the corresponding rotation matrices.

Solution: The total Jones matrix is $\mathbf{J}(\omega)=\mathbf{J}_3\mathbf{J}_2(\omega)\mathbf{J}_1$. The intensity transfer function (2.306) is $\mathbf{H}_m(\omega_m)=(1/2)\mathbf{e}_{in}^+\begin{pmatrix}(\mathbf{J}_3\mathbf{J}_2(\omega)\mathbf{J}_1)^+(\mathbf{J}_3\mathbf{J}_2(\omega+\omega_m)\mathbf{J}_1)\\ +(\mathbf{J}_3\mathbf{J}_2(\omega-\omega_m)\mathbf{J}_1)^+(\mathbf{J}_3\mathbf{J}_2(\omega)\mathbf{J}_1)\end{pmatrix}\mathbf{e}_{in}=\mathbf{e}_{in}^+\mathbf{J}_1^+\mathbf{J}_2(\omega_m)\mathbf{J}_1\mathbf{e}_{in}$. The input Jones vector of the DGD section is $\mathbf{e}_{in,2}=\mathbf{J}_1\mathbf{e}_{in}$. It follows

$$\begin{aligned}\mathbf{H}_m(\omega_m)&=\mathbf{e}_{in,2}^+\mathbf{J}_2(\omega_m)\mathbf{e}_{in,2}=\left|e_{in,2,1}\right|^2 e^{j\omega_m\tau/2}+\left|e_{in,2,2}\right|^2 e^{-j\omega_m\tau/2}\\ &=\cos\omega_m\tau/2+jS_{in,2,1}\sin\omega_m\tau/2\end{aligned}.$$

The input Stokes vector of the DGD section is $\mathbf{S}_{in,2}=\mathbf{G}_1\mathbf{S}_{in}$ where $\mathbf{G}_1$ is the rotation matrix representing $\mathbf{J}_1$ and $\mathbf{S}_{in}$ is the normalized Stokes vector which corresponds to $\mathbf{e}_{in}$.

According to (2.312), the local PMD vector of the DGD section is $\mathbf{\Omega}_2 = \tau[1 \; 0 \; 0]^T$, or $\mathbf{\Omega}_{2,n} = [1 \; 0 \; 0]^T$ when normalized. The PMD vectors of the frequency-independent polarization transformers are zero. The overall, input referred PMD vector is therefore $\tilde{\mathbf{\Omega}} = \mathbf{G}_1^T \mathbf{\Omega}_2$, its normalized counterpart is $\tilde{\mathbf{\Omega}}_n = \mathbf{G}_1^T \mathbf{\Omega}_{2,n}$, i.e. the first column of $\mathbf{G}_1^T$. The first component of $\mathbf{S}_{in,2}$ is $S_{in,2,1} = [1 \; 0 \; 0]\mathbf{S}_{in,2} = \mathbf{\Omega}_{2,n}^T \mathbf{G}_1 \mathbf{S}_{in} = \left(\mathbf{G}_1^T \mathbf{\Omega}_{2,n}\right)^T \mathbf{S}_{in} = \tilde{\mathbf{\Omega}}_n^T \mathbf{S}_{in}$. The final result is

$$\mathbf{H}_m(\omega_m) = \cos\omega_m \tau/2 + j\tilde{\mathbf{\Omega}}_n^T \mathbf{S}_{in} \sin\omega_m \tau/2 . \tag{2.317}$$

The difference to (2.309), (2.311) is due to the fact that (2.308) is based on a first-order approximation. The result (2.317) holds not only for this specific retarder chain but for any cascade of frequency-independent retarders and one DGD section because its Jones matrix can always be expressed as $\mathbf{J}_3\mathbf{J}_2(\omega)\mathbf{J}_1$.

Using the rotation matrices we find $\mathbf{G}(\omega) = \mathbf{G}_3\mathbf{G}_2(\omega)\mathbf{G}_1$, $\frac{\mathbf{G}(\omega)}{d\omega} = \mathbf{G}_3 \frac{\mathbf{G}_2(\omega)}{d\omega}\mathbf{G}_1$, $\mathbf{G}_2(\omega) = \begin{bmatrix} 1 & 0 & 0 \\ 0 & \cos\omega\tau & -\sin\omega\tau \\ 0 & \sin\omega\tau & \cos\omega\tau \end{bmatrix}$, $\frac{d\mathbf{G}_2(\omega)}{d\omega} = \tau\begin{bmatrix} 0 & 0 & 0 \\ 0 & -\sin\omega\tau & -\cos\omega\tau \\ 0 & \cos\omega\tau & -\sin\omega\tau \end{bmatrix}$. We solve (2.315) $\mathbf{G}_3 \frac{d\mathbf{G}_2(\omega)}{d\omega}\mathbf{G}_1\tilde{\mathbf{\Omega}}_n \overset{!}{=} \mathbf{0}$ and obtain $\frac{d\mathbf{G}_2(\omega)}{d\omega}\left(\mathbf{G}_1\tilde{\mathbf{\Omega}}_n\right) \overset{!}{=} \mathbf{0}$, $\mathbf{G}_1\tilde{\mathbf{\Omega}}_n = [1 \; 0 \; 0]^T$. Like before this means that the input-referred PSP $\tilde{\mathbf{\Omega}}_n = \mathbf{G}_1^T[1 \; 0 \; 0]^T$ is the first column of $\mathbf{G}_1^T$. If both PSPs are excited with equal powers then the first element of $\mathbf{G}_1\mathbf{S}_m$ must be zero. For the example $\mathbf{G}_1\mathbf{S}_m = [0 \; 1 \; 0]^T$ we obtain $\frac{d\mathbf{G}(\omega)}{d\omega}\mathbf{S}_m = \mathbf{G}_3\tau[0 \; -\sin\omega\tau \; \cos\omega\tau]^T$, $\left|\frac{d\mathbf{G}(\omega)}{d\omega}\mathbf{S}_m\right| = |\tau|$, thereby confirming (2.316).

Problem: Derive the small-signal intensity transfer function $H_m(\omega_m)$ at the angular carrier frequency ω_0 of a medium of length L with chromatic dispersion. Its phase constant can be expressed by the truncated Taylor series $\beta(\omega) = \beta(\omega_0) + (\omega - \omega_0)\beta'(\omega_0) + (1/2)(\omega - \omega_0)^2 \beta''(\omega_0)$.

Solution: The optical transfer function is $H(\omega) = e^{-j\beta(\omega)L}$. Using (2.300) and $\mathbf{E}_{out}(\omega) = H(\omega)\mathbf{E}_{in}(\omega)$, the output electrical field is $\mathbf{E}_{out}(t) = \left(e^{-j\beta(\omega_0)L} + (a/4)e^{j(\omega_m t - \beta(\omega_0+\omega_m)L)} + (a/4)e^{j(-\omega_m t - \beta(\omega_0-\omega_m)L)}\right)e^{j\omega_0 t}\mathbf{e}_{in}$
$= \left(1 + (a/2)\cos(\omega_m(t - \beta'(\omega_0)L))e^{-(j/2)\omega_m^2\beta''(\omega_0)L}\right)e^{j(\omega_0(t-(\beta(\omega_0)/\omega_0)L))}\mathbf{e}_{in}$.

We identify the phase velocity $v_{ph} = \omega_0/\beta(\omega_0)$ at which the carrier travels and the group velocity $v_{ph} = 1/\beta'(\omega_0)$ at which the modulation travels. For $|a| << 1$ the output intensity is $I_{out}(t) = |\mathbf{E}_{out}(t)|^2 = 1 + a\cos(\omega_m(t - \beta'(\omega_0)L))\cos\left((1/2)\omega_m^2\beta''(\omega_0)L\right)$. Due to

$(I_{out}(t)-1)/a = \mathrm{Re}\left(e^{j\omega_m t} e^{-j\beta'(\omega_0)L} \cos\left((1/2)\omega_m^2 \beta''(\omega_0)L\right)\right)$ the (scalar) small signal intensity transfer function is

$$H_m(\omega_m) = e^{-j\beta'(\omega_0)L} \cos\left((1/2)\omega_m^2 \beta''(\omega_0)L\right). \tag{2.318}$$

For rising modulation frequency its magnitude decreases, reaching a first zero at $(1/2)\omega_m^2 |\beta''(\omega_0)|L = \pi/2$ or $\omega_m = \sqrt{\pi/(|\beta''(\omega_0)|L)}$, with $D = -\frac{2\pi c}{\lambda^2}\beta''$ (2.159). Chromatic dispersion aggravates proportional to L and to the square of ω_m ! See also the Problem on p. 16.

Some properties of eigenmodes and principal states-of-polarizations are summarized in Table 2.3.

Table 2.3 Properties of eigenmodes and principal states-of-polarization

Eigenmodes	*Principal states-of-polarization (PSPs)*
Output polarization equals input polarization if eigenmode is launched and/or if, in a retarder, the retardation equals an integer multiple of 2π.	Minimum and maximum group delays occur for fast and slow PSP, respectively. Output polarization does not vary linearly with frequency (when displayed on the Poincaré sphere) if PSP is launched. Output PSPs generally differ from input PSPs.
Lossless medium → orthogonal eigenmodes	Lossless medium → orthogonal PSPs
Adding a retarder before or behind a fiber can change the eigenmodes.	Adding a frequency-independent retarder before (behind) a fiber can change only the input-referred (output-referred) PSPs.
For a transmission medium the PSPs become the eigenmodes if the local PSPs, equal to the local eigenmodes, are constant along the transmission medium.	

2.4 Linear Electrooptic Effect

2.4.1 Phase Modulation

So far we have considered that the electromagnetic displacements are strictly proportional to the respective fields. A closer investigation reveals nonlinear behavior. While $\mathbf{B} = \mu_0 \mathbf{H}$ holds for optical media the dielectric displacement must be written as

$$\mathbf{D} = \varepsilon_0 \left(\left(\mathbf{1} + \chi^{(1)} \right) \mathbf{E} + \chi^{(2)} : \mathbf{EE} + \chi^{(3)} \vdots \mathbf{EEE} + \ldots \right). \tag{2.319}$$

The term proportional to $\mathbf{E}$ describes linear behavior including anisotropy, while the others describe quadratic and cubic dependencies. Matrix $\left(\mathbf{1} + \chi^{(1)} \right) = \varepsilon_r$ is a tensor of rank 2, since matrix elements can be arranged 2-dimensionally. Second and third order susceptibilities $\chi^{(2)}$, $\chi^{(3)}$ are tensors of ranks 3 and 4 and their elements can be arranged 3- or 4-dimensionally, respectively. The number of vertically stacked multiplication dots is explained by the following expression which shows how the vector elements of $\mathbf{D}$ are calculated from those of $\mathbf{E}$,

$$D_i = \varepsilon_0 \left(E_i + \sum_j \chi_{ij}^{(1)} E_j + \sum_j \sum_k \chi_{ijk}^{(2)} E_j E_k + \sum_j \sum_k \sum_l \chi_{ijkl}^{(3)} E_j E_k E_l + \ldots \right). \tag{2.320}$$

In crystals $\chi^{(2)}$ usually has non-zero elements, and the associated term delivers more important contributions to D than that with $\chi^{(3)}$. In contrast, silica is amorphous so that $\chi^{(2)}$ vanishes and $\chi^{(3)}$ is the most important nonlinearity. Returning to crystals and $\chi^{(2)}$, $\mathbf{D}$ has terms proportional to all combinations $E_j E_k$. Let for example be $E_3 = E_{DC} + E_0 \cos \omega t$, $j = k = 3$, where E_{DC} is an electrostatic field and $E_0 \cos \omega t$ an optical field. Then $\mathbf{D}$ contains terms proportional to

$$E_3^2 = \left(E_{DC}^2 + \frac{1}{2} E_0^2 \right) + 2 E_{DC} E_0 \cos \omega t + \frac{1}{2} E_0^2 \cos 2\omega t. \tag{2.321}$$

The first term on the right hand side causes a static dielectric displacement, the third a contribution toward optical frequency doubling and the second is proportional to both the static field and the optical field. In other words, it is a contribution to $\mathbf{D}$ that is linear in the optical field and can be controlled by the electrostatic field. This linear electrooptic effect is also called Pockels effect (Friedrich Pockels, 1906). Expressed in terms of our previously developed terminology it deforms the index ellipsoid. In the main coordinate system the index ellipsoid equation for a lossless reciprocal medium can be written as [1]

$$\begin{gathered}X^2\left(\frac{1}{n_1^2}+\sum_{j=1}^{3} r_{1j}E_j\right)+Y^2\left(\frac{1}{n_2^2}+\sum_{j=1}^{3} r_{2j}E_j\right)+Z^2\left(\frac{1}{n_3^2}+\sum_{j=1}^{3} r_{3j}E_j\right)\\ +2YZ\sum_{j=1}^{3} r_{4j}E_j+2ZX\sum_{j=1}^{3} r_{5j}E_j+2XY\sum_{j=1}^{3} r_{6j}E_j=1\end{gathered}, \quad (2.322)$$

where indexes 1, 2, 3 mean x, y, z and the field components E_j are the electrostatic ones. The linear electrooptic coefficients r_{ij} can be arranged in a second rank tensor (matrix) with 6 lines and 3 columns. This is simpler than using the 27 elements of $\chi^{(2)}$.

We pick Lithium Niobate ($LiNbO_3$) as one example that is important for optical communication. This transparent, synthetic, non-conductive crystal belongs to the symmetry group 3m. It is uniaxial with $n_X = n_Y = n_o \approx 2.21283$, $n_Z = n_e \approx 2.13739$. The electrooptic tensor contains mostly zeros and among the eight non-zero elements there are equal ones, which leaves only four independent coefficients,

$$\left[r_{ij}\right]=\begin{bmatrix}0 & r_{12} & r_{13}\\ 0 & -r_{12} & r_{13}\\ 0 & 0 & r_{33}\\ 0 & r_{51} & 0\\ r_{51} & 0 & 0\\ r_{12} & 0 & 0\end{bmatrix}. \quad (2.323)$$

At 1550 nm wavelength it holds $r_{33} \approx 30.8\,\text{pm/V}$, $r_{51} = r_{42} \approx 26\,\text{pm/V}$, $r_{13} = r_{23} \approx 8.6\,\text{pm/V}$, $r_{12} = -r_{22} = r_{61} \approx -3.4\,\text{pm/V}$. Consider the phase modulator of Fig. 2.17. The crystal is cut in the crystallographic Z direction. The cut axis is denominated according to the sawing process; it is perpendicular to the crystal surface. Along the Y crystal axis, at the surface, there are two waveguides 1, 2 formed by Titanium dioxide (TiO_2). It is fabricated by previous indiffusion at ~1080°C of a Ti stripe which is deposited photolithographically on the crystal surface. At the crystal surface there is also a buffer layer, for example SiO_2. On top of it, above the waveguides, there are gold electrodes between which a voltage U is applied. The fundamental modes of the integrated waveguides correspond to the HE_{11} modes of the optical fiber. However, usually they are denoted as TE (parallel to the surface, here the X crystal axis) and TM modes (perpendicular to the surface, here the Z crystal axis), like in a slab waveguide.

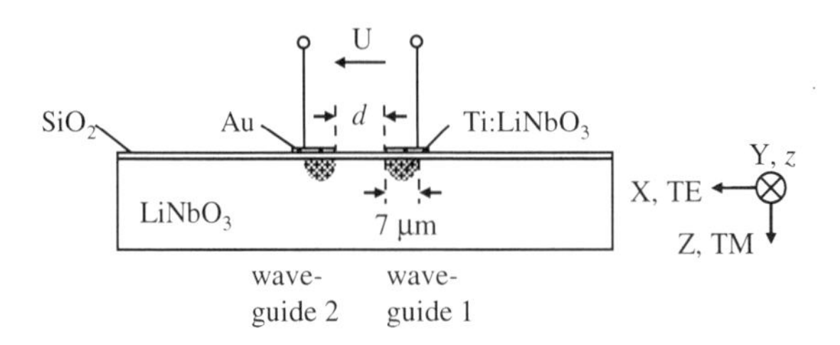

Fig. 2.17 Cross-section of Mach-Zehnder modulator in Z-cut $LiNbO_3$. The differential phase modulation in an interferometer serves for amplitude, polarity and power modulation

The electrostatic field has X and Z components E_1, E_3. The last column of (2.323) with $r_{43} = r_{53} = r_{63} = 0$ shows that in (2.322) the main axes of the index ellipsoid do not change directions, only lengths. For a TM wave, polarized in Z direction (= 3) as one eigenmode, only the term $\left(\frac{1}{n_3^2} + \sum_{j=1}^{3} r_{3j} E_j\right)$ is therefore relevant, where $n_3 = n_Z = n_e$. Since $E_2 = 0$ and $r_{31} = 0$ holds, the term simplifies to $\left(\frac{1}{n_e^2} + r_{33} E_3\right)$. We take it to the power $-1/2$ to obtain the relevant refractive index,

$$n = \left(\frac{1}{n_e^2} + r_{33} E_3\right)^{-1/2} \approx n_e - \frac{n_e^3}{2} r_{33} E_3 . \tag{2.324}$$

The electrostatic field equals $E_3 = \Gamma U/d$, where $0 < \Gamma < 1$ is an overlap factor which takes into account that the relevant electrostatic field is weaker than that caused by a voltage U between two parallel plates having the same distance as the electrode gap d. So, the refractive index difference introduced by the voltage U in waveguide 1 equals

$$\Delta n = -\frac{n_e^3}{2} r_{33} \Gamma \frac{U}{d} \tag{2.325}$$

whereas for waveguide 2 it equals $-\Delta n$ because the Z component of the electrostatic field is oppositely poled there. Phase modulation as caused in one of the two waveguides is possible, though there are not many applications where this is needed. However, the phase difference $-2\Delta n L/\lambda$ between two waveguides 1 and 2 having the length L serves to modulate the optical field amplitude and polarity in Mach-Zehnder modulators.

The chirp factor of a device such as a modulator driven by voltage U, is defined as $\alpha_H = \frac{\partial n/\partial U}{\partial n_i/\partial U}$, where n is the real part and n_i is the negative imaginary (absorptive) part of the complex refractive index. Since the transfer function is $H(U) = e^{-j\beta_0(n-jn_i)L} = e^{-\beta_0 n_i(1+j\alpha_H)L}$, a corresponding modulator driven by a quadratically time-dependent n_i can readily generate chirped Gaussian pulses (2.86). For electroabsorption modulators the chirp factor is usually positive and fairly independent of U, on the order of +1...+3, which makes them difficult to use in the presence of chromatic dispersion. With the optical field $\underline{E}$ being proportional to the transfer function $H(U)$ we find the chirp factor definitions

$$\alpha_H = \frac{\partial n/\partial U}{\partial n_i/\partial U} = |\underline{E}|\frac{\partial\varphi/\partial U}{\partial|\underline{E}|/\partial U} = 2P\frac{\partial\varphi/\partial U}{\partial P/\partial U}. \tag{2.326}$$

The last expression holds due to $P = \frac{1}{2}|\underline{E}|^2$, $\frac{1}{P}\frac{\partial P}{\partial U} = \frac{1}{P}\frac{\partial P}{\partial|\underline{E}|}\frac{\partial|\underline{E}|}{\partial U} = \frac{2}{|\underline{E}|}\frac{\partial|\underline{E}|}{\partial U}$.

The Mach-Zehnder modulator, which is an interferometer with electrically controllable phase difference, is sketched as seen from the top in the left part of Fig. 3.27 (p. 210). The field transfer function of the input Y fork is $\frac{1}{\sqrt{2}}\begin{bmatrix}1\\1\end{bmatrix}$. Due to reciprocity the transfer function of the Y combiner at the output is just its transpose, $\frac{1}{\sqrt{2}}[1 \;\; 1]$. If the phase shifts imposed in waveguides 1 and 2 are $\varphi_{1,2}$ then the total modulator transfer function is $H(U) = \frac{1}{2}\left(e^{j\varphi_1} + e^{j\varphi_2}\right) = e^{j\varphi}\cos(\Delta\varphi/2)$ with mean phase shift $\varphi = \frac{1}{2}(\varphi_1 + \varphi_2)$ (= 0) and differential phase shift $\Delta\varphi = (\varphi_1 - \varphi_2)$ (= $-2\Delta nL/\lambda$). The mean phase φ is independent of U, and $\alpha_H = 0$ holds therefore.

Problem: If the left electrode 1 in Fig. 2.17 is chosen very broad, so that it serves as a ground electrode, and if another ground electrode is placed at the right side of electrode 2, not all electric field lines will not go through waveguide 2. Instead of $\varphi_2/\varphi_1 = -1$ it then holds $\varphi_2/\varphi_1 = K \approx -0.5$. Calculate the chirp factor generally and at the point $|H(U)|^2 = \frac{1}{2}$, i.e. the operation point of an intensity modulator.

Solution: With $P = |H(U)|^2 = \frac{1}{2}(1+\cos\Delta\varphi)$ we calculate $\frac{\partial\varphi}{\partial U} = \frac{\partial\Delta\varphi}{\partial\varphi_1}\frac{\partial\varphi_1}{\partial U} = \frac{1}{2}(1+K)\frac{\partial\varphi_1}{\partial U}$,

$\frac{1}{2P}\frac{\partial P}{\partial U} = \frac{1}{2P}\frac{\partial P}{\partial\Delta\varphi}\frac{\partial\Delta\varphi}{\partial\varphi_1}\frac{\partial\varphi_1}{\partial U} = \frac{1}{1+\cos\Delta\varphi}\frac{-\sin\Delta\varphi}{2}(1-K)\frac{\partial\varphi_1}{\partial U}$,

$$\alpha_H = -\frac{1+\cos\Delta\varphi}{\sin\Delta\varphi}\frac{1+K}{1-K}. \quad (2.327)$$

Here the chirp factor depends on the operation point or applied voltage, other than in (2.86)! For $|H(U)|^2 = \frac{1}{2}$ one finds $\Delta\varphi = \pm\pi/2$, $\alpha_H = \mp 2\frac{1+K}{1-K} \approx \mp 0.7$. The chirp sign depends on the chosen modulator slope (normal or inverted). At the operation point $\Delta\varphi = \pi/2$, the resulting $\alpha_H \approx -0.7$ mitigates the effect of SSMF fiber dispersion somewhat.

2.4.2 Soleil-Babinet Compensator

Another integrated electrooptic component is a Soleil-Babinet compensator in X-cut, Z-propagation $LiNbO_3$ which may be used for unlimited, endless polarization control (Fig. 2.18). The X-Y plane is perpendicular to the Z propagation axis, so that (2.322) simplifies into

$$X^2\left(\frac{1}{n_o^2} + \sum_{j=1}^{3} r_{1j}E_j\right) + Y^2\left(\frac{1}{n_o^2} + \sum_{j=1}^{3} r_{2j}E_j\right) + 2XY\sum_{j=1}^{3} r_{6j}E_j = 1. \quad (2.328)$$

Considering $E_3 = 0$ and (2.323) the ellipse equation

$$X^2\left(\frac{1}{n_o^2} + r_{12}E_2\right) + Y^2\left(\frac{1}{n_o^2} - r_{12}E_2\right) + 2XYr_{12}E_1 = 1. \quad (2.329)$$

results. Voltage U_1 causes a field which is X-directed on average, $E_1 = \Gamma_1 U_1/d$. The average field generated by voltage U_2 is Y-directed, $E_2 = \Gamma_2 U_2/d$. The waveguide has a length L.

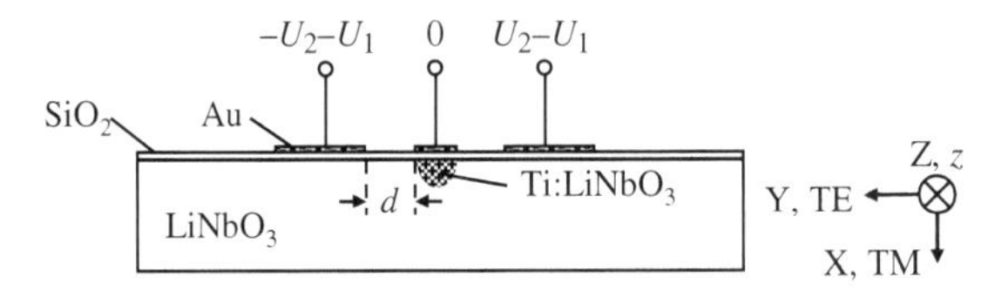

Fig. 2.18 Cross-section of electrooptic Soleil-Babinet compensator for endless polarization control in X-cut, Z-propagation $LiNbO_3$

Problem: Find normalized Stokes eigenvectors and retardation of this retarder.

Solution: The ellipse equation can be written as a Hermitian form, $\begin{bmatrix} X \\ Y \end{bmatrix}^+ \begin{bmatrix} n_o^{-2} + r_{12}E_2 & r_{12}E_1 \\ r_{12}E_1 & n_o^{-2} - r_{12}E_2 \end{bmatrix} \begin{bmatrix} X \\ Y \end{bmatrix} = 1$. The eigenvalues and the first of the two orthogonal eigenvectors are $\lambda_{1,2} = n_o^{-2} \pm r_{12}\sqrt{E_2^2 + E_1^2}$, $\begin{bmatrix} X_1 \\ Y_1 \end{bmatrix} = \begin{bmatrix} -E_1 \\ E_2 - \sqrt{E_2^2 + E_2^2} \end{bmatrix}$. Its normalized Stokes vector $\mathbf{V}$ is given in (2.330). The refractive indexes are $n_{1,2} = \lambda_{1,2}^{-1/2} \approx n_o\left(1 \mp \left(n_o^2/2\right)r_{12}\sqrt{E_2^2 + E_1^2}\right)$. The retardation is $\delta = (n_2 - n_1)L$.

The normalized Stokes vector of one of the eigenmodes is

$$\mathbf{V} = \frac{1}{\sqrt{E_2^2 + E_1^2}} \begin{bmatrix} E_2 \\ E_1 \\ 0 \end{bmatrix}. \tag{2.330}$$

The retardation of the second one, $-\mathbf{V}$, with respect to the $\mathbf{V}$ equals

$$\delta = n_o^3 r_{12}\sqrt{E_2^2 + E_1^2}\,L. \tag{2.331}$$

We set $\pi = n_o^3 r_{12}\Gamma_2\left(U_{2,\pi}/d\right)L = n_o^3 r_{12}\Gamma_1\left(U_{1,\pi}/d\right)L$ to define the voltages $U_{2,\pi}$, $U_{1,\pi}$ which, if applied alone, generate a retardation equal to π. This results in

$$\mathbf{V} = \frac{\pi}{\delta}\begin{bmatrix} U_2/U_{2,\pi} \\ U_1/U_{1,\pi} \\ 0 \end{bmatrix} = \begin{bmatrix} \cos 2\alpha \\ \sin 2\alpha \\ 0 \end{bmatrix}, \tag{2.332}$$

$$\delta = \pi\sqrt{\left(U_2/U_{2,\pi}\right)^2 + \left(U_1/U_{1,\pi}\right)^2}, \tag{2.333}$$

where α is the azimuth angle of the linearly polarized eigenmode (2.330). It is convenient to write

$$\mathbf{V} = \frac{1}{\sqrt{\left(U_2/U_{2,\pi}\right)^2 + \left(U_1/U_{1,\pi}\right)^2}}\begin{bmatrix} U_2/U_{2,\pi} \\ U_1/U_{1,\pi} \\ 0 \end{bmatrix}. \tag{2.334}$$

Note that even in the absence of voltages there is a TE-TM birefringence due to the absence of point symmetry of the waveguide cross section. This must be taken into account by an offset voltage that is added to U_2. But for simplicity we do not take this into account in the equations. The retarder is an electrooptic waveplate with adjustable retardation and azimuth angle. It works like a bulk-optic Soleil-Babinet compensator (SBC). An SBC is composed of two birefringent wedges. These are inserted from both sides into the optical path so that they form one waveplate if one ignores the slanted slot between them. Sliding the wedges against each other changes the thickness of this plate, hence the optical path lengths and the retardation. The whole arrangement can be turned around the propagation axis so that the azimuth angle of the waveplate is also adjustable.

Jones and Müller matrices of the SBC are given by (2.241), (2.242). Or else, (2.240) yields

$$\mathrm{SBC}(\delta,2\alpha)=\mathbf{G}=\begin{bmatrix} V_1^2+V_2^2\cos\delta & V_1V_2(1-\cos\delta) & V_2\sin\delta \\ V_1V_2(1-\cos\delta) & V_2^2+V_1^2\cos\delta & -V_1\sin\delta \\ -V_2\sin\delta & V_1\sin\delta & \cos\delta \end{bmatrix}. \tag{2.335}$$

With circular input polarization, the output polarization becomes

$$\mathbf{S}=\mathrm{SBC}(\delta,2\alpha)\begin{bmatrix}0\\0\\1\end{bmatrix}=\begin{bmatrix}V_2\sin\delta\\-V_1\sin\delta\\\cos\delta\end{bmatrix}. \tag{2.336}$$

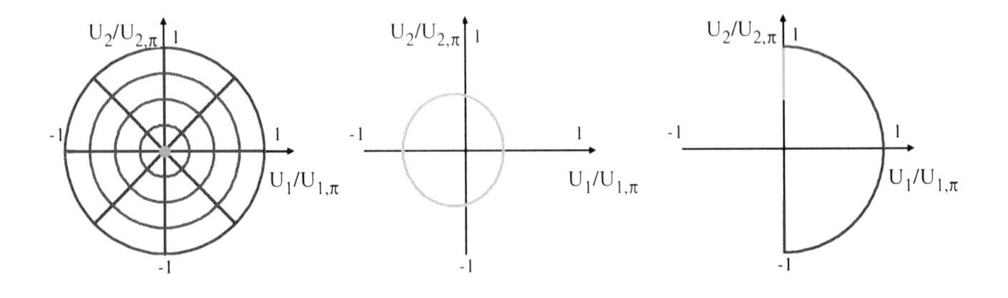

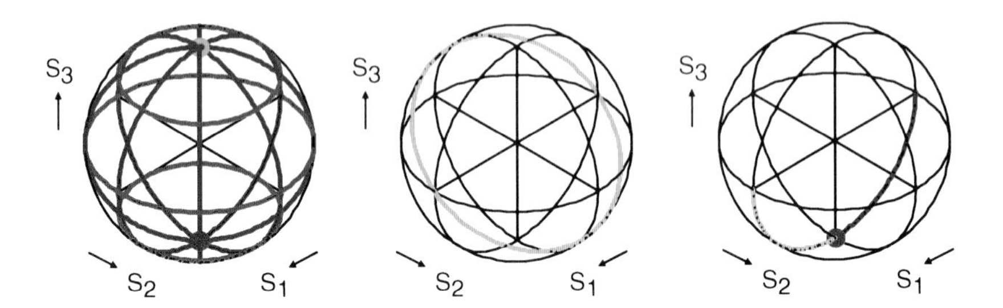

Fig. 2.19 Azimuthal equidistant projecctions of normalized voltage plane (top) to respective output polarization on Poincaré sphere surface (bottom) for circular input polarization (green spot on left Poincaré sphere). The circumference of the unit circle of the voltage plane is mapped into the second, orthogonal circular output polarization. This allows to prevent voltage overflows by limiting the output voltages within the unit circle.

Fig. 2.19 shows trajectories in the normalized voltage plane and associated output polarization trajectories. The whole Poincaré sphere is covered, by an azimuthal equidistant projection from the voltage plane to the sphere surface (left). Normally any required Poincaré sphere trajectories are achieved by trajectories inside the unit circle of normalized voltages, for example the particular Poincaré sphere great circle shown in the middle. A problem occurs for trajectories passing the $\begin{bmatrix}0 & 0 & -1\end{bmatrix}^T$ pole of the sphere. Under normal circumstances tracking will make the control voltages leave the unit circle. At least if this continues the component will be destroyed by overvoltage. Else the retardation must be diminuished by 2π. The latter operation makes the output polarization rotate on one complete great circle, and it needs some time, on the order of ns to ms depending on implementation. But if polarization control is needed for optical information transmission it must be interruption-free, "endless". One could not afford the loss of, say, $10^3...10^6$ data symbols due to a polarization control reset. A suitable strategy to make this polarization control endless is to limit the control voltages within the unit circle [11]. As the output polarization passes the $\begin{bmatrix}0 & 0 & -1\end{bmatrix}^T$ pole the SBC azimuth will or must be changed until a further output polarization change can result in a reduced distance from the origin of the voltage plane.

We have found an endless electrooptic polarization transformer that can transform circular input polarization into any output polarization. Looking at the third line of (2.335), an input polarization with the normalized Stokes vector $\mathbf{S} = \begin{bmatrix}S_1 & S_2 & S_3\end{bmatrix}^T$ is transformed into an output polarization $\begin{bmatrix}0 & 0 & 1\end{bmatrix}^T$ if we choose $-V_2 \sin\delta = S_1$, $V_1 \sin\delta = S_2$, $\cos\delta = S_3$. So, the SBC allows endless polarization control also in the opposite sense. A very similar component is described on p. 107.

Experimentally, endless polarization control with good quality is possible up to speeds of 56 krad/s on the Poincaré sphere [12–17].

2.5 Mode Coupling

Many optical components and effects can be described and explained by the mode coupling theory: Anisotropic components including components with PMD and polarization-maintaining fibers (PMFs), directional couplers, gratings and hence also most single-mode lasers, electrooptical and acoustooptical components, nonlinear effects. Mode coupling [1] means the conversion of power from one mode into another, and vice versa. The power is just a quantity derived from the electric fields. In many cases not only the magnitude but also the phase of the coupled field is of importance.

The integration of coupled differential equations for the fields corresponds to the multiplication of a field vector onto many matrices. For codirectional mode coupling these are transfer matrices or Jones matrices, for counterdirectional mode coupling these are transmission matrices.

2.5.1 Mode Orthogonality

In a solenoidal, homogeneous medium there be various electromagnetical fields μ, ν with $\mu, \nu \in \{1, 2, ..., n\}$. These fields be generated by sources outside the volume under consideration. One multiplies $\text{curl}\,\underline{\mathbf{H}}_\mu = j\omega\underline{\varepsilon}\underline{\mathbf{E}}_\mu$ in a scalar multiplication by $\underline{\mathbf{E}}_\nu$ and $\text{curl}\,\underline{\mathbf{E}}_\nu = -j\omega\underline{\mu}\underline{\mathbf{H}}_\nu$ by $\underline{\mathbf{H}}_\mu$, subtracts the resulting equations and obtains with $\text{div}(\underline{\mathbf{A}} \times \underline{\mathbf{B}}) = \underline{\mathbf{B}} \cdot \text{curl}\,\underline{\mathbf{A}} - \underline{\mathbf{A}} \cdot \text{curl}\,\underline{\mathbf{B}}$

$$\begin{aligned} -\text{div}\left(\underline{\mathbf{E}}_\nu \times \underline{\mathbf{H}}_\mu\right) &= \underline{\mathbf{E}}_\nu \cdot \text{curl}\,\underline{\mathbf{H}}_\mu - \underline{\mathbf{H}}_\mu \cdot \text{curl}\,\underline{\mathbf{E}}_\nu \\ &= j\omega\left(\underline{\mathbf{E}}_\nu \cdot \underline{\varepsilon}\underline{\mathbf{E}}_\mu + \underline{\mathbf{H}}_\mu \cdot \underline{\mu}\underline{\mathbf{H}}_\nu\right) \end{aligned} . \quad (2.337)$$

We exchange the indices μ, ν and subtract the resulting equation from (2.337), to obtain

$$\begin{aligned} &\text{div}\left(\underline{\mathbf{E}}_\nu \times \underline{\mathbf{H}}_\mu - \underline{\mathbf{E}}_\mu \times \underline{\mathbf{H}}_\nu\right) \\ &= j\omega\left(\underline{\mathbf{E}}_\nu \cdot \left(\underline{\varepsilon}\underline{\mathbf{E}}_\mu\right) - \underline{\mathbf{E}}_\mu \cdot \left(\underline{\varepsilon}\underline{\mathbf{E}}_\nu\right) + \underline{\mathbf{H}}_\mu \cdot \left(\underline{\mu}\underline{\mathbf{H}}_\nu\right) - \underline{\mathbf{H}}_\nu \cdot \left(\underline{\mu}\underline{\mathbf{H}}_\mu\right)\right) \end{aligned} . \quad (2.338)$$

The 4 summands at the right hand side are scalars. They can be written as matrix products. The 1st and 4th summand are transposed and allow to rewrite the right hand side as $j\omega\left(\underline{\mathbf{E}}_\mu^T\left(\underline{\varepsilon}^T - \underline{\varepsilon}\right)\underline{\mathbf{E}}_\nu + \underline{\mathbf{H}}_\mu^T\left(\underline{\mu} - \underline{\mu}^T\right)\underline{\mathbf{H}}_\nu\right)$. For symmetric material tensors it follows the

$$\text{div}\left(\underline{\mathbf{E}}_\nu \times \underline{\mathbf{H}}_\mu - \underline{\mathbf{E}}_\mu \times \underline{\mathbf{H}}_\nu\right) = 0 \qquad \textit{Lorentz reciprocity theorem} \quad (2.339)$$

for solenoidal media in differential form. The divergence is integrated over a volume and the volume integral is, using Gauß's integral theorem, replaced by a closed surface integral. It results the

$$\oint\!\!\!\oint \left(\underline{\mathbf{E}}_\nu \times \underline{\mathbf{H}}_\mu - \underline{\mathbf{E}}_\mu \times \underline{\mathbf{H}}_\nu\right) \cdot d\mathbf{A} = 0 \quad \textit{Lorentz reciprocity theorem} \quad (2.340)$$

for solenoidal media in integral form. Before we derive from it a mode orthogonality condition we discuss its implications:

Most optical media are reciprocal and have symmetric dielectricity tensors. In the exercise lossless anisotropic media will be considered, among these one with Faraday rotation and a non-symmetrical

$$\underline{\varepsilon} = \varepsilon_0 \begin{bmatrix} \varepsilon_r & -jG_0 & 0 \\ jG_0 & \varepsilon_r & 0 \\ 0 & 0 & \varepsilon_r \end{bmatrix}. \tag{2.341}$$

The eigenmodes are circular here. The azimuth of incident linearly polarized light is rotated. Since $\underline{\varepsilon}$ is Hermitian the medium is lossless, but it is non-reciprocal. The tensor depends on the direction of a static magnetic field. For a given magnetic field $\underline{\varepsilon}$ is constant in any given coordinate system. In an exercise it has been discussed that such a YIG crystal can be used to build an optical isolator. The direction dependence of signal transmissivity in non-reciprocal media is possible only because one direction is designated by an external effect, here by the static magnetic field.

In addition to these there are *optically active* media with elliptically or circularly polarized eigenmodes. For a given propagation direction these (nearly) lossless media have a Hermitian, non-symmetrical $\underline{\varepsilon}$ tensor. If the medium shall also be reciprocal one finds a contradiction because a non-vanishing, even and odd symmetric imaginary part of the $\underline{\varepsilon}$ tensor is not possible. In optically active media the magnetic field which is responsible for the gyration is not external and static but belongs to the wave itself. Therefore the tensor elements depend on the propagation direction. In other words, in the term $\underline{\varepsilon}\underline{\mathbf{E}}$, $\underline{\varepsilon}$ depends on $\underline{\mathbf{E}}$, which is not the case for linearly birefringent nor for Faraday media. However, Maxwell's equations require material constants and therefore need not cover such special cases in their general form. If one defines a vector function $\underline{\mathbf{D}} = f(\underline{\mathbf{E}})$ rather than the linear relation $\underline{\mathbf{D}} = \underline{\varepsilon}\underline{\mathbf{E}}$ then optically active media can be included without contradiction. So, the Lorentz reciprocity theorem does not "prevent" optically active media from being reciprocal. Indeed they are reciprocal. Their molecules are chiral. E.g., carbon with four different ligands forms chiral molecules. These have a handedness, like a screw. The polarization rotation direction is specified by the handedness of the screw.

The *Sagnac* effect, which results from the general relativity theory, is non-reciprocal. If a system has a velocity then the time a lightwave needs to pass through a medium depends on the direction. In a fiber loop which is passed by a wave from both sides the relative phase of one with respect to the other wave changes if the measurement apparatus including fiber loop is being rotated. This allows to construct fiberoptic gyroscopes.

Summarizing: Non-reciprocity results in a direction-dependence of propagation properties. In reciprocal, optically active media a material-"constant", the $\underline{\varepsilon}$ tensor, is not a constant.

We derive now the mode orthogonality in a medium with symmetric material tensors. Let the waves μ,ν be guided modes with the fields

$$\underline{\mathbf{E}}_i = \underline{\mathbf{E}}_{i0}(x,y)e^{j(\omega t-\underline{k}_i z)} \qquad \underline{\mathbf{H}}_i = \underline{\mathbf{H}}_{i0}(x,y)e^{j(\omega t-\underline{k}_i z)} \qquad i=\mu,\nu \tag{2.342}$$

and propagate in the positive z direction. Insertion into $\operatorname{div}(\underline{\mathbf{E}}_\nu \times \underline{\mathbf{H}}_\mu - \underline{\mathbf{E}}_\mu \times \underline{\mathbf{H}}_\nu)=0$ (2.339) results in

$$\operatorname{div}_t(\underline{\mathbf{E}}_\nu \times \underline{\mathbf{H}}_\mu - \underline{\mathbf{E}}_\mu \times \underline{\mathbf{H}}_\nu) - j\mathbf{e}_z \cdot (\underline{k}_\mu + \underline{k}_\nu)(\underline{\mathbf{E}}_\nu \times \underline{\mathbf{H}}_\mu - \underline{\mathbf{E}}_\mu \times \underline{\mathbf{H}}_\nu) = 0 . \tag{2.343}$$

This equation is integrated over the whole x-y plane. The operator of the left term is a 2-dimenisonal, transversal divergence. The surface integral over the transversal divergence is transformed by a 2-dimensional Gauß integral theorem (instead of the usual 3-dimensional one) into a 1-dimensional closed loop integral (instead of the usual 2-dimensional surface integral) over the argument of the transversal divergence,

$$\begin{aligned}&\oint(\underline{\mathbf{E}}_\nu \times \underline{\mathbf{H}}_\mu - \underline{\mathbf{E}}_\mu \times \underline{\mathbf{H}}_\nu)\cdot \mathbf{e}_n dl \\ &- j(\underline{k}_\mu + \underline{k}_\nu)\iint(\underline{\mathbf{E}}_\nu \times \underline{\mathbf{H}}_\mu - \underline{\mathbf{E}}_\mu \times \underline{\mathbf{H}}_\nu)\cdot \mathbf{e}_z dxdy = 0\end{aligned} \tag{2.344}$$

Here $\mathbf{e}_n$ is a unit vector in the x-y plane which is perpendicular to the integration path and points outward. The closed-loop integral equals zero because the fields decay roughly exponentially in a large distance from the z axis. This way eqn. (2.343) reduces to

$$(\underline{k}_\mu + \underline{k}_\nu)\iint(\underline{\mathbf{E}}_\nu \times \underline{\mathbf{H}}_\mu - \underline{\mathbf{E}}_\mu \times \underline{\mathbf{H}}_\nu)\cdot \mathbf{e}_z\, dxdy = 0 . \tag{2.345}$$

Only the transversal fields (index t) contribute, since only the z components of the outer products are taken into account in the scalar multiplication with $\mathbf{e}_z$. This can be shown by application of $\mathbf{u}\cdot(\mathbf{v}\times\mathbf{w}) = \mathbf{v}\cdot(\mathbf{w}\times\mathbf{u})$; for example $(\underline{\mathbf{E}}_\nu \times \underline{\mathbf{H}}_\mu)\cdot \mathbf{e}_z = (\underline{\mathbf{H}}_\mu \times \mathbf{e}_z)\cdot \underline{\mathbf{E}}_\nu = (\underline{\mathbf{H}}_{\mu,t} \times \mathbf{e}_z)\cdot \underline{\mathbf{E}}_\nu = (\underline{\mathbf{E}}_\nu \times \underline{\mathbf{H}}_{\mu,t})\cdot \mathbf{e}_z$. Now the fields (2.342) are inserted. The common phasor is canceled and one obtains

$$(\underline{k}_\mu + \underline{k}_\nu)\iint(\underline{\mathbf{E}}_{\nu 0} \times \underline{\mathbf{H}}_{\mu 0} - \underline{\mathbf{E}}_{\mu 0} \times \underline{\mathbf{H}}_{\nu 0})\cdot \mathbf{e}_z\, dxdy = 0 . \tag{2.346}$$

It is now convenient to assign mode indices with opposite signs to modes which differ only by the sign of the propagation constant. In addition to the propagation constant $\underline{k}_{-\nu} = -\underline{k}_\nu$ and the mode index also the longitudinal electric and the transversal magnetic field change their signs, see also (2.354). Since only the transversal components have an effect here, the analog to (2.346) for the modes $\mu,-\nu$ is

$$(\underline{k}_\mu - \underline{k}_\nu)\iint(\underline{\mathbf{E}}_{\nu 0} \times \underline{\mathbf{H}}_{\mu 0} + \underline{\mathbf{E}}_{\mu 0} \times \underline{\mathbf{H}}_{\nu 0})\cdot \mathbf{e}_z\, dxdy = 0 . \tag{2.347}$$

For $|\mu| \neq |\nu|$ we may divide here and in (2.346) by the difference and sum of propagation constants, respectively. The resulting equations are added to a mode orthogonality relation (under the condition $\nu \neq -\mu$)

$$\iint \left(\underline{\mathbf{E}}_{\nu 0} \times \underline{\mathbf{H}}_{\mu 0}\right) \cdot \mathbf{e}_z \, dxdy = 0 \qquad \text{in } \underline{\text{reciprocal}} \text{ media.} \tag{2.348}$$

This holds also if instead of the field amplitudes the z- and t-dependent fields are inserted. It is general and holds, e.g., for eigenmodes of anisotropic media with symmetric $\underline{\varepsilon}$, slab waveguides, cylindrical dielectric waveguides.

In solenoidal and lossless media similar mode orthogonality relations can be derived: Multiplication of $\operatorname{curl} \underline{\mathbf{H}}_{\mu}^{*} = -j\omega \underline{\varepsilon}^{*} \underline{\mathbf{E}}_{\mu}^{*}$ with $\underline{\mathbf{E}}_{\nu}$, $\operatorname{curl} \underline{\mathbf{E}}_{\nu} = -j\omega \underline{\mu} \underline{\mathbf{H}}_{\nu}$ with $\underline{\mathbf{H}}_{\mu}^{*}$, results in

$$-\operatorname{div}\left(\underline{\mathbf{E}}_{\nu} \times \underline{\mathbf{H}}_{\mu}^{*}\right) = j\omega\left(-\underline{\mathbf{E}}_{\nu} \cdot \underline{\varepsilon}^{*} \underline{\mathbf{E}}_{\mu}^{*} + \underline{\mathbf{H}}_{\mu}^{*} \cdot \underline{\mu} \underline{\mathbf{H}}_{\nu}\right). \tag{2.349}$$

For $\mu = \nu$ this is Poynting's theorem for anisotropic media. One exchanges the indices in (2.349), takes the complex conjugate and adds it to the original eqn. (2.349). For Hermitian material tensors (= lossless media) it results

$$\operatorname{div}\left(\underline{\mathbf{E}}_{\nu} \times \underline{\mathbf{H}}_{\mu}^{*} + \underline{\mathbf{E}}_{\mu}^{*} \times \underline{\mathbf{H}}_{\nu}\right) = 0. \tag{2.350}$$

From this one obtains a mode orthogonality relation (under the condition $\nu \neq -\mu$)

$$\iint \left(\underline{\mathbf{E}}_{\nu 0} \times \underline{\mathbf{H}}_{\mu 0}^{*}\right) \cdot \mathbf{e}_z \, dxdy = 0 \qquad \text{for } \underline{\text{lossless}} \text{ media.} \tag{2.351}$$

In the following we stay in a lossless medium with a real characteristic impedance. With the restriction $\nu \neq -\mu$, and combined with Poynting's theorem in integral form one obtains

$$\left|\frac{1}{2} \iint \left(\underline{\mathbf{E}}_{\mu} \times \underline{\mathbf{H}}_{\nu}^{*}\right) \cdot d\mathbf{A}\right| = \left|\frac{1}{2} \iint \left(\underline{\mathbf{E}}_{t,\mu} \times \underline{\mathbf{H}}_{t,\nu}^{*}\right) \cdot d\mathbf{A}\right| = \delta_{|\mu||\nu|} \sqrt{P_{\mu} P_{\nu}}\,. \tag{2.352}$$

Next the fields are normalized using

$$\begin{aligned} \underline{\mathbf{E}}(\mathbf{r}, t) &= \underline{E} Z_F^{1/2} \underline{\mathbf{E}}^{(1/m)}(x, y) e^{j(\omega t - \beta z)} \\ \underline{\mathbf{H}}(\mathbf{r}, t) &= \underline{E} Z_F^{-1/2} \underline{\mathbf{H}}^{(1/m)}(x, y) e^{j(\omega t - \beta z)} \end{aligned}. \tag{2.353}$$

For counterdirectional waves $\mu, -\mu$ the signs of one longitudinal field component, say $\underline{\mathbf{E}}_z^{(1/m)}$, and of the other transversal component, here $\underline{\mathbf{H}}_t^{(1/m)}$, must be inverted,

$$\underline{\mathbf{E}}_{t,-\mu}^{(1/m)} = \underline{\mathbf{E}}_{t,\mu}^{(1/m)}, \qquad \underline{\mathbf{E}}_{z,-\mu}^{(1/m)} = -\underline{\mathbf{E}}_{z,\mu}^{(1/m)},$$
$$\underline{\mathbf{H}}_{t,-\mu}^{(1/m)} = -\underline{\mathbf{H}}_{t,\mu}^{(1/m)}, \qquad \underline{\mathbf{H}}_{z,-\mu}^{(1/m)} = \underline{\mathbf{H}}_{z,\mu}^{(1/m)}. \tag{2.354}$$

This can be shown as follows: A direction change means $\beta \to -\beta$. In (2.143)–(2.146) the substitutions $\beta \to -\beta$, $\underline{A} \to -\underline{A}$, $\underline{C} \to -\underline{C}$ result in $\underline{E}_z e^{-j\beta z} \to -\underline{E}_z e^{j\beta z}$, $\underline{E}_r e^{-j\beta z} \to \underline{E}_r e^{j\beta z}$, $\underline{E}_\varphi e^{-j\beta z} \to \underline{E}_\varphi e^{j\beta z}$, $\underline{H}_z e^{-j\beta z} \to \underline{H}_z e^{j\beta z}$, $\underline{H}_r e^{-j\beta z} \to -\underline{H}_r e^{j\beta z}$, $\underline{H}_\varphi e^{-j\beta z} \to -\underline{H}_\varphi e^{j\beta z}$.

$\underline{E}$ is a wave quantity (as used together with scattering parameters) and has the unit $\sqrt{W}$. Since Z_F has the unit Ω the units of the normalized fields or structure functions $\underline{\mathbf{E}}^{(1/m)}$, $\underline{\mathbf{H}}^{(1/m)}$ are $1/m$. The phase of $\underline{E}$ is arbitrary, the magnitude is defined by

$$\iint \left(\underline{\mathbf{E}}_\mu^{(1/m)} \times \underline{\mathbf{H}}_\nu^{(1/m)*} \right) \cdot d\mathbf{A} = \iint \left(\underline{\mathbf{E}}_{t,\mu}^{(1/m)} \times \underline{\mathbf{H}}_{t,\nu}^{(1/m)*} \right) \cdot d\mathbf{A} = \delta_{|\mu||\nu|} \operatorname{sgn}(\nu). \tag{2.355}$$

Let the permeability μ be scalar, constant and real, hence

$$Z_F = \frac{\omega\mu}{|\beta|} = \frac{|\beta|}{\omega\varepsilon_{eff}}. \tag{2.356}$$

Due to (2.352), (2.355) it holds

$$\frac{1}{2} \underline{E}_\mu \underline{E}_{-\mu}^* = \sqrt{P_\mu P_{-\mu}} \qquad \frac{1}{2} |\underline{E}|^2 = P. \tag{2.357}$$

From $\operatorname{curl} \underline{\mathbf{E}} = -j\omega\mu \underline{\mathbf{H}}$ it follows due to (2.353)

$$\underline{\mathbf{H}}^{(1/m)} = e^{j\beta z} \frac{jZ_F}{\omega\mu} \operatorname{curl}\left(\underline{\mathbf{E}}^{(1/m)} e^{-j\beta z} \right) = e^{j\beta z} \frac{j}{|\beta|} \operatorname{curl}\left(\underline{\mathbf{E}}^{(1/m)} e^{-j\beta z} \right). \tag{2.358}$$

The structure functions of the electric fields of TE and TEM waves can be assumed to be real without loss of generality; this is always possible by choosing a suitable phase of $\underline{E}$. In order to be able to use the following derivation also for other waves we continue with complex numbers. For TE waves, e.g., $\underline{\mathbf{E}}^{(1/m)}(x,y) = \underline{\mathbf{E}}_t^{(1/m)}(x,y) = E_x(x,y)\mathbf{e}_x + E_y(x,y)\mathbf{e}_y$, one applies the curl operator to derive

$$(\operatorname{curl} \underline{\mathbf{E}})_t = -j\beta (\mathbf{e}_z \times \underline{\mathbf{E}}_t). \tag{2.359}$$

From (2.358), (2.359) it follows due to $\operatorname{sgn}(\beta) = \operatorname{sgn}(\nu)$

$$\underline{\mathbf{H}}_{t,\nu}^{(1/m)} \operatorname{sgn}(\nu) = e^{j\beta z} \frac{j}{\beta} \left(\operatorname{curl} \underline{\mathbf{E}}_\nu^{(1/m)} e^{-j\beta z} \right)_t = \left(\mathbf{e}_z \times \underline{\mathbf{E}}_{t,\nu}^{(1/m)} \right), \tag{2.360}$$

$$\begin{aligned}\delta_{|\mu||\nu|} &= \iint \left(\underline{\mathbf{E}}_{t,\mu}^{(1/m)} \times \underline{\mathbf{H}}_{t,\nu}^{(1/m)*} \operatorname{sgn}(\nu) \right) \cdot d\mathbf{A} \\ &= \iint \left(\underline{\mathbf{E}}_{t,\mu}^{(1/m)} \times \left(\mathbf{e}_z \times \underline{\mathbf{E}}_{t,\nu}^{(1/m)*} \right) \right) \cdot d\mathbf{A} \\ &= \iint \left(\mathbf{e}_z \left(\underline{\mathbf{E}}_{t,\mu}^{(1/m)} \cdot \underline{\mathbf{E}}_{t,\nu}^{(1/m)*} \right) - \underline{\mathbf{E}}_{t,\nu}^{(1/m)*} \left(\underline{\mathbf{E}}_{t,\mu}^{(1/m)} \cdot \mathbf{e}_z \right) \right) \cdot \mathbf{e}_z dA\end{aligned} \quad . \tag{2.361}$$

In the last step $d\mathbf{A} = \mathbf{e}_z dA$ and $\mathbf{u} \times (\mathbf{v} \times \mathbf{w}) = \mathbf{v}(\mathbf{u} \cdot \mathbf{w}) - \mathbf{w}(\mathbf{u} \cdot \mathbf{v})$ was used. $\underline{\mathbf{E}}_{t,\mu}^{(1/m)}$ and $\mathbf{e}_z$ are perpendicular. The result

$$\iint \underline{\mathbf{E}}_{t,\mu}^{(1/m)} \cdot \underline{\mathbf{E}}_{t,\nu}^{(1/m)*} dA = \iint \underline{\mathbf{E}}_{t,\mu}^{(1/m)+} \underline{\mathbf{E}}_{t,\nu}^{(1/m)} dA = \delta_{|\mu||\nu|} \tag{2.362}$$

is a mode orthogonality relation for the transversal electrical fields. It also holds between orthogonal polarizations. The result holds for arbitrarily polarized TE waves and homogeneous plane waves as a special case thereof. If one allows for arbitrary phases for the waves then the modulus must be taken.

2.5.2 Mode Coupling Theory

We consider quasi-monochromatic signals. Modes (= eigenmodes) have the property of propagating undisturbed. In a system with mode coupling there is always a perturbation which couples the modes among each other. An exact solution is found if one can calculate the new modes of the perturbed system. An approximation is offered by the mode coupling theory [1]. Variable complex amplitudes are assigned to the modes of the unperturbed system. The accuracy is impaired by the fact that the modes of the unperturbed system form no orthogonal mode basis of the perturbed system.

The wave equation in a weakly inhomogeneous, solenoidal, lossless, and dielectrically perturbed medium is

$$\Delta \underline{\mathbf{E}} = -\omega^2 \mu \underline{\varepsilon} \underline{\mathbf{E}} = -\omega^2 \mu \left(\underline{\varepsilon}_n \underline{\mathbf{E}} + \underline{\varepsilon}_p \underline{\mathbf{E}} \right) \qquad \underline{\varepsilon} = \underline{\varepsilon}_n + \underline{\varepsilon}_p \, . \tag{2.363}$$

The dielectricity constant/tensor $\underline{\varepsilon}_n$ in the unperturbed medium is chosen such that phase constants $\pm\beta$ result in $\pm z$ directions. $\underline{\varepsilon}_p$ is the perturbation. While neglecting radiation modes we define a system of orthogonal modes,

$$\underline{\mathbf{E}}(\mathbf{r},t) = \sum_{i \neq 0} \underline{E}_i(z) Z_{Fi}^{1/2} \underline{\mathbf{E}}_i^{(1/m)}(x,y) e^{j(\omega t - \beta_i z)} \, . \tag{2.364}$$

Per definition the $\underline{E}_i$ are normally constant. In the approximation of the mode coupling theory a slow longitudinal position dependence $\underline{E}_i(z)$ is permitted. The phase constants $\beta_i = \beta_0 n_{eff,i} \operatorname{sgn} i$ with positive effective refractive indices $n_{eff,i} = n_{eff,-i}$ and hence also the characteristic field impedances $Z_{Fi} = Z_{F0} / n_{eff,i}$ are taken as constants. Insertion into the wave equation results in

$$\sum_{i\neq 0}\left(\begin{pmatrix}\left(\frac{\partial^2}{\partial x^2}+\frac{\partial^2}{\partial y^2}-\beta_i^2+\omega^2\mu\underline{\varepsilon}\right)\underline{E}_i(z)\\ -2j\beta_i\frac{\partial \underline{E}_i(z)}{\partial z}+\frac{\partial^2 \underline{E}_i(z)}{\partial z^2}\end{pmatrix}\cdot Z_{Fi}^{1/2}\underline{\mathbf{E}}_i^{(1/m)}e^{j(\omega t-\beta_i z)}\right)=0\,. \quad (2.365)$$

The i-th eigenmode of the unperturbed system fulfills

$$\left(\frac{\partial^2}{\partial x^2}+\frac{\partial^2}{\partial y^2}-\beta_i^2+\omega^2\mu\underline{\varepsilon}_{n,i}(x,y)\right)\underline{\mathbf{E}}_i^{(1/m)}(x,y)=0\,, \quad (2.366)$$

and since the wave amplitude changes only slowly,

$$\left|2\beta_i\frac{\partial \underline{E}_i(z)}{\partial z}\right| >> \left|\frac{\partial^2 \underline{E}_i(z)}{\partial z^2}\right|. \quad (2.367)$$

Insertion into (2.365) results in a system of linear coupled first-order differential equations

$$\sum_{i\neq 0}\left(\left(\omega^2\mu Z_{Fi}^{1/2}\underline{\varepsilon}_{p,i}\underline{E}_i(z)-2j\beta_i Z_{Fi}^{1/2}\frac{d\underline{E}_i(z)}{dz}\right)\underline{\mathbf{E}}_i^{(1/m)}e^{j(\omega t-\beta_i z)}\right)=0\,. \quad (2.368)$$

The sum remains unchanged if each summand is multiplied by $(j/2)\beta_0^{-1}Z_{F0}^{-1/2}=(j/2)\beta_0\omega^{-2}\mu^{-1}\varepsilon_0^{-1}Z_{F0}^{-1/2}$,

$$\begin{aligned}&\sum_{i\neq 0}\left(\operatorname{sgn} i\cdot n_{eff,i}^{1/2}\frac{d\underline{E}_i(z)}{dz}\underline{\mathbf{E}}_i^{(1/m)}e^{j(\omega t-\beta_i z)}\right)\\ &=\sum_{i\neq 0}\left(-\frac{j\beta_0}{2\varepsilon_0 n_{eff,i}^{1/2}}\underline{E}_i(z)\underline{\varepsilon}_{p,i}\underline{\mathbf{E}}_i^{(1/m)}e^{j(\omega t-\beta_i z)}\right)\end{aligned}. \quad (2.369)$$

Here $\beta_i=\beta_0 n_{eff,i}\operatorname{sgn} i$ has been inserted, and amplitudes and their derivatives have been separated. The cross product with $n_{eff,s}^{-1/2}\underline{\mathbf{H}}_s^{(1/m)*}$ is formed on both sides. The respective products are multiplied by the area element $d\mathbf{A}$ and are then integrated over the cross section,

$$\begin{aligned}&\frac{d\underline{E}_s}{dz}e^{j(\omega t-\beta_s z)}-\frac{d\underline{E}_{-s}}{dz}e^{j(\omega t+\beta_s z)}\\ &=-\frac{j\beta_0}{2\varepsilon_0}\sum_{i\neq 0}\underline{E}_i(z)n_{eff,i}^{-1/2}n_{eff,s}^{-1/2}e^{j(\omega t-\beta_i z)}\iint\left(\left(\underline{\varepsilon}_{p,i}\underline{\mathbf{E}}_i^{(1/m)}\right)\times\underline{\mathbf{H}}_s^{(1/m)*}\right)\cdot d\mathbf{A}\end{aligned}. \quad (2.370)$$

At the left hand side (2.355) has been inserted. Which summands vanish and which disappear at the right hand side of (2.370) depends on the particular case. The equation can be evaluated numerically or analytically. We also derive its equivalent for the electric fields of TE and TEM waves. One obtains it by multiplication of the transversal part of (2.369) by $n_{eff,s}^{-1/2}\underline{\mathbf{E}}_{t,s}^{(1/m)+}$ from left and subsequent integration over the cross section, or direct conversion of (2.370) using (2.360),

$$\begin{aligned}&\frac{d\underline{E}_s}{dz}e^{j(\omega t-\beta_s z)}-\frac{d\underline{E}_{-s}}{dz}e^{j(\omega t+\beta_s z)}\\&=-\frac{j\beta_0}{2\varepsilon_0}\sum_{i\neq 0}\underline{E}_i n_{eff,i}^{-1/2}n_{eff,s}^{-1/2}e^{j(\omega t-\beta_i z)}\iint\underline{\mathbf{E}}_{t,s}^{(1/m)+}\underline{\varepsilon}_{p,t,i}\underline{\mathbf{E}}_{t,i}^{(1/m)}dA\end{aligned}\quad (2.371)$$

In weakly inhomogeneous waveguides the fundamental modes are almost TEM modes because the longitudinal fields are very weak. For simplicity we therefore use (2.371) in the following.

To obtain the refractive index tensor $\underline{\mathbf{n}}$ from $\underline{\varepsilon}/\varepsilon_0=\underline{\mathbf{n}}^2$ the square root of a matrix must be taken. To this purpose one diagonalizes $\underline{\varepsilon}/\varepsilon_0=\underline{\mathbf{A}}\,\mathbf{diag}(\underline{\lambda}_k)\underline{\mathbf{A}}^{-1}$. Then it holds $\underline{\mathbf{n}}=\underline{\mathbf{A}}\,\mathbf{diag}\left(\underline{\lambda}_k{}^{1/2}\right)\underline{\mathbf{A}}^{-1}$. Among the two possible signs, the one which yields $\mathrm{Re}\left(\underline{\lambda}_k{}^{1/2}\right)>0$ must be chosen here. Otherwise negative refractive indices could result. From $\left(\underline{\varepsilon}_n+\underline{\varepsilon}_p\right)/\varepsilon_0=\left(\underline{\mathbf{n}}_n+\underline{\mathbf{n}}_p\right)^2\approx\underline{\mathbf{n}}_n^2+\underline{\mathbf{n}}_n\underline{\mathbf{n}}_p+\underline{\mathbf{n}}_p\underline{\mathbf{n}}_n$ it follows

$$\underline{\varepsilon}_p/\varepsilon_0=\underline{\mathbf{n}}_n\underline{\mathbf{n}}_p+\underline{\mathbf{n}}_p\underline{\mathbf{n}}_n\,. \quad (2.372)$$

If $\underline{\mathbf{n}}_n$ is a scalar or if $\underline{\mathbf{n}}_n$ and $\underline{\mathbf{n}}_p$ are diagonal matrices it holds

$$\underline{\varepsilon}_p/\varepsilon_0=2\underline{\mathbf{n}}_p\underline{\mathbf{n}}_n\,. \quad (2.373)$$

2.5.3 Codirectional Coupling in Anisotropic Waveguide

We consider an anisotropic waveguide in the (nearly) transversal fundamental mode which supports two mutually orthogonal polarizations. We may for example choose x and y polarizations as the coordinate system but any other pair of orthogonal eigenmodes of the isotropic (unperturbed) waveguide would also be fine. There they possess the same phase constant $\beta=\beta_0 n_{eff}$. Backward propagating modes are (almost) not excited, as will later be seen in the context of counterdirectional mode coupling. Therefore (2.371) becomes a system of two coupled differential equations

$$\begin{aligned}\frac{d\underline{E}_x}{dz}&=-\frac{j\beta_0}{2\varepsilon_0 n_{eff}}\begin{pmatrix}\underline{E}_x\iint\underline{\mathbf{E}}_{t,x}^{(1/m)+}\underline{\varepsilon}_{p,t}\underline{\mathbf{E}}_{t,x}^{(1/m)}dA\\+\underline{E}_y\iint\underline{\mathbf{E}}_{t,x}^{(1/m)+}\underline{\varepsilon}_{p,t}\underline{\mathbf{E}}_{t,y}^{(1/m)}dA\end{pmatrix}\\\frac{d\underline{E}_y}{dz}&=-\frac{j\beta_0}{2\varepsilon_0 n_{eff}}\begin{pmatrix}\underline{E}_x\iint\underline{\mathbf{E}}_{t,y}^{(1/m)+}\underline{\varepsilon}_{p,t}\underline{\mathbf{E}}_{x}^{(1/m)}dA\\+\underline{E}_y\iint\underline{\mathbf{E}}_{t,y}^{(1/m)+}\underline{\varepsilon}_{p,t}\underline{\mathbf{E}}_{t,y}^{(1/m)}dA\end{pmatrix}\end{aligned}\quad (2.374)$$

If a dielectric waveguide has not the circular cross section which was used to define the eigenmodes of the unperturbed system but an elliptical one then the perturbation can be taken into account by a transversally position-dependent $\underline{\varepsilon}_p$. If both ellipse main axes coincide with modes 1,2 (which are assumed to be x, y) then it (approximately) results

$$\begin{aligned}&\iint \underline{\mathbf{E}}_{t,x}^{(1/m)+} \underline{\varepsilon}_{p,t} \underline{\mathbf{E}}_{t,x}^{(1/m)} dA = -\iint \underline{\mathbf{E}}_{t,y}^{(1/m)+} \underline{\varepsilon}_{p,t} \underline{\mathbf{E}}_{t,y}^{(1/m)} dA = \text{real}, \neq 0 \\ &\iint \underline{\mathbf{E}}_{t,x}^{(1/m)+} \underline{\varepsilon}_{p,t} \underline{\mathbf{E}}_{t,y}^{(1/m)} dA = \iint \underline{\mathbf{E}}_{t,y}^{(1/m)+} \underline{\varepsilon}_{p,t} \underline{\mathbf{E}}_{t,x}^{(1/m)} dA = 0\end{aligned}, \tag{2.375}$$

so that there are two independent differential equations, without mode coupling. The same holds for a transversal dielectricity tensor

$$\begin{aligned}&\underline{\varepsilon}_t = \begin{bmatrix} \varepsilon_x & 0 \\ 0 & \varepsilon_y \end{bmatrix} \qquad \underline{\varepsilon}_{n,t} = \begin{bmatrix} \bar{\varepsilon} & 0 \\ 0 & \bar{\varepsilon} \end{bmatrix} \qquad \underline{\varepsilon}_{p,t} = \begin{bmatrix} -\Delta\varepsilon_1/2 & 0 \\ 0 & \Delta\varepsilon_1/2 \end{bmatrix}, \\ &\bar{\varepsilon} = (\varepsilon_x + \varepsilon_y)/2 \qquad \Delta\varepsilon_1 = \varepsilon_y - \varepsilon_x\end{aligned} \tag{2.376}$$

i.e, for a birefringence between x and y. Stress-induced birefringence of the waveguide, by lateral pressure in x or pull in y direction, or by bending around the x or y axis causes such x-y birefringence. The fields of the circular core waveguide have been taken here although the core is non-circular, which is an approximation.

If the ellipse main axes coincide with ±45° then we find

$$\begin{aligned}&\iint \underline{\mathbf{E}}_{t,x}^{(1/m)+} \underline{\varepsilon}_{p} \underline{\mathbf{E}}_{x}^{(1/m)} dA = \iint \underline{\mathbf{E}}_{t,y}^{(1/m)+} \underline{\varepsilon}_{p} \underline{\mathbf{E}}_{y}^{(1/m)} dA = 0 \\ &\iint \underline{\mathbf{E}}_{t,x}^{(1/m)+} \underline{\varepsilon}_{p} \underline{\mathbf{E}}_{y}^{(1/m)} dA = \iint \underline{\mathbf{E}}_{t,y}^{(1/m)+} \underline{\varepsilon}_{p} \underline{\mathbf{E}}_{x}^{(1/m)} dA = \text{real}, \neq 0\end{aligned}. \tag{2.377}$$

The same holds if the medium with dielectricity tensor (2.376) is rotated about the z axis by 45°, according to $\begin{bmatrix} 1/\sqrt{2} & -1/\sqrt{2} \\ 1/\sqrt{2} & 1/\sqrt{2} \end{bmatrix} \underline{\varepsilon}_t \begin{bmatrix} 1/\sqrt{2} & 1/\sqrt{2} \\ -1/\sqrt{2} & 1/\sqrt{2} \end{bmatrix}$. This results in new

$$\underline{\varepsilon}_t = \begin{bmatrix} \bar{\varepsilon} & -\Delta\varepsilon_2/2 \\ -\Delta\varepsilon_2/2 & \bar{\varepsilon} \end{bmatrix} \qquad \underline{\varepsilon}_{p,t} = \begin{bmatrix} 0 & -\Delta\varepsilon_2/2 \\ -\Delta\varepsilon_2/2 & 0 \end{bmatrix}, \tag{2.378}$$

here with $\Delta\varepsilon_2 = \Delta\varepsilon_1$, and $\underline{\varepsilon}_{n,t}$ according to (2.376). If the waveguide is twisted it holds

$$\begin{aligned}&\iint \underline{\mathbf{E}}_{t,x}^{(1/m)+} \underline{\varepsilon}_{p,t} \underline{\mathbf{E}}_{t,x}^{(1/m)} dA = \iint \underline{\mathbf{E}}_{t,y}^{(1/m)+} \underline{\varepsilon}_{p,t} \underline{\mathbf{E}}_{t,y}^{(1/m)} dA = 0 \\ &\iint \underline{\mathbf{E}}_{t,x}^{(1/m)+} \underline{\varepsilon}_{p,t} \underline{\mathbf{E}}_{t,y}^{(1/m)} dA = -\iint \underline{\mathbf{E}}_{t,y}^{(1/m)+} \underline{\varepsilon}_{p,t} \underline{\mathbf{E}}_{t,x}^{(1/m)} dA = \text{imaginary}, \neq 0\end{aligned}. \tag{2.379}$$

The twist corresponds to reciprocal circular birefringence. Since we consider only one propagation direction it may be described by a dielectricity tensor which is the same as that for non-reciprocal circular birefringence,

$$\underline{\varepsilon}_t = \begin{bmatrix} \bar{\varepsilon} & -j\Delta\varepsilon_3/2 \\ j\Delta\varepsilon_3/2 & \bar{\varepsilon} \end{bmatrix} \qquad \underline{\varepsilon}_{p,t} = \begin{bmatrix} 0 & -j\Delta\varepsilon_3/2 \\ j\Delta\varepsilon_3/2 & 0 \end{bmatrix}, \tag{2.380}$$

with $\Delta\varepsilon_3 = 2\varepsilon_0 G_0$ and $\underline{\varepsilon}_{n,t}$ according to (2.376). In a real optical waveguide these effects occur usually jointly so that (2.374) may be written under the assumption of TEM waves as the differential equation

$$\frac{d\underline{\mathbf{E}}}{dz} = j\frac{\beta_0}{2\varepsilon_0 n_{eff}} \begin{bmatrix} \Delta\varepsilon_1/2 & \Delta\varepsilon_2/2 + j\Delta\varepsilon_3/2 \\ \Delta\varepsilon_2/2 - j\Delta\varepsilon_3/2 & -\Delta\varepsilon_1/2 \end{bmatrix} \underline{\mathbf{E}} \tag{2.381}$$

of a Jones vector $\underline{\mathbf{E}} = \begin{bmatrix} \underline{E}_x & \underline{E}_y \end{bmatrix}^T$. With

$$\Delta\beta_k = \beta_0 \Delta n_k = \frac{\beta_0}{2\varepsilon_0 n_{eff}} \Delta\varepsilon_k \qquad (k = 1,2,3) \tag{2.382}$$

(2.381) may be written as

$$\frac{d\underline{\mathbf{E}}}{dz} = j\begin{bmatrix} \Delta\beta_1/2 & \Delta\beta_2/2 + j\Delta\beta_3/2 \\ \Delta\beta_2/2 - j\Delta\beta_3/2 & -\Delta\beta_1/2 \end{bmatrix} \underline{\mathbf{E}}. \tag{2.383}$$

If the perturbation is constant over z this is solved by an eigenvalue search,

$$\begin{gathered} \underline{\mathbf{E}} = \underline{\mathbf{E}}_{\pm} e^{j\lambda_{\pm} z} \\ \begin{vmatrix} \Delta\beta_1/2 - \lambda & \Delta\beta_2/2 + j\Delta\beta_3/2 \\ \Delta\beta_2/2 - j\Delta\beta_3/2 & -\Delta\beta_1/2 - \lambda \end{vmatrix} = 0 \\ \lambda_{\pm} = \pm\frac{1}{2}\sqrt{\Delta\beta_1^2 + \Delta\beta_2^2 + \Delta\beta_3^2} \\ \underline{\mathbf{E}}_{\pm} = \begin{bmatrix} \pm\sqrt{\Delta\beta_1^2 + \Delta\beta_2^2 + \Delta\beta_3^2} + \Delta\beta_1 \\ \Delta\beta_2 - j\Delta\beta_3 \end{bmatrix} \end{gathered} \tag{2.384}$$

If the eigenvectors are normalized one obtains

$$\begin{gathered} \underline{\mathbf{E}}_+ = \frac{1}{\sqrt{2(1+V_1)}} \begin{bmatrix} 1+V_1 \\ V_2 - jV_3 \end{bmatrix} \qquad \underline{\mathbf{E}}_- = \frac{1}{\sqrt{2(1+V_1)}} \begin{bmatrix} -V_2 - jV_3 \\ 1+V_1 \end{bmatrix} \\ V_k = \frac{\Delta\beta_k}{\sqrt{\Delta\beta_1^2 + \Delta\beta_2^2 + \Delta\beta_3^2}} = \frac{\Delta\varepsilon_k}{\sqrt{\Delta\varepsilon_1^2 + \Delta\varepsilon_2^2 + \Delta\varepsilon_3^2}} \qquad k = 1,2,3. \\ \mathbf{S}_{\pm} = \pm\begin{bmatrix} V_1 & V_2 & V_3 \end{bmatrix}^T \end{gathered} \tag{2.385}$$

Here $\mathbf{S}_\pm$ are the normalized Stokes vectors corresponding to $\underline{\mathbf{E}}_\pm$. Such a waveguide, with a length L, is an elliptical retarder with retardation

$$\delta = (\lambda_+ - \lambda_-)L = \sqrt{\Delta\beta_1^2 + \Delta\beta_2^2 + \Delta\beta_3^2}\, L = \Delta\beta L\,, \tag{2.386}$$

and Jones matrix (2.239). Instead of (2.383) one may write a differential equation for the normalized Stokes vectors. From the matrix equation $\mathbf{S}(z+dz) = \mathbf{GS}(z)$ of an elliptical retarder (2.240) having a length dz and a retardation $\delta = \Delta\beta dz$ it follows for $dz \to 0$

$$\mathbf{S}(z+dz) - \mathbf{S}(z) = (\mathbf{G} - \mathbf{1})\mathbf{S}(z) \approx \begin{bmatrix} 0 & -V_3 & V_2 \\ V_3 & 0 & -V_1 \\ -V_2 & V_1 & 0 \end{bmatrix} \Delta\beta dz \mathbf{S}\,, \tag{2.387}$$

$$\frac{d\mathbf{S}}{dz} = \begin{bmatrix} 0 & -V_3 & V_2 \\ V_3 & 0 & -V_1 \\ -V_2 & V_1 & 0 \end{bmatrix} \Delta\beta \mathbf{S} = \mathbf{S}_+ \Delta\beta \times \mathbf{S} = \begin{bmatrix} \Delta\beta_1 \\ \Delta\beta_2 \\ \Delta\beta_3 \end{bmatrix} \times \mathbf{S}\,. \tag{2.388}$$

On the Poincaré sphere this is the already-known rotation of the input polarization about the axis $\mathbf{S}_+$.

If for example the x-y birefringence dominates ($|\Delta\beta_1| >> |\Delta\beta_{2,3}|$) the eigenmodes and principal states-of-polarization are nearly x and y polarizations. A practical application of this observation are polarization-maintaining optical fibers which are fabricated with a strong x-y birefringence. Without loss of generality we assume a constant $\Delta\beta_1$. Using

$$\underline{\mathbf{E}} = \underline{\mathbf{N}}\underline{\tilde{\mathbf{E}}} = \begin{bmatrix} e^{j(\Delta\beta_1/2)z} & 0 \\ 0 & e^{-j(\Delta\beta_1/2)z} \end{bmatrix} \underline{\tilde{\mathbf{E}}} \tag{2.389}$$

one obtains from (2.383)

$$\begin{aligned} \frac{d(\underline{\mathbf{N}}\underline{\tilde{\mathbf{E}}})}{dz} &= j\begin{bmatrix} \Delta\beta_1/2 & 0 \\ 0 & -\Delta\beta_1/2 \end{bmatrix} \underline{\mathbf{N}}\underline{\tilde{\mathbf{E}}} + \underline{\mathbf{N}}\frac{d\underline{\tilde{\mathbf{E}}}}{dz} \\ &= j\begin{bmatrix} \Delta\beta_1/2 & \Delta\beta_2/2 + j\Delta\beta_3/2 \\ \Delta\beta_2/2 - j\Delta\beta_3/2 & -\Delta\beta_1/2 \end{bmatrix} \underline{\mathbf{N}}\underline{\tilde{\mathbf{E}}} \end{aligned}, \tag{2.390}$$

$$\frac{d\underline{\tilde{\mathbf{E}}}}{dz} = j\begin{bmatrix} 0 & \underline{K} \\ \underline{K}^* & 0 \end{bmatrix} \underline{\tilde{\mathbf{E}}} \qquad \underline{K}(z) = e^{-j\Delta\beta_1 z}(\Delta\beta_2(z)/2 + j\Delta\beta_3(z)/2)\,. \tag{2.391}$$

We excite only one mode at the input, for example $\underline{\tilde{\mathbf{E}}}(0) = \begin{bmatrix} 1 \\ 0 \end{bmatrix}$. Under the condition $\left|\int_0^z \underline{K}(\zeta)d\zeta\right| << 1$ the output field is $\underline{\tilde{\mathbf{E}}}(z) \approx \begin{bmatrix} 1 \\ j\int_0^z \underline{K}^*(\zeta)d\zeta \end{bmatrix}$. The power fraction coupled into the other mode is $\sim\left|\int_0^z \underline{K}(\zeta)d\zeta\right|^2$. Now, $\int_0^z \underline{K}(\zeta)d\zeta$ is proportional to the spatial Fourier transform of the ±45° and circular birefringence. Even if $\Delta\beta_2(z)$, $\Delta\beta_3(z)$ can not be neglected their spatial frequency spectra will have a pronounced lowpass characteristic, for example because the bending direction of a fiber is constant over several cm, and does not change every mm or so between 45° and −45° directions.

A waveguide piece with a length equal to a *beat length* possesses a retardation of 2π. Technologically one reach beat lengths

$$\Lambda = 2\pi/\Delta\beta_1 = \lambda_0/\Delta n_1 \qquad (\lambda_0 = \text{wavelength in vacuum}) \qquad (2.392)$$

down to about 2 mm in optical fibers. The spectra of $\Delta\beta_2(z)$, $\Delta\beta_3(z)$ therefore have quite low amplitudes at the spatial frequency Λ^{-1}, and the polarization is very well preserved. One can generate linear birefringence in optical fibers either by an elliptical rather than circular core cross section or by a lateral mechanical stress. More common is the latter method, see Fig. 2.20.

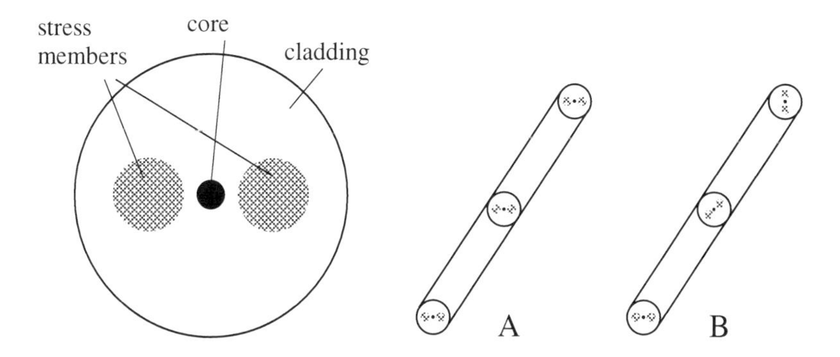

Fig. 2.20 Polarization-maintaining optical fiber: cross section, untwisted (A), twisted (B).

The glass composition and the thermal expansion coefficients of the two stress-generating lateral zones, the stress members, differ from those of the core. Once the stress-free melt has solidified, and is cooled down to the operation temperature it generates a mechanical stress along the straight line connecting the two zones. This results in a length- and frequency[9]-proportional retardation $\delta = \Delta\beta_1 z = \frac{\omega}{c}\Delta n_1 z$ between x and y polarizations, i.e., polarization mode dispersion. The differential group delay per length unit is

[9] It is not exactly proportional to frequency. The phase velocity difference generally differs from the group velocity difference.

$$\frac{d\tau}{dz} = \frac{\partial^2 \delta}{\partial \omega \partial z} = \frac{\Delta n_1}{c}. \tag{2.393}$$

For $\Lambda = 2\,\text{mm}$, $\lambda = 1550\,\text{nm}$ a very high value $d\tau/dz = 2{,}5\,\text{ps/m}$ results. Yet the signals are transmitted without distortion if one takes care to excite only one mode at the fiber input. Since this is not a stochastic but a deterministic fiber perturbation τ is proportional to the length, not to its square root. Polarization-maintaining fibers (PMF) are not only more expensive than standard fibers but also exhibit a higher attenuation. In the field more problems can arise if PMFs are inadvertedly broken and then re-spliced under a small misalignment angle once or several times.

If the PMF is not laid straight (A in Fig. 2.20) but bent or twisted (B) this results in $\Delta\beta_2, \Delta\beta_3 \neq 0$. These effects are suppressed as described before. Gradual onset and offset of the perturbation, which is quite usual due to fiber and cable elasticity, can contribute to a very good suppression. If an input PSP of the PMF is excited, i.e., one of the stress main axes, then the output polarization corresponds also to the respective stress main axes, i.e., to one of the output PSPs. This holds with good accuracy even if the fiber is twisted.

Problem: A straight PMF with a beat length Λ = 4 mm and a length L has a very large twist rate of 10 turns per meter. The stress members follow the twist, but in silica the polarization is only rotated by roughly the 0.08fold of the twist angle. How large is the worst-case polarization extinction at the fiber output if the input polarization is a PSP?

Solution: The beat length yields $\Delta\beta_1 = 2\pi/\Lambda \approx 1571/\text{m}$. The physical twist of the polarization with respect to the stress members is $(0.08-1)\cdot 2\pi$ per fiber turn. The circular retardation is twice this value. So $\Delta\beta_3 = 2\cdot(0.08-1)\cdot 2\pi \cdot 10/\text{m} \approx -116/\text{m}$. For weak coupling, the coupled power fraction equals $\left|\int_0^L K(\zeta)d\zeta\right|^2 = \left|\int_0^z e^{-j\Delta\beta_1 z} j\Delta\beta_3/2 d\zeta\right|^2 = \left(\frac{\Delta\beta_3}{2\Delta\beta_1}\left(e^{-j\Delta\beta_1 L}-1\right)\right)^2$ $\leq (\Delta\beta_3/\Delta\beta_1)^2$. The worst polarization extinction equals $-10\log\left((\Delta\beta_3/\Delta\beta_1)^2\right) = 22.6\,\text{dB}$, which justifies the assumption of weak coupling.

2.5.4 Codirectional Coupling of Two Waveguides

Codirectional mode coupling with constant coupling coefficients occurs in optical directional couplers. The codirectional waves 1 and 2 propagate in the unperturbed waveguides with the refractive index profiles $n_1(x,y)$ and $n_2(x,y)$, respectively (Fig. 2.21). The mode fields of the unperturbed waveguides stay nearly unaltered when the perturbation is added.

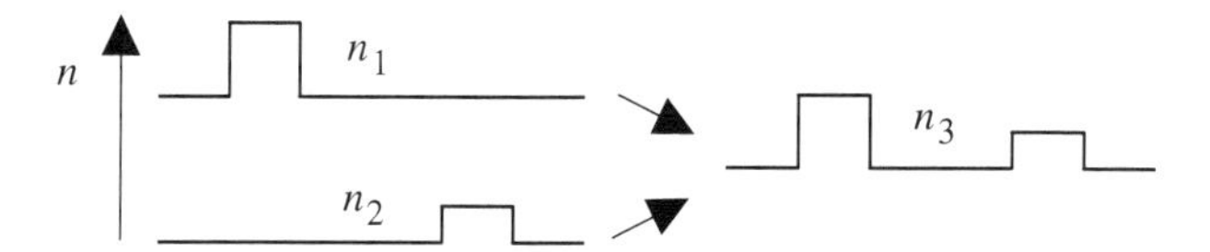

Fig. 2.21 System of two coupled waveguides, refractive index profile across both fiber cores

The perturbations of the dielectricity constant are due to the

$$\varepsilon_{p,1} = \varepsilon_0 \left(n_3^2(x,y) - n_1^2(x,y) \right) \qquad \varepsilon_{p,2} = \varepsilon_0 \left(n_3^2(x,y) - n_2^2(x,y) \right) \tag{2.394}$$

For TEM waves (approximation) a multiplication from the left side of the transversal part of (2.369) by $n_{eff,i}^{-1/2} \underline{\mathbf{E}}_{t,i}^{(1/m)+} e^{-j(\omega t - \beta_i z)}$ ($i = 1,2$) and subsequent integration over the whole cross section results in

$$\begin{aligned}
&\frac{d\underline{E}_1}{dz} + \frac{d\underline{E}_2}{dz} n_{eff,1}^{-1/2} n_{eff,2}^{1/2} e^{-j(\beta_2 - \beta_1)z} \iint \underline{\mathbf{E}}_{t,1}^{(1/m)+} \underline{\mathbf{E}}_2^{(1/m)} dA \\
&= -\frac{j\beta_0}{2\varepsilon_0} \begin{pmatrix} \underline{E}_1 n_{eff,1}^{-1} \iint \underline{\mathbf{E}}_{t,1}^{(1/m)+} \varepsilon_{p,1} \underline{\mathbf{E}}_1^{(1/m)} dA \\ + \underline{E}_2 n_{eff,1}^{-1/2} n_{eff,2}^{-1/2} e^{-j(\beta_2 - \beta_1)z} \iint \underline{\mathbf{E}}_{t,1}^{(1/m)+} \varepsilon_{p,2} \underline{\mathbf{E}}_2^{(1/m)} dA \end{pmatrix} \\
&\frac{d\underline{E}_1}{dz} n_{eff,1}^{1/2} n_{eff,2}^{-1/2} e^{-j(\beta_1 - \beta_2)z} \iint \underline{\mathbf{E}}_{t,2}^{(1/m)+} \underline{\mathbf{E}}_1^{(1/m)} dA + \frac{d\underline{E}_2}{dz} \\
&= -\frac{j\beta_0}{2\varepsilon_0} \begin{pmatrix} \underline{E}_1 n_{eff,1}^{-1/2} n_{eff,2}^{-1/2} e^{-j(\beta_1 - \beta_2)z} \iint \underline{\mathbf{E}}_{t,2}^{(1/m)+} \varepsilon_{p,1} \underline{\mathbf{E}}_1^{(1/m)} dA \\ + \underline{E}_2 n_{eff,2}^{-1} \iint \underline{\mathbf{E}}_{t,2}^{(1/m)+} \varepsilon_{p,2} \underline{\mathbf{E}}_2^{(1/m)} dA \end{pmatrix}
\end{aligned} \tag{2.395}$$

Attention: The various modes of <u>one</u> waveguide are orthogonal but we use here the modes of two different waveguides with the refractive index profiles $n_1(x,y)$ and $n_2(x,y)$, respectively, which can not coexist. Therefore $\underline{\mathbf{E}}_{t,1}^{(1/m)}$ and $\underline{\mathbf{E}}_{t,2}^{(1/m)}$ are not orthogonal. Exceptionally it holds $\iint \underline{\mathbf{E}}_{t,2}^{(1/m)+} \underline{\mathbf{E}}_{t,1}^{(1/m)} dA \neq 0$, so that on each left hand sides there is a second summand beside $d\underline{E}_1/dz$ and $d\underline{E}_2/dz$, respectively. However, these can be neglected – see also Fig. 2.22. Eqn. (2.395) can then be written in matrix form as

$$\frac{d\underline{\mathbf{E}}}{dz} = -j\underline{\underline{\mathbf{H}}}\underline{\mathbf{E}} = -j \begin{bmatrix} M_1 & \kappa_{12} \\ \kappa_{21} & M_2 \end{bmatrix} \underline{\mathbf{E}} \qquad \underline{\mathbf{E}} = \begin{bmatrix} \underline{E}_1 \\ \underline{E}_2 \end{bmatrix}, \tag{2.396}$$

$$\begin{aligned}
M_i &= \frac{\beta_0}{2\varepsilon_0} n_{eff,i}^{-1} \iint \underline{\mathbf{E}}_{t,i}^{(1/m)+} \varepsilon_{p,i} \underline{\mathbf{E}}_{t,i}^{(1/m)} dA \qquad i = 1,2 \qquad k = 3 - i \\
\kappa_{ik} &= \frac{\beta_0}{2\varepsilon_0} n_{eff,i}^{-1/2} n_{eff,k}^{-1/2} e^{-j(\beta_k - \beta_i)z} \iint \underline{\mathbf{E}}_{t,i}^{(1/m)+} \varepsilon_{p,k} \underline{\mathbf{E}}_{t,k}^{(1/m)} dA
\end{aligned} \tag{2.397}$$

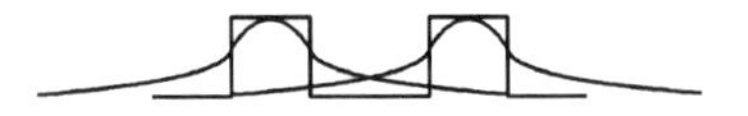

Fig. 2.22 Representation of the functions which occur in the overlap integrals

For the analytical treatment it is now convenient to set

$$\frac{d\underline{\mathbf{E}}}{dz} = -j\begin{bmatrix} M_1 & \underline{\kappa} \\ \underline{\kappa}^* & M_2 \end{bmatrix}\underline{\mathbf{E}} \quad \text{with} \quad \underline{\kappa} = \left(\kappa_{12} + \kappa_{21}^*\right)/2\,, \tag{2.398}$$

because this enforces the codirectional mode coupling to be lossless,

$$\frac{d}{dz}\left(\left|\underline{E}_1\right|^2 + \left|\underline{E}_2\right|^2\right) = 0\,. \tag{2.399}$$

The solution of (2.398)

$$\begin{aligned}
&\underline{\mathbf{E}} = \underline{\mathbf{E}}_{1,2}e^{j\lambda_{1,2}z} \qquad \begin{vmatrix} M_1+\lambda & \underline{\kappa} \\ \underline{\kappa}^* & M_2+\lambda \end{vmatrix} = 0 \qquad \lambda_{1,2} = -\overline{M} \pm \Psi \\
&\overline{M} = \frac{M_1+M_2}{2} \qquad \Delta\beta = M_1 - M_2 \qquad \Psi = \sqrt{(\Delta\beta/2)^2 + |\underline{\kappa}|^2} \\
&\underline{\mathbf{E}}_{1,2} = \frac{1}{\sqrt{2\Psi(\Psi \pm 2(\Delta\beta/2))}}\begin{bmatrix} -\underline{\kappa} \\ (\Delta\beta/2) \pm \Psi \end{bmatrix} \qquad \text{or} \\
&\underline{\mathbf{E}}_{1,2} = \frac{1}{\sqrt{2\Psi(\Psi \mp 2(\Delta\beta/2))}}\begin{bmatrix} (\Delta\beta/2) \mp \Psi \\ \underline{\kappa}^* \end{bmatrix}
\end{aligned} \tag{2.400}$$

is similar to that of (2.383). We write the transfer matrix $\underline{\mathbf{J}}$,

$$\begin{aligned}
&\underline{\mathbf{E}}(z) = \underline{\mathbf{J}}\underline{\mathbf{E}}(0) \qquad \underline{\mathbf{E}}_e = \left[\underline{\mathbf{E}}_1 \quad \underline{\mathbf{E}}_2\right] \\
&\underline{\mathbf{J}} = e^{-j\overline{M}z}\underline{\mathbf{E}}_e\begin{bmatrix} e^{j\Psi z} & 0 \\ 0 & e^{-j\Psi z} \end{bmatrix}\underline{\mathbf{E}}_e^{-1} \\
&= e^{-j\overline{M}z}\begin{bmatrix} \cos(\Psi z) - j\frac{\Delta\beta/2}{\Psi}\sin(\Psi z) & -j\frac{\underline{\kappa}}{\Psi}\sin(\Psi z) \\ -j\frac{\underline{\kappa}^*}{\Psi}\sin(\Psi z) & \cos(\Psi z) + j\frac{\Delta\beta/2}{\Psi}\sin(\Psi z) \end{bmatrix}
\end{aligned} \tag{2.401}$$

Here the additional terms $e^{-j\beta_i z}$ of (2.364) have to be dropped because they are part of $\underline{\mathbf{J}}$. A comparison with (2.239) shows that the coupler behaves like a retarder. Only the field vectors do not describe x and y polarizations of one waveguide but two equally polarized modes of waveguides with the refractive index profiles $n_1(x,y)$, $n_2(x,y)$, respectively.

If a signal is coupled into one of the waveguides it generally appears at both waveguide outputs. The power transfer coupling factor into the second waveguide is

$$\frac{P_2(z)}{P_1(0)} = \left|\underline{J}_{21}\right|^2 = \frac{|\underline{\kappa}|^2}{\Psi^2}\sin^2(\Psi z)\,. \tag{2.402}$$

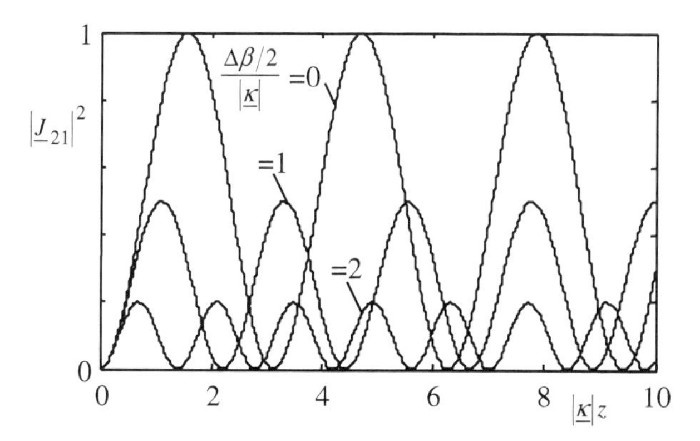

Fig. 2.23 Power transfer coupling factor $|\underline{J}_{21}|^2$ as a function of $|\underline{\kappa}|z$ for various $\Delta\beta$

In Fig. 2.23 this function is plotted. Coupling is periodical. Between two subsequent maxima of the coupled wave the coupled wave amplitude changes its sign. Complete coupling is possible only if the phase mismatch $\Delta\beta$ between the two waveguides disappears. In that case, $\Delta\beta = 0$, $M_1 = M_2 = \overline{M}$,

$$\underline{\mathbf{J}} = e^{-j\overline{M}z} \begin{bmatrix} \cos(|\underline{\kappa}|z) & -je^{j\arg(\underline{\kappa})}\sin(|\underline{\kappa}|z) \\ -je^{-j\arg(\underline{\kappa})}\sin(|\underline{\kappa}|z) & \cos(|\underline{\kappa}|z) \end{bmatrix} \tag{2.403}$$

holds. For TE and TEM waves $\underline{\kappa} = \kappa > 0$ is real and positive. For small z the coupled wave lags 90° in phase with respect to the residual wave.

For comparison we derive the directional coupler transfer matrix now directly from the field modes of a system of coupled waveguides. The coupled waveguides form together a new waveguide in which in the case of TE waves the fundamental mode (TE_0 or H_0) and the next higher mode (TE_1 or H_1) are sustained (Fig. 2.24).

Fig. 2.24 TE_0 mode (left) and TE_1 mode (right) in a system of coupled waveguides

If TE_0 and TE_1 modes occur with equal amplitudes and phases they add to form nearly the field of the TE_0 mode in one of the waveguides alone. The difference of these fields corresponds to the TE_0 mode in the other waveguide alone. Therefore we can write

$$\begin{bmatrix} \underline{E}_1 \\ \underline{E}_2 \end{bmatrix} = \frac{1}{\sqrt{2}} \begin{bmatrix} 1 & 1 \\ 1 & -1 \end{bmatrix} \begin{bmatrix} \underline{E}_{TE_0} \\ \underline{E}_{TE_1} \end{bmatrix}. \tag{2.404}$$

In the right vector there are the modes of the waveguide system. They propagate according to

$$\begin{bmatrix} \underline{E}_{TE_0}(z) \\ \underline{E}_{TE_1}(z) \end{bmatrix} = \begin{bmatrix} e^{-j\beta_{TE_0} z} & 0 \\ 0 & e^{-j\beta_{TE_1} z} \end{bmatrix} \begin{bmatrix} \underline{E}_{TE_0}(0) \\ \underline{E}_{TE_1}(0) \end{bmatrix} \qquad \beta_{TE_0} > \beta_{TE_1} . \tag{2.405}$$

In agreement with (2.403) it follows

$$\begin{aligned} \underline{\mathbf{J}} &= \frac{1}{\sqrt{2}} \begin{bmatrix} 1 & 1 \\ 1 & -1 \end{bmatrix} \begin{bmatrix} e^{-j\beta_{TE_0} z} & 0 \\ 0 & e^{-j\beta_{TE_1} z} \end{bmatrix} \left(\frac{1}{\sqrt{2}} \begin{bmatrix} 1 & 1 \\ 1 & -1 \end{bmatrix} \right)^{-1} \\ &= e^{-j\overline{\beta} z} \begin{bmatrix} \cos(\kappa z) & -j\sin(\kappa z) \\ -j\sin(\kappa z) & \cos(\kappa z) \end{bmatrix} \\ \overline{\beta} &= \frac{\beta_{TE_0} + \beta_{TE_1}}{2} \qquad \kappa = \frac{\beta_{TE_0} - \beta_{TE_1}}{2} > 0 \end{aligned} \tag{2.406}$$

The relation $\kappa > 0$ holds because the field of the higher-order mode reaches further into the cladding than that of the fundamental mode. (As the mode number rises, u increases while $v = \sqrt{V^2 - u^2}$ and $\beta = \sqrt{(2v/d)^2 + \beta_2^2}$ decrease.).

Problem: An infinite number of coupled waveguides, where coupling occurs only between neighbor waveguides, fulfill the coupled differential equations $\frac{dE_n}{dz} = -j\kappa E_{n-1} - j\kappa E_{n+1}$. If only the 0-th waveguide is excited, $E_i(0) = \delta_{i0}$, the solution is $E_n(z) = (-j)^n J_n(2\kappa z)$.

Problem: Design 3×3 fiber couplers, using 3 identical waveguides of lengths l with cores placed at the three corners of a regular triangle. (1) Find the phase shift between coupled waves (e.g., 1–5) and through-waves (e.g., 1–4) for any value of κl .(2) Find κl and field transfer matrix $\mathbf{T}$ for a short coupler which has equal output powers at all waveguides if light is connected to only one input waveguide. (3) Repeat this for the case that only 1/5 of the light power coupled into one waveguide remains there.

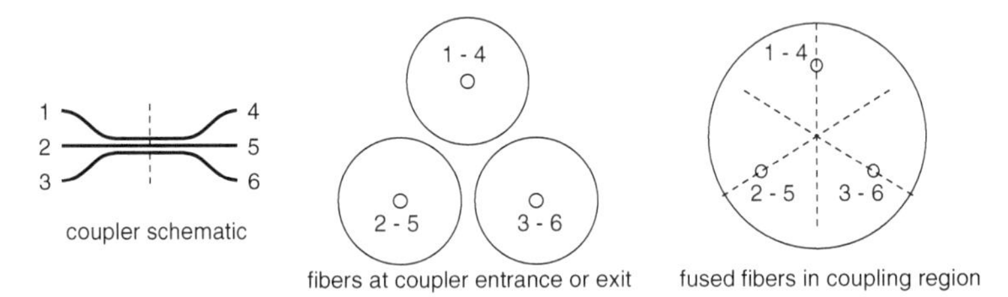

Solution: (1) Due to the sketched symmetry planes, coupling between neighbor waveguides is identical. The fibers also have identical propagation constants, without phase mismatch.

In extension of the foregoing we find $\frac{d\mathbf{E}}{dz} = -j\mathbf{N}\mathbf{E}$ with $\mathbf{N} = \begin{bmatrix} M & \kappa & \kappa \\ \kappa & M & \kappa \\ \kappa & \kappa & M \end{bmatrix}$. For $\mathbf{E} = \mathbf{E}_0 e^{-j\beta z}$ it results $\frac{d\mathbf{E}}{dz} = -j\beta\mathbf{E}$. Symmetry allows to guess a suitable set of (forcibly) orthogonal eigenmodes $\mathbf{E}_1 = \frac{1}{\sqrt{3}}\begin{bmatrix}1\\1\\1\end{bmatrix}$, $\mathbf{E}_2 = \frac{1}{\sqrt{2}}\begin{bmatrix}0\\1\\-1\end{bmatrix}$ and $\mathbf{E}_3 = \frac{1}{\sqrt{6}}\begin{bmatrix}2\\-1\\-1\end{bmatrix}$ of the symmetric matrix $\mathbf{N}$. For these, which are inserted for $\mathbf{E}_0$, we find the eigenvalues $\beta_1 = M + 2\kappa$, $\beta_{2,3} = M - \kappa$. Since $\beta_{2,3}$ are degenerate, any pair of orthogonal linear combinations of $\mathbf{E}_{2,3}$ would also be eigenvectors. (Alternatively we start by solving the 3rd-order eigenvalue equation and then compute the eigenvectors.) The eigenvector matrix $\mathbf{A} = [\mathbf{E}_1 \quad \mathbf{E}_2 \quad \mathbf{E}_3]$ is orthogonal. Its inverse is $\mathbf{A}^{-1} = \mathbf{A}^{-T}$. This results in a unitary coupler transfer matrix

$$\mathbf{T} = \mathbf{A}e^{-jMl}\begin{bmatrix} e^{-j2\kappa l} & 0 & 0 \\ 0 & e^{j\kappa l} & 0 \\ 0 & 0 & e^{j\kappa l}\end{bmatrix}\mathbf{A}^T,$$

$$\mathbf{T} = \left[T_{ij}\right] = e^{-j(M+2\kappa)l}\,\frac{1}{3}\begin{bmatrix} 1+2e^{j3\kappa l} & 1-e^{j3\kappa l} & 1-e^{j3\kappa l} \\ 1-e^{j3\kappa l} & 1+2e^{j3\kappa l} & 1-e^{j3\kappa l} \\ 1-e^{j3\kappa l} & 1-e^{j3\kappa l} & 1+2e^{j3\kappa l}\end{bmatrix}. \tag{2.407}$$

The phase shift of coupled vs. uncoupled waves is

$$\varphi_c = \arg(T_{21}) - \arg(T_{11}) = \arg\left(\left(1-e^{j3\kappa l}\right)/\left(1+2e^{j3\kappa l}\right)\right) = -\pi + \arctan\frac{3\sin(3\kappa l)}{1-\cos(3\kappa l)}. \tag{2.408}$$

The remaining fraction of light power launched into any one particular waveguide is $|T_{11}|^2 = \left|\left(1+2e^{j3\kappa l}\right)/3\right|^2 = (5+4\cos(3\kappa l))/9$.

(2) Each power transfer factor $|T_{ij}|^2$ must be 1/3, since the power incident at one waveguide alone must be equally distributed to all outputs. Condition $|T_{11}|^2 = 1/3$ leads to $\cos(3\kappa l) = -1/2$, $\kappa l = 2\pi/9$, $\varphi_c = -2\pi/3$ and

$$\mathbf{T} = e^{j(-Ml+\pi/18)}\frac{j}{\sqrt{3}}\begin{bmatrix} 1 & e^{-j2\pi/3} & e^{-j2\pi/3} \\ e^{-j2\pi/3} & 1 & e^{-j2\pi/3} \\ e^{-j2\pi/3} & e^{-j2\pi/3} & 1\end{bmatrix}. \tag{2.409}$$

(3) If 1/5 of the launched power is to remain in the same waveguide it must hold $|T_{11}|^2 = 1/5$. This leads to $\cos(3\kappa l) = -4/5$, $\sin(3\kappa l) = 3/5$, $\kappa l = (1/3)\arccos(-4/5) \approx 0.8327$, $\varphi_c = -3\pi/4$,

$$\mathbf{T} = e^{j\xi}\frac{1}{\sqrt{5}}\begin{bmatrix} 1 & -1-j & -1-j \\ -1-j & 1 & -1-j \\ -1-j & -1-j & 1\end{bmatrix} \quad \xi = -Ml - (2/3)\arccos(-4/5) - \pi + \arctan 2. \tag{2.410}$$

Such couplers can be used for simultaneous in-phase and quadrature interference detection, for example in direct-detection DQPSK receivers or in coherent intradyne receivers.

2.5.5 Periodic Codirectional Coupling

If the wave vectors of both modes to be coupled are identical or almost identical then codirectional mode coupling with constant coupling coefficients is possible. For large wave vector differences only little power can be coupled because already after a short length the coupled power is being coupled back. This problem can be solved by forming the coupling periodical. The grating vector of the perturbation must compensate the wave vector difference.

An example is TE-TM mode coupling in crystals with birefringent waveguides. In Lithium Niobate ($LiNbO_3$) with X cut and Y propagation direction the mode coupling can be controlled electrooptically (Fig. 2.25). Along the Y crystal axis, at the surface, there is a waveguide (TiO_2). On top of a SiO_2 buffer layer at the surface there are comb-shaped gold electrodes, a meandering ground electrode and two signal electrodes, the combs of which interlace with the combs of the ground electrode. They receive voltages U_1, U_2. The fundamental modes of the integrated waveguide, corresponding to the HE_{11} modes of the optical fiber, are usually denoted as TE (parallel to the surface, here the Z crystal axis) and TM modes (perpendicular to the surface, here the X crystal axis). The polished end faces of the crystal are slanted, for example at 8°. Reflections between chip and a fiber coupled to it (under a $\arcsin\left((1/2)(n_o + n_e)\cdot \sin 8°/n_{SiO2}\right) \approx 11.8°$ angle) therefore occur outside the acceptance angle of the respective waveguide and are radiated away, and detrimental multipath reflections are avoided.

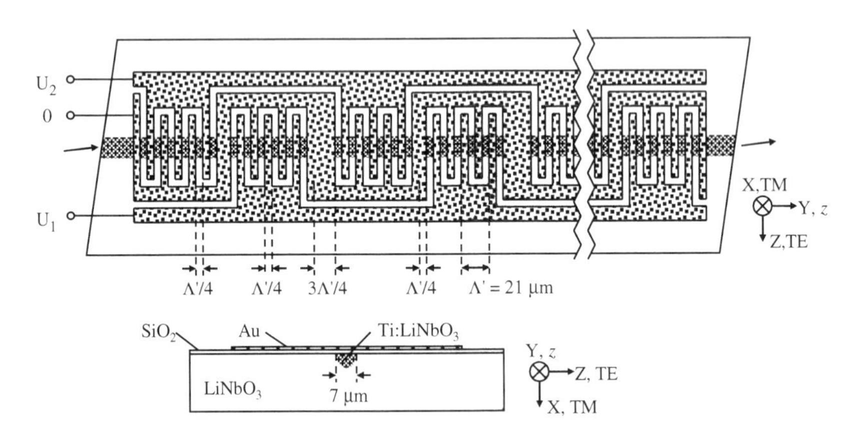

Fig. 2.25 Schematic drawing of a TE-TM mode converter in X-cut, Y-propagation $LiNbO_3$, top and cross-sectional view. The dimensions hold for λ = 1550 nm.

All electrode combs possess a spatial period Λ'. Comb teeth and their gaps are therefore each about $\Lambda'/4$ broad. Subsequent combs of one signal electrode have separations equal to an integer multiple of Λ'. Ideally the electrode period Λ' equals the beat length $\Lambda = \lambda/|n_e - n_o|$ between TE and TM waves. The quadrature electrode (voltage U_2) is displaced in propagation direction with respect to the in-phase electrode (voltage U_1) by $\Lambda'/4$. This is obtained by a broadening of the mass electrode when it passes between the combs of different signal electrodes, by alterately $\Lambda'/4$ and $3\Lambda'/4$. The voltages generate vertical (X-directed) periodic electrostatic field components in the waveguide. Due to the electrooptic coefficient r_{51} they slightly rotate the index ellipsoid so that its principal axes run no longer in X and Z directions. In the vicinity of the comb the vertical electric field and the index ellipsoid rotation angle α vary roughly proportional to $\cos(2\pi z/\Lambda')$, i.e., along the Y axis. It holds $|\alpha| << 1$. The rotation in the X-Z coordinate system by α corresponds to a rotation by $-\alpha$ in the TE-TM coordinate system. There the transversal part of the refractive index tensor is

$$\mathbf{n}_t = \begin{bmatrix} \cos\alpha & \sin\alpha \\ -\sin\alpha & \cos\alpha \end{bmatrix} \begin{bmatrix} n_e & 0 \\ 0 & n_o \end{bmatrix} \begin{bmatrix} \cos\alpha & -\sin\alpha \\ \sin\alpha & \cos\alpha \end{bmatrix} \approx \begin{bmatrix} n_e & (n_o - n_e)\alpha \\ (n_o - n_e)\alpha & n_o \end{bmatrix}. \quad (2.411)$$

When it is separated as

$$\mathbf{n}_t = \mathbf{n}_{n,t} + \mathbf{n}_{p,t} \qquad \mathbf{n}_{n,t} = \begin{bmatrix} n_e & 0 \\ 0 & n_o \end{bmatrix} \qquad \mathbf{n}_{p,t} = (n_o - n_e)\alpha \begin{bmatrix} 0 & 1 \\ 1 & 0 \end{bmatrix} \quad (2.412)$$

then according to (2.372) it holds

$$\underline{\boldsymbol{\varepsilon}}_p/\varepsilon_0 = (n_o - n_e)(n_o + n_e)\alpha \begin{bmatrix} 0 & 1 \\ 1 & 0 \end{bmatrix}. \quad (2.413)$$

Since the comb teeth are a lot longer than the waveguide is wide we may use a separation ansatz $\alpha = f(x, y)g(z)$. This means, the transversal integration of the mode fields and the longitudinal integration for the genuine solution of the coupled mode equations can be conducted separately.

Instead of (2.371) one now obtains

$$\frac{d\underline{E}_{TE}}{dz} = -\frac{j\beta_0}{2\varepsilon_0}\left(\begin{array}{l} \underline{E}_{TE} n_{eff,TE}^{-1} \iint \underline{\mathbf{E}}_{t,TE}^{(1/m)+} \underline{\boldsymbol{\varepsilon}}_p \underline{\mathbf{E}}_{TE}^{(1/m)} dA \\ + \underline{E}_{TM} n_{eff,TE}^{-1/2} n_{eff,TM}^{-1/2} e^{j(\beta_{TE}-\beta_{TM})z} \iint \underline{\mathbf{E}}_{t,TE}^{(1/m)+} \underline{\boldsymbol{\varepsilon}}_p \underline{\mathbf{E}}_{TM}^{(1/m)} dA \end{array} \right)$$

$$\frac{d\underline{E}_{TM}}{dz} = -\frac{j\beta_0}{2\varepsilon_0}\left(\begin{array}{l} \underline{E}_{TE} n_{eff,TE}^{-1/2} n_{eff,TM}^{-1/2} e^{-j(\beta_{TE}-\beta_{TM})z} \iint \underline{\mathbf{E}}_{t,TM}^{(1/m)+} \underline{\boldsymbol{\varepsilon}}_p \underline{\mathbf{E}}_{TE}^{(1/m)} dA \\ \underline{E}_{TM} n_{eff,TM}^{-1} \iint \underline{\mathbf{E}}_{t,TM}^{(1/m)+} \underline{\boldsymbol{\varepsilon}}_p \underline{\mathbf{E}}_{TM}^{(1/m)} dA \end{array} \right)$$

$$(2.414)$$

Inserting $n_{eff,TE} \approx n_e$, $n_{eff,TM} \approx n_o$ and (2.413) results in

$$\begin{aligned} \frac{d\underline{E}_{TE}}{dz} &= -\frac{j\beta_0}{2\varepsilon_0 n_e^{1/2} n_o^{1/2}} \underline{E}_{TM} e^{j(\beta_{TE}-\beta_{TM})z} \iint \underline{\mathbf{E}}_{t,TE}^{(1/m)+} \underline{\boldsymbol{\varepsilon}}_p \underline{\mathbf{E}}_{TM}^{(1/m)} dA \\ \frac{d\underline{E}_{TM}}{dz} &= -\frac{j\beta_0}{2\varepsilon_0 n_e^{1/2} n_o^{1/2}} \underline{E}_{TE} e^{-j(\beta_{TE}-\beta_{TM})z} \iint \underline{\mathbf{E}}_{t,TM}^{(1/m)+} \underline{\boldsymbol{\varepsilon}}_p \underline{\mathbf{E}}_{TE}^{(1/m)} dA \end{aligned} . \quad (2.415)$$

For the first electrode it holds $g(z) \propto U_1 \cos \beta_p z$, for the second $g(z) \propto -U_2 \sin \beta_p z$, because it is displaced with respect to the first one by $\Lambda'/4$. Here $\beta_p = 2\pi/\Lambda'$ is the phase constant of the perturbation. In order to take into account the fact that only one, not both signal electrodes can be present above a particular section of the waveguide we substitute

$$\frac{\beta_0}{2\varepsilon_0 n_e^{1/2} n_o^{1/2}} \iint \underline{\mathbf{E}}_{t,TE}^{(1/m)+} \underline{\boldsymbol{\varepsilon}}_p \underline{\mathbf{E}}_{TM}^{(1/m)} dA = 2\kappa' \left(U_1(z) \cos \beta_p z - U_2(z) \sin \beta_p z \right). \quad (2.416)$$

Here each of the two position-dependent voltages $U_i(z)$ ($i = 1,2$) is alternately equal to the applied voltage U_i (where the corresponding comb is there), or zero (where the other comb is there). It holds

$$\begin{aligned} & 2e^{jv(\beta_{TE}-\beta_{TM})z} \left(U_1(z) \cos \beta_p z - U_2(z) \sin \beta_p z \right) \\ &= U_1(z) \left(e^{jv((\beta_{TE}-\beta_{TM})+v\beta_p)z} + e^{jv((\beta_{TE}-\beta_{TM})-v\beta_p)z} \right) \qquad (v = \pm 1). \quad (2.417) \\ &+ jU_2(z) \left(e^{jv((\beta_{TE}-\beta_{TM})+v\beta_p)z} - e^{jv((\beta_{TE}-\beta_{TM})-v\beta_p)z} \right) \end{aligned}$$

Two out of four exponential terms at the right hand side may be neglected in the integration because they oscillate quite rapidly as a function of z and do not contribute significantly. We define the phase mismatch by

$$\Delta\beta = \beta_{TE} - \beta_{TM} \pm \beta_p \text{ with } |\Delta\beta| = \left\| \beta_{TE} - \beta_{TM} \right| - \left| \beta_p \right\| . \quad (2.418)$$

The sign is chosen so that $\Delta\beta$ has the smallest possible absolute value. This allows to replace expression (2.417) by $\left(U_1(z) \pm jvU_2(z) \right) e^{jv\Delta\beta z}$. Alternately only one of the voltages can be applied. The spatially averaged voltages are $\overline{U_i(z)} = U_i/2$ (if the broadening of the mass electrode by alternately $\Lambda'/4$ and $3\Lambda'/4$ is neglected). So, (2.417) takes the form $\left(\overline{U_1(z)} \pm jv\overline{U_2(z)} \right) e^{jv\Delta\beta z}$, since the combs of the two signal electrodes also follow one after the other also with a fairly high spatial frequency. In total one obtains

$$\frac{d\underline{\mathbf{E}}}{dz} = -j\begin{bmatrix} 0 & \underline{\kappa} e^{j\Delta\beta z} \\ \underline{\kappa}^* e^{-j\Delta\beta z} & 0 \end{bmatrix}\underline{\mathbf{E}} \qquad \underline{\mathbf{E}} = \begin{bmatrix} \underline{E}_{TE} \\ \underline{E}_{TM} \end{bmatrix}. \\ \underline{\kappa} = \kappa'\left(\overline{U_1(z)} \pm j\overline{U_2(z)}\right) = \frac{\kappa'}{2}\left(U_1 \pm jU_2\right) \tag{2.419}$$

A substitution (inverse to that which led to (2.391)) allows to transform (2.419) into a system of coupled differential equations with constant coefficients,

$$\underline{\tilde{\mathbf{E}}} = \begin{bmatrix} e^{-j(\Delta\beta/2)z} & 0 \\ 0 & e^{j(\Delta\beta/2)z} \end{bmatrix}\underline{\mathbf{E}} \Rightarrow \frac{d\underline{\tilde{\mathbf{E}}}}{dz} = -j\begin{bmatrix} \Delta\beta/2 & \underline{\kappa} \\ \underline{\kappa}^* & -\Delta\beta/2 \end{bmatrix}\underline{\tilde{\mathbf{E}}}. \tag{2.420}$$

A comparison with (2.398) yields a Jones matrix (2.401) with $\overline{M} = 0$ for $\underline{\tilde{\mathbf{E}}}$. However, $\underline{\tilde{\mathbf{E}}}$ is not the physically occurring Jones vector. $\underline{\mathbf{E}}$ (2.364) is closer to reality. In order to take the various propagation constants into account in a component which begins in the coordinate origin and ends at z, the Jones matrix (2.401) must be multiplied from the left side by $\begin{bmatrix} e^{j(-\beta_{TE}+\Delta\beta/2)z} & 0 \\ 0 & e^{j(-\beta_{TM}-\Delta\beta/2)z} \end{bmatrix}$. The right term in the parentheses inverts (2.420), the left one performs (2.364). Using $\Delta\beta = \beta_{TE} - \beta_{TM} \pm \beta_p$ one obtains the true Jones matrix

$$e^{-j\overline{\beta}z}\begin{bmatrix} \begin{pmatrix} \cos\Psi z \\ -j\dfrac{\Delta\beta/2}{\Psi}\sin\Psi z \end{pmatrix} e^{\pm j(\beta_p/2)z} & -j\dfrac{\underline{\kappa}}{\Psi}\sin\Psi z\, e^{\pm j(\beta_p/2)z} \\ -j\dfrac{\underline{\kappa}^*}{\Psi}\sin\Psi z\, e^{\mp j(\beta_p/2)z} & \begin{pmatrix} \cos\Psi z \\ +j\dfrac{\Delta\beta/2}{\Psi}\sin\Psi z \end{pmatrix} e^{\mp j(\beta_p/2)z} \end{bmatrix} \tag{2.421}$$

with $\overline{\beta} = (\beta_{TE} + \beta_{TM})/2$. Most interesting is the case of vanishing phase mismatch, $\Delta\beta = 0$. If one considers a component section with a length z equal to the m-fold of an electrode period Λ' then $e^{\pm j(\beta_p/2)z} = (-1)^m$ holds and the Jones matrix is simplified into

$$e^{-j\overline{\beta}z}(-1)^m\begin{bmatrix} \cos|\underline{\kappa}|z & -je^{j\arg\underline{\kappa}}\sin|\underline{\kappa}|z \\ -je^{-j\arg\underline{\kappa}}\sin|\underline{\kappa}|z & \cos|\underline{\kappa}|z \end{bmatrix}. \tag{2.422}$$

According to (2.239) it belongs to a retarder with retardation $\delta = 2|\underline{\kappa}|z \sim \sqrt{U_1^2 + U_2^2}$ and an eigenmode $\dfrac{1}{\sqrt{2}}\begin{bmatrix} 1 \\ -e^{-j\arg\underline{\kappa}} \end{bmatrix}$

$=\frac{1}{\sqrt{2}}\left[\begin{array}{c}1\\ (-U_1+jU_2)/\sqrt{U_1^2+U_2^2}\end{array}\right]$ (for $z>0$: the fast one). The eigenmodes may be varied on the S_2 - S_3 great circle of the Poincaré sphere at will. This is a mode converter where in-phase and quadrature mode coupling may be freely chosen.

This component exhibits also strong PMD. With $n_o=2.21283$, $n_e=2.13739$ an extremely high differential group delay per unit length results, $d\tau/dz=|n_o-n_e|/c\approx 250\,\mathrm{ps/m}$. If several such mode converters, with individually choosable voltages, are cascaded on a $LiNbO_3$ substrate (maximum possible length: 100 mm at present) a fairly ideal PMD equalizer is obtained [18].

Problem: A mode converter with a Jones matrix (2.422) and a retardation equal to π is driven by the voltages $U_1=U_0\cos\Omega t$, $U_2=\pm U_0\sin\Omega t$. The optical input signal has TE polarization. Which properties does the output signal have?

Solution: The multiplication of (2.422) by the vector $[1\ \ 0]^T$ results in an output signal which is proportional to $-je^{-j\arg\underline{\kappa}}[0\ \ 1]^T=-j\frac{U_1-jU_2}{\sqrt{U_1^2+U_2^2}}[0\ \ 1]^T=-je^{\mp j\Omega t}[0\ \ 1]^T$. The output signal of this time-variable retarder is TM-polarized, and its angular frequency is shifted with respect to that of the input signal by Ω upwards or downwards [19]! If one chooses a TM input polarization then the output polarization is TE, and the frequency shift occurs in opposite direction.

In the case of a longitudinally variable $\underline{\kappa}(z)$ no closed-form expression exists for the solution of (2.419). However, for small $\int_0^L|\underline{\kappa}(z)|dz$ and $\underline{E}_{TE}(0)\neq 0$, $\underline{E}_{TM}(0)=0$ there exists the approximate solution

$$\begin{aligned}&\underline{E}_{TE}(z)\approx\underline{E}_{TE}(0)\qquad \frac{d\underline{E}_{TM}}{dz}\approx -j\underline{\kappa}^*(z)\underline{E}_{TE}(0)e^{-j\Delta\beta z}\\ &\underline{E}_{TM}(L)=-j\underline{E}_{TE}(0)\int_0^L\underline{\kappa}^*(z)e^{-j\Delta\beta z}dz\end{aligned}\quad. \tag{2.423}$$

The conversion factor is proportional to the spatial Fourier transform of the (complex conjugate) coupling coefficient.

2.5.6 *Periodic Counterdirectional Coupling*

Periodic counterdirectional mode coupling occurs in an optical Bragg grating. For simplicity let us consider TEM waves. Let the medium be isotropic, and exhibit a rectangular refractive index modulation with a period that is roughly half a wavelength. In forward and in backward direction the waveguide is perturbed by

$$\varepsilon_p(\mathbf{r})=\frac{\varepsilon_p}{2}\mathrm{sgn}(\sin(\eta z+\chi))=-j\frac{\varepsilon_p}{\pi}\sum_{l\ \mathrm{odd}}l^{-1}e^{jl(\eta z+\chi)}\qquad \eta\approx 2\beta\,. \tag{2.424}$$

Let $\beta = \beta_s$. If (2.371) is multiplied on both sides by $e^{-j(\omega t - \beta z)}$ and integrated over z then at the left side only the first term and at the right side only the term $i = -s$, $l = -1$ contribute significantly to the result because all other terms have a strong longitudinal periodicity. One roughly obtains

$$\begin{aligned} \frac{d\underline{E}_s}{dz} &= -\frac{j\beta_0}{2\varepsilon_0 n_{eff}} \underline{E}_{-s} e^{j(2\beta-\eta)z} e^{-j\chi}(-j)\frac{\varepsilon_p}{\pi(-1)} \iint \underline{\mathbf{E}}_{t,s}^{(1/m)+} \underline{\mathbf{E}}_{t,-s}^{(1/m)} dA \\ &= -j\underline{\kappa}\underline{E}_{-s} e^{j\delta z} \qquad \underline{\kappa} = \frac{j\beta_0}{2\varepsilon_0 n_{eff}} \frac{\varepsilon_p}{\pi} e^{-j\chi} \qquad \delta = 2\beta - \eta \end{aligned} \quad . \tag{2.425}$$

Because $\varepsilon_p(\mathbf{r})$ depends only on z it has been pulled out from the integral, which therefore simplifies to 1. If one multiplies (2.371) instead by $e^{-j(\omega t + \beta z)}$, then only the second term at the left side and the term $i = s$, $l = 1$ at the right side must be considered. In summary, this results in

$$\frac{d\underline{E}_+}{dz} = -j\underline{\kappa}\underline{E}_- e^{j\delta z} \qquad \frac{d\underline{E}_-}{dz} = j\underline{\kappa}^* \underline{E}_+ e^{-j\delta z} \tag{2.426}$$

with $\underline{E}_+ = \underline{E}_s$, $\underline{E}_- = \underline{E}_{-s}$. The grating periodicity provides *phase matching* between forward and backward propagating waves. This is brought about by the fundamental of the grating function, which therefore is a 1st-order grating. For other grating periodicities other Fourier coefficients and possibly other grating orders may result. The total forward propagating power is $(1/2)\left(|\underline{E}_+|^2 - |\underline{E}_-|^2\right)$. The counterdirectional mode coupling is lossless because we can show

$$\frac{d}{dz}\left(|\underline{E}_+|^2 - |\underline{E}_-|^2\right) = 0 . \tag{2.427}$$

Let a wave be incident in forward direction into a grating of length L with the amplitude $\underline{E}_+(0)$. No wave is fed in from the back side, $\underline{E}_-(L) = 0$. The general solution of (2.426) occurs similar to that of (2.419) and yields

$$\begin{aligned} \underline{E}_+(z) &= \underline{E}_+(0) \frac{\Theta \cosh \Theta(L - z) + j(\delta/2) \sinh \Theta(L - z)}{\Theta \cosh \Theta L + j(\delta/2) \sinh \Theta L} e^{j(\delta/2)z} \\ \underline{E}_-(z) &= \underline{E}_+(0) \frac{-j\underline{\kappa}^* \sinh \Theta(L - z)}{\Theta \cosh \Theta L + j(\delta/2) \sinh \Theta L} e^{-j(\delta/2)z} \\ \Theta &= \sqrt{|\kappa|^2 - (\delta/2)^2} \end{aligned} \quad . \tag{2.428}$$

This allows to plot Fig. 2.26, in agreement with (2.427). At the ends the waves

$$\begin{aligned} \underline{E}_+(L) &= \underline{\tau}\underline{E}_+(0) \qquad & \underline{\tau} &= \frac{\Theta}{\Theta\cosh\Theta L + j(\delta/2)\sinh\Theta L} e^{j(\delta/2)L} \\ \underline{E}_-(0) &= \underline{\rho}\underline{E}_+(0) \qquad & \underline{\rho} &= \frac{-j\underline{\kappa}^*\sinh\Theta L}{\Theta\cosh\Theta L + j(\delta/2)\sinh\Theta L} \end{aligned} \tag{2.429}$$

exit, see Fig. 2.27.

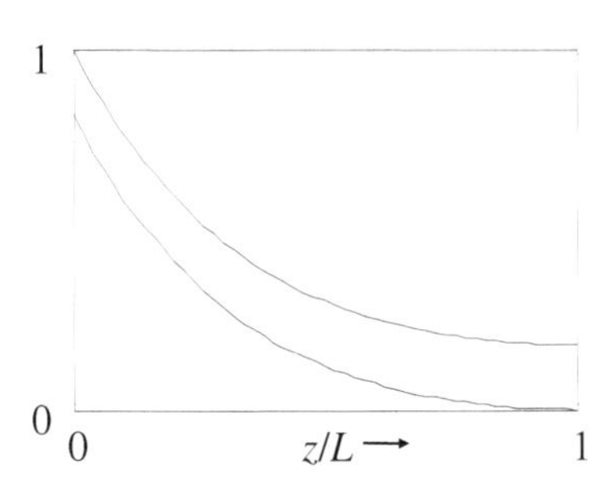

Fig. 2.26 Position-dependent intensities $\left|\underline{E}_+(z)/\underline{E}_+(0)\right|^2$ (top) and $\left|\underline{E}_-(z)/\underline{E}_+(0)\right|^2$ (bottom) for $|\underline{\kappa}|L = 1{,}5$, $\delta = 0$

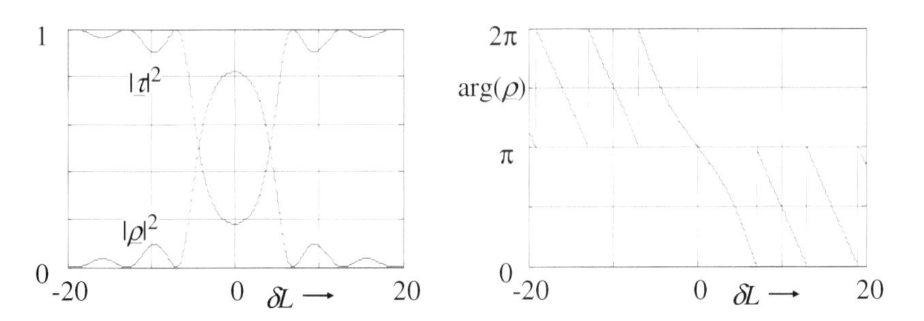

Fig. 2.27 Backscattered intensity $|\underline{\rho}|^2$, transmitted intensity $|\underline{\tau}|^2$ and phase angle $\arg(\underline{\rho})$ of the reflection factor, each as a function of the normalized phase mismatch, for $|\underline{\kappa}|L = 1{,}5$

In order to avoid imaginary arguments of the hyperbolic functions at $|\delta/2| > |\underline{\kappa}|$ one may instead use trigonometric functions,

$$\begin{aligned} \underline{\tau} &= \frac{\Phi}{\Phi\cos\Phi L + j(\delta/2)\sin\Phi L} e^{j(\delta/2)L} \\ \underline{\rho} &= \frac{-j\underline{\kappa}^*\sin\Phi L}{\Phi\cos\Phi L + j(\delta/2)\sin\Phi L} \qquad \Phi = \sqrt{(\delta/2)^2 - |\underline{\kappa}|^2} \end{aligned} . \tag{2.430}$$

The reflection factor $\underline{\rho}$ possesses zeros at

$$\delta = \pm 2\sqrt{|\underline{\kappa}|^2 + (n\pi/L)^2} \qquad n = 1,2,3,\ldots \quad . \tag{2.431}$$

Similar to (2.423), a closed form solution of (2.426) for a longitudinally variable $\kappa(z)$ exists under the condition of a small $\int_0^L |\underline{\kappa}(z)| dz$ and $\underline{E}_+(0) \neq 0$, $\underline{E}_-(L) = 0$.

$$\begin{aligned} &\underline{E}_+(z) \approx \underline{E}_+(0) \qquad \frac{d\underline{E}_-}{dz} \approx j\underline{\kappa}^*(z)\underline{E}_+(0)e^{-j\delta z} \\ &\underline{E}_-(0) = j\underline{E}_+(0)\int_0^L \underline{\kappa}^*(z)e^{-j\delta z}dz \end{aligned} \quad . \tag{2.432}$$

The reflection factor is again proportional to the spatial Fourier transform of the (complex conjugate) coupling coefficient. For large $|\underline{\kappa}|$ the reflection spectrum becomes very broad. This is because the first grating periods are already sufficient to achieve full reflection.

2.6 Mode Coupling for Dispersion Compensation

We will first introduce a standard description of mode coupling and the so-called differential group delay profile (DGD), a powerful tool for a geometrical understanding of PMD. Then possible implementations of polarization mode and chromatic dispersion compensators will be discussed.

2.6.1 Mode Coupling and Differential Group Delay Profiles

In order to be able to compensate polarization mode dispersion (PMD) a suitable equalizer structure must be found. We will derive the necessary properties of polarization transformers and differential group delay (DGD) sections [10]. Transmission fiber span and equalizer are assumed to consist of lossless optical retarders, and isotropic propagation delays are neglected.

The most general retarder is an endless elliptical retarder (ER) as described by (2.239), (2.240). One of its eigenmodes, the fast one for a positive retardation δ, has the normalized Stokes vector

$$\begin{bmatrix} V_1 \\ V_2 \\ V_3 \end{bmatrix} = \begin{bmatrix} \cos 2\vartheta \cos 2\varepsilon \\ \sin 2\vartheta \cos 2\varepsilon \\ \sin 2\varepsilon \end{bmatrix} \qquad (V_1^2 + V_2^2 + V_3^2 = 1), \tag{2.433}$$

with azimuth angle ϑ and ellipticity angle ε. Relative strengths of 0°, 45° and circular birefringence are given by V_1, V_2 and V_3, respectively. These as well as the retardation $\delta = -\infty...\infty$ can vary endlessly. Angles $2\vartheta = -\infty...\infty$ and $2\varepsilon = -\pi/2...\pi/2$ allow for endless variations of the eigenmodes.

Both Jones and Müller matrices of optical components can be multiplied in reversed order of light propagation to describe their concatenation. This means the following expressions (2.434)–(2.443) are valid for Jones matrices (2.239) and 3×3 rotation matrices (2.240). We use the nomenclature

$$\begin{aligned}
&\text{(endless) phase shifter } (2\vartheta = 0,\ 2\varepsilon = 0): && \mathrm{PS}(\delta = -\infty...\infty),\\
&\text{(finite) mode converter } (2\vartheta = \pi/2,\ 2\varepsilon = 0): && \mathrm{MC}(\delta = 0...\pi),\\
&\text{(endless) Soleil-Babinet compensator } (2\varepsilon = 0): && \mathrm{SBC}(\delta = 0...\pi, 2\vartheta = -\infty...\infty),\\
&\text{(endless) Soleil-Babinet analog } (2\vartheta = \pi/2): && \mathrm{SBA}(\delta = 0...\pi, 2\varepsilon = -\infty...\infty),
\end{aligned} \tag{2.434}$$

to define some special cases of an (endless) elliptical retarder ER. We call the last example a Soleil-Babinet analog because it is related to the familiar Soleil-Babinet compensator, a rotatable waveplate of adjustable retardation, by cyclical shifts of rows and columns of the rotation matrix. An SBA is described by either of

$$\mathrm{SBA}(\delta,\psi)=\begin{bmatrix}\cos\delta/2 & je^{j\psi}\sin\delta/2\\ je^{-j\psi}\sin\delta/2 & \cos\delta/2\end{bmatrix}\quad \text{Jones matrix,}$$

$$\mathrm{SBA}(\delta,\psi)=\begin{bmatrix}\cos\delta & -\sin\psi\sin\delta & \cos\psi\sin\delta\\ \sin\psi\sin\delta & \cos^2\psi+\sin^2\psi\cos\delta & \cos\psi\sin\psi(1-\cos\delta)\\ -\cos\psi\sin\delta & \cos\psi\sin\psi(1-\cos\delta) & \sin^2\psi+\cos^2\psi\cos\delta\end{bmatrix}\quad \text{rotation matrix.} \tag{2.435}$$

A rotating SBC ($2\vartheta=\Omega t$) converts circular input polarization partly or fully into its orthogonal including a frequency shift by Ω. A "rotating" SBA ($2\varepsilon=\Omega t$) does the same thing to an x-polarized input signal. An SBA can endlessly transform x polarization at its input into any output polarization or vice versa. An SBA can be replaced by a phase shifter, a mode converter, and another phase shifter of opposite retardation:

$$\mathrm{SBA}(\varphi,2\varepsilon)=\mathrm{PS}(2\varepsilon)\mathrm{MC}(\varphi)\mathrm{PS}(-2\varepsilon) \tag{2.436}$$

An ER can be replaced by (or expressed as) a sequence of a quarterwave plate, a halfwave plate, and another quarterwave plate, each of them endlessly rotatable [20], or by an endlessly rotatable halfwave plate and an SBC. Alternatively, an ER can be replaced by a phase shifter, a mode converter, and another phase shifter [21]. It may be shown that the mode converter needs only a finite retardation in the range $0...\pi$. This means the ER can also be substituted by an SBA, preceded or followed by one PS,

$$\begin{aligned}\mathrm{ER} &= \mathrm{PS}(\varphi_3)\mathrm{MC}(\varphi_2)\mathrm{PS}(\varphi_1)\\ &= \mathrm{PS}(\varphi_3+\varphi_1)\mathrm{SBA}(\varphi_2,-\varphi_1)=\mathrm{SBA}(\varphi_2,\varphi_3)\mathrm{PS}(\varphi_1+\varphi_3)\end{aligned}. \tag{2.437}$$

Comparison of the last two expressions shows that a phase shifter may be "pushed" from left to right through an SBA if the SBA orientation angle is increased by the transferred phase shift.

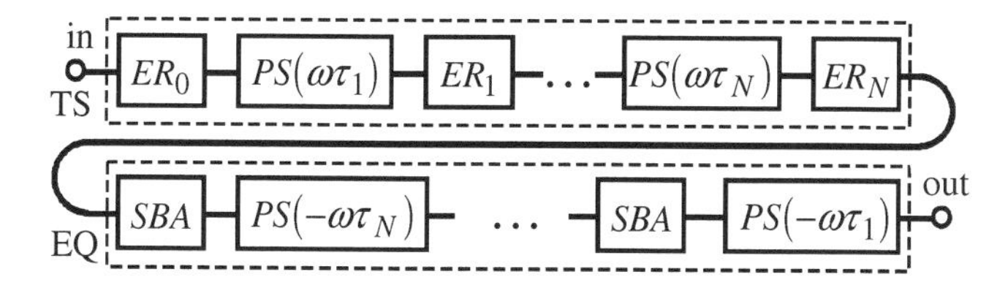

Fig. 2.28 Transmission fiber span (TS) with elliptical retarders (ER) and DGD sections. Suitable equalizer (EQ) with Soleil-Babinet analogs (SBA; see text) and DGD sections, each of which may be accompanied by arbitrary differential phase shifts. © 1999 IEEE.

At an angular optical frequency ω we represent a fiber section having a differential group delay τ by a phase shifter $\mathrm{PS}(\omega\tau)$. Its eigenmodes and principal states-of-polarization (PSP) coincide. The whole transmission span TS may consist of an infinite sequence of alternating ERs and DGD sections. In practice, we limit ourselves to N DGD sections and $N+1$ ERs which are passed by the light signal in ascending order of index i. This is shown in Fig. 2.28. The equalizer (EQ) will be explained later. The matrix describing the transmission fiber span is

$$\begin{aligned} \mathrm{TS} &= \prod_{i=N}^{1} \left(\mathrm{ER}_i \mathrm{PS}(\omega\tau_i)\right)\mathrm{ER}_0 \\ &= \prod_{i=N}^{1} \left(\mathrm{SBA}(\varphi_{2i},\varphi_{3i})\mathrm{PS}(\varphi_{1i}+\varphi_{3i}+\omega\tau_i)\right)\mathrm{ER}_0 \end{aligned} \quad . \qquad (2.438)$$

The product symbols with a stop index 1 that is lower than the start index N have to be written from left to right in descending order of index i. Adjacent phase shifts or shifters may be interchanged. All frequency-independent phase shifts in the second expression can therefore be transferred to the right through preceding SBAs according to (2.437),

$$\begin{aligned} \mathrm{TS} = &\prod_{i=N}^{1} \left(\mathrm{SBA}\left(\varphi_{2,i}, \varphi_{3i} + \sum_{k=i+1}^{N} (\varphi_{1k}+\varphi_{3k}) \right) \mathrm{PS}(\omega\tau_i) \right) \\ &\cdot \mathrm{PS}\left(\sum_{i=1}^{N} (\varphi_{1i}+\varphi_{3i}) \right) \mathrm{ER}_0 \end{aligned} \quad . \qquad (2.439)$$

The retarders in the second line can be replaced by another ER.

Quite generally the summands $\tilde{\mathbf{\Omega}}_i$ of the overall PMD vector (2.313) of a system can be plotted individually in what may be called the *differential group delay profile*. The input polarization of the first retarder can arbitrarily be set to be horizontal if we multiply its rotation matrix from the right side by a rotation matrix which will transform horizontal into the true retarder input polarization. It is useful to indicate this standardized input polarization by an arrow in $\begin{bmatrix}1 & 0 & 0\end{bmatrix}^T$ direction which ends at the origin. From there on the tail of $\tilde{\mathbf{\Omega}}_i$ joins the head of $\tilde{\mathbf{\Omega}}_{i-1}$ and so on. It is always allowed to add a DGD section with zero DGD and a locally horizontal fast PSP at the end of the retarder cascade. Since its PMD vector would have zero length an arrow is plotted instead in the direction of the input-referred PMD vector. This is useful if the retarder cascade is terminated by a frequency-independent retarder. If $\mathbf{G}$ is the total rotation matrix an input polarization $\mathbf{G}^T\begin{bmatrix}1 & 0 & 0\end{bmatrix}^T$ would be needed to generate a $\begin{bmatrix}1 & 0 & 0\end{bmatrix}^T$ output polarization. So the direction $\mathbf{G}^T\begin{bmatrix}1 & 0 & 0\end{bmatrix}^T$ of that output arrow indicates the input polarization which would be necessary to locally hit the fast PSP of that fictitiuos DGD section with zero DGD. The DGD sections have constant DGD τ, so the phase shift between the two PSPs is proportional to $\omega\tau$. Each of the DGD profile rods defined by $\tilde{\mathbf{\Omega}}_i$ therefore twists along its axis if the optical frequency ω is varied. Fig. 2.29 (see legend) shows a number of DGD profiles and associated simulated received eye diagrams. The distortions of Fig. 2.29 e) can be detected by an electrical slope steepness detector [22, 23].

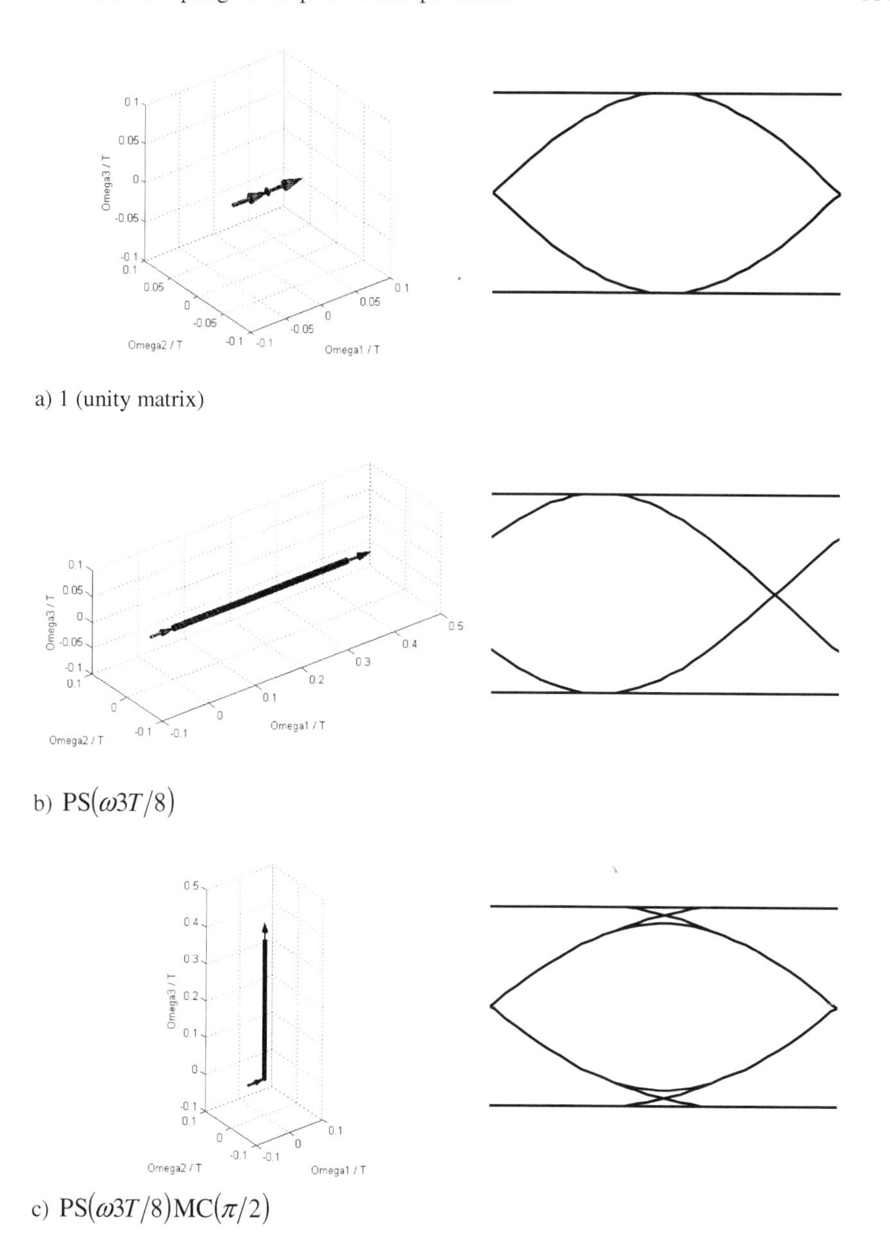

Fig. 2.29 DGD profiles (left), corresponding simulated NRZ eye diagrams in the receiver (right), and matrix of the investigated retarder chain (below DGD profile) using abbreviations (2.434). Each DGD profile is plotted 3 times: at the center angular frequency ω where we have assumed here that the differential phase shift of each DGD section is an integer of 2π, and with frequency offsets of $\pm 1/(2T)$ where T is the bit duration. The plots at the three different frequencies become distinguishable only from subdiagram d) on.

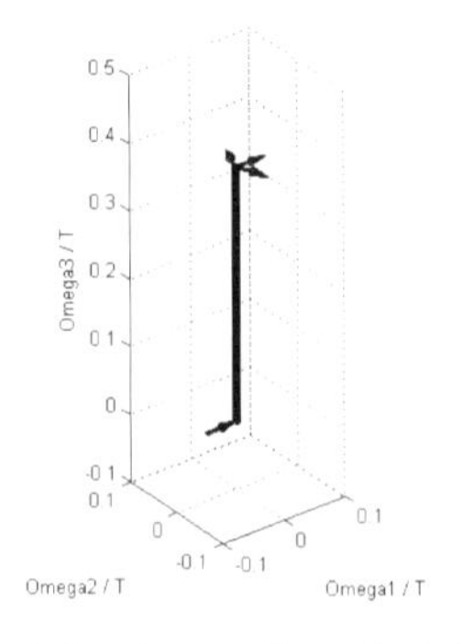

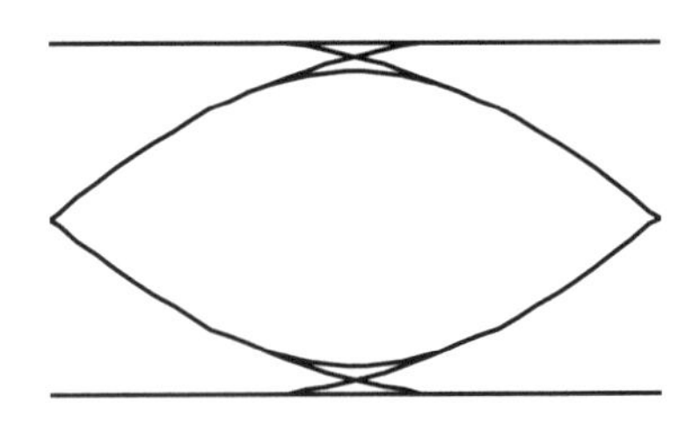

d) $\text{MC}(-\pi/2)\text{PS}(\omega 3T/8)\text{MC}(\pi/2)$

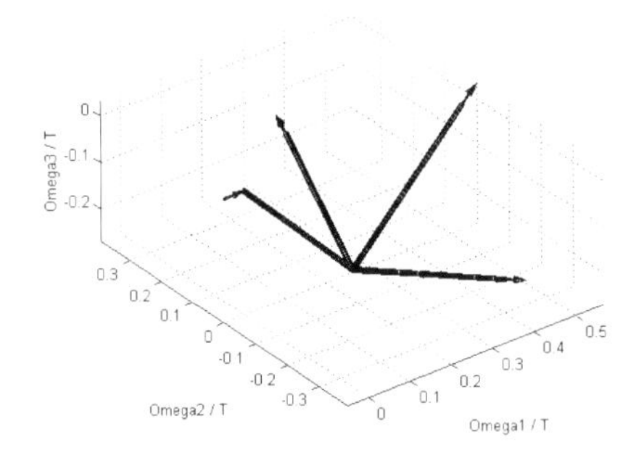

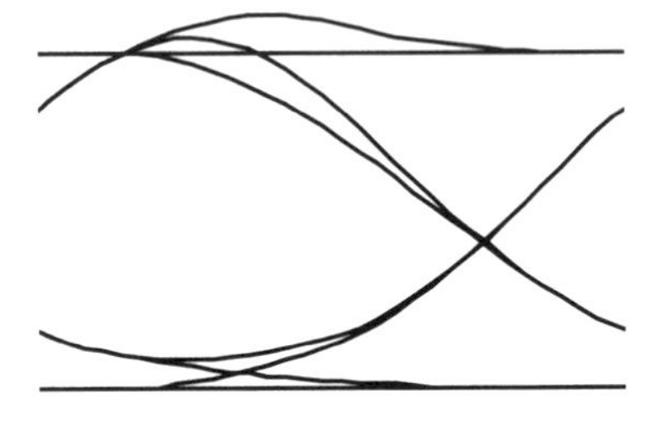

e) $\text{PS}(\omega 3T/8)\text{MC}(\pi/2)\text{PS}(\omega 3T/8)\text{MC}(-\pi/4)$

Fig. 2.29 (*continued*)

2.6.2 Polarization Mode Dispersion Compensation

A perfect PMD equalizer (EQ or EQ′) has to mirror the PMD profile of the transmission span [10]. Each DGD section of the transmission span will have an oppositely directed, direct neighbor of the equalizer. The principle can be understood from Figs. 2.28 and 2.30.

Fig. 2.30 PMD profile of transmission span (solid) and perfect equalizer (dashed, dotted) in the 3-dimensional normalized Stokes or PMD vector space. © 1999 IEEE.

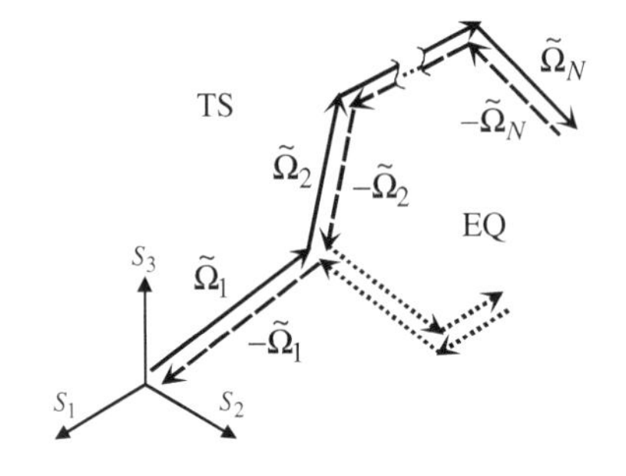

In a particular equalizer implementation EQ' may consist of the same sequence of SBAs and DGD sections as the TS, but with reversed order. The signs of SBA and PS retardations are inverted which can also be accomplished by orthogonal orientations.

$$EQ' = \prod_{i=1}^{N}\left(PS(-\omega\tau_i)SBA\left(-\varphi_{2,i},\varphi_{3i} + \sum_{k=i+1}^{N}(\varphi_{1k} + \varphi_{3k})\right)\right) \tag{2.440}$$

The phase delay $-\omega\tau_i$ of a DGD section may be of the order of many, many 2π. It is therefore practically impossible to avoid frequency-independent, possibly endless offset retardations $\varphi_{4i} = -\infty...\infty$ in each DGD section. However, these can be taken care of by moving them through subsequent SBAs, i.e. to the left, according to (2.437). In this generalized implementation with additional degrees-of-freedom the equalizer (EQ) is described by

$$\begin{aligned} EQ &= PS\left(\sum_{k=1}^{N}\varphi_{4k}\right)EQ' \\ &= \prod_{i=1}^{N}\left(PS(-\omega\tau_i + \varphi_{4i})SBA\left(-\varphi_{2i},\varphi_{3i} + \sum_{k=i+1}^{N}(\varphi_{4k} + \varphi_{1k} + \varphi_{3k})\right)\right) \end{aligned} . \tag{2.441}$$

As desired, the concatenation of TS and EQ (or EQ') results in a frequency-independent ER that does not exhibit any PMD,

$$EQ \cdot TS = PS\left(\sum_{i=1}^{N}(\varphi_{4i} + \varphi_{1i} + \varphi_{3i})\right)ER_0 = ER . \tag{2.442}$$

Although suggested by (2.441), the polarization transformers in the EQ need not necessarily be SBAs. We may substitute $\varphi_{4i} = \varphi'_{4i} + \varphi''_{4i}$ and consider $PS(\varphi'_{4i})$ to be part of a DGD section $PS(-\omega\tau_i + \varphi'_{4i})$ while $PS(\varphi''_{4i})$ belongs to the preceding polarization transformer. The *i*th polarization transformer, now described by

$$PS(\varphi''_{4i})SBA\left(-\varphi_{2i},\varphi_{3i} + \sum_{k=i+1}^{N}(\varphi'_{4k} + \varphi''_{4k} + \varphi_{1k} + \varphi_{3k})\right), \tag{2.443}$$

is a general ER (2.437) if its three variables, one for the PS and two for the SBA, are independent. However, to minimize efforts the designer should let φ''_{4i} depend on the two SBA parameters at will, thereby retaining only two independent variables. Such a polarization transformer may sometimes be easier to realize than a true SBA. Like the SBA it has the property of being able to **endlessly transform any input polarization into a PSP of the following DGD section**. Firstly, *any* polarization transformer capable of such operation may be used here, since the PS in (2.443) constitutes the difference between the SBA, which is needed to fulfill the required functionality, and the most general case, an ER. Secondly, it is indeed a PSP we need to consider here, not just an eigenmode. To see this we place another, arbitrary retarder and its inverse between $\mathrm{PS}(\varphi''_{4i})$ and $\mathrm{PS}(-\omega\tau_i+\varphi'_{4i})$. One of them is considered to belong to $\mathrm{PS}(-\omega\tau_i+\varphi'_{4i})$ where it transforms PSPs (but modifies EMs in a different way), while the other is part of the polarization transformer (2.443) and will transform horizontal and vertical polarizations which are available at the output of $\mathrm{PS}(\varphi''_{4i})$ into the transformed PSPs.

Eqn. (2.443) can also be implemented by a phase shifter placed between two mode converters, all with finite retardations [11]. The matrix $\mathrm{MC}(0...\pi)\mathrm{PS}(0...2\pi)\mathrm{MC}(0...\pi)$ is a suitable description. In this particular example, which requires a proper control algrorithm, φ''_{4i} in the equivalent expression (2.443) is sometimes a step function of the SBA orientation angle. This matter can complicate control considerably because φ''_{4i} appears in the orientation angles of all subsequent SBAs of the PMD compensator. On the other hand, practical difficulties are minimized if φ''_{4i} can be chosen to be constant.

Problem: A reciprocal polarization transformer which can be represented by the matrix $\mathrm{PS}(\psi_3)\mathrm{SBA}(\psi_1,\psi_2)$ with $\psi_3=f(\psi_1,\psi_2)$ can in forward direction endlessly transform any input polarization into x or y output polarization. (1) Show that in backward direction it can endlessly transform x or y polarization into any polarization. (2) Under which condition can it, in forward direction, also endlessly transform x or y input polarization into any output polarization? (3) A reciprocal polarization transformer is now given by $\mathrm{MC}(\delta_2)\mathrm{PS}(\delta_1)$. Can you represent it as $\mathrm{PS}(\psi_3)\mathrm{SBA}(\psi_1,\psi_2)$ with $\psi_3=f(\psi_1,\psi_2)$? (4) Can it in backward direction endlessly transform x or y polarization into any polarization? (5) Can it, in forward direction, endlessly transform x or y input polarization into any output polarization?

Solution: In the following we refer to Jones matrices. For reciprocity the Jones matrix in backward direction is the transpose of that in forward direction.

(1) For the backward direction we obtain $(\mathrm{PS}(\psi_3)\mathrm{SBA}(\psi_1,\psi_2))^T=\mathrm{SBA}(\psi_1,\psi_2)^T\,\mathrm{PS}(\psi_3)^T$ $=\mathrm{SBA}(\psi_1,-\psi_2)\mathrm{PS}(\psi_3)$; see also (2.252). $\mathrm{PS}(\psi_3)$ leaves x and y polarizations unchanged, and $\mathrm{SBA}(\psi_1,\psi_4)$ with $\psi_4=-\psi_2=-\infty...\infty$ has all necessary properties to be able to endlessly transform this x or y polarization into any polarization. Alternative, more complicated proof: The inverse of $\mathrm{SBA}(\psi_1,-\psi_2)\mathrm{PS}(\psi_3)$ must be able to endlessly transform any polarization into x or y polarization. That is indeed the case because we can write $(\mathrm{SBA}(\psi_1,-\psi_2)\mathrm{PS}(\psi_3))^{-1}=\mathrm{PS}(\psi_3)^{-1}\mathrm{SBA}(\psi_1,-\psi_2)^{-1}=\mathrm{PS}(-\psi_3)\mathrm{SBA}(-\psi_1,-\psi_2)$ $=\mathrm{PS}(\psi_6)\mathrm{SBA}(\psi_1,\psi_5)$ with $\psi_5=\pi-\psi_2=-\infty...\infty$ and $\psi_6=-\psi_3=g(\psi_1,\psi_5)$.

(2) With the assumed reciprocity the task is identical to endlessly transforming the arbitrary polarization of a backward-propagating signal at the output into x or y polarization at the input. The Jones matrix in backward direction $\mathrm{SBA}(\psi_1,-\psi_2)\mathrm{PS}(\psi_3)$ can according to (2.437) be written as $\mathrm{PS}(\psi_3)\mathrm{SBA}(\psi_1,-\psi_2-\psi_3)$. The task is fulfilled if $\mathrm{SBA}(\psi_1,\psi_7)$ with $\psi_7=-\psi_2-\psi_3$ is an SBA. So the question is whether $\psi_7=-\psi_2-f(\psi_1,\psi_2)$ can be continuously and strictly

monotonically varied in the range $-\infty ... +\infty$ by varying the only available independent variable ψ_2. It must therefore hold $0 < \partial\psi_7/\partial\psi_2 < \infty$ or $-\infty < \partial\psi_7/\partial\psi_2 < 0$. Using $\partial\psi_7/\partial\psi_2 = -1 - \partial\psi_3/\partial\psi_2$ it follows $-\infty < \partial\psi_3/\partial\psi_2 < -1$ or $-1 < \partial\psi_3/\partial\psi_2 < \infty$. The second expression is more easily fulfilled than the first, in particular by $\psi_3 = \text{const.}$.

(3) It holds $\mathrm{PS}(\psi_3)\mathrm{SBA}(\psi_1,\psi_2) = \begin{bmatrix} e^{j\psi_3/2}\cos\psi_1/2 & je^{j(\psi_2+\psi_3/2)}\sin\psi_1/2 \\ je^{-j(\psi_2+\psi_3/2)}\sin\psi_1/2 & e^{-j\psi_3/2}\cos\psi_1/2 \end{bmatrix}$ and $\mathrm{MC}(\delta_2)\mathrm{PS}(\delta_1) = \begin{bmatrix} e^{j\delta_1/2}\cos\delta_2/2 & je^{-j\delta_1/2}\sin\delta_2/2 \\ je^{j\delta_1/2}\sin\delta_2/2 & e^{-j\delta_1/2}\cos\delta_2/2 \end{bmatrix}$. An element comparison yields successively $\delta_2 = \psi_1$, $\psi_3/2 = \delta_1/2 = -\psi_2 - \psi_3/2$, $\psi_3 = -\psi_2$, $\delta_1 = -\psi_2$. Yes, the required δ_1, δ_2, ψ_3 are indeed given as functions of ψ_1, ψ_2. The device can in forward direction endlessly transform any input polarization into x or y output polarization. Much easier to understand, $\mathrm{PS}(\delta_1)$ transforms the input polarization to a position on the S_2 - S_3 great circle from where the subsequent $\mathrm{MC}(\delta_2)$ can transform it into x and y polarization.

(4) We combine solutions (3) and (1) and find: yes.

(5) We need to fulfill either $-\infty < \partial\psi_3/\partial\psi_2 < -1$ or $-1 < \partial\psi_3/\partial\psi_2 < \infty$. But in (3) we have found $\psi_3 = -\psi_2$, which means $\partial\psi_3/\partial\psi_2 = -1$. The answer is no. Much easier to understand, $\mathrm{PS}(\delta_1)$ leaves x and y input polarizations unchanged, and the subsequent $\mathrm{MC}(\delta_2)$ can therefore reach only half the S_2 - S_3 great circle but not the whole Poincaré sphere.

Fig. 2.30 shows the DGD profile of the concatenation of TS and a perfect EQ fulfilling (2.442). All PMD vectors are canceled by opposed adjacent ones because PMD is compensated not just to 1st order (which merely requires the vector sum and thereby the overall PMD vector to vanish) but completely (assuming that all existing frequency dependence has been covered by vectors Ω_i). As exemplified by the dotted arrows, an excess of total DGD of the EQ over that of the TS is not of concern if some adjacent compensator sections are made to cancel each other. Less perfect PMD vector cancelling normally indicates a PMD penalty. Provided the signal happens to coincide with a PSP it may be transmitted without distortion if merely 1st-order PMD persists after the EQ.

An important practical question is whether a PMD compensator should have fixed or variable DGD sections, and there is a straight answer to it [10]. It may be calculated that fixed differential group delays can cause detrimental side maxima of compensator performance. An immediate argument in favor of a variable DGD compensator is therefore that side maxima vanish. This is useful and important for one-section equalizers which, however, leave 2nd- and higher-order PMD uncompensated.

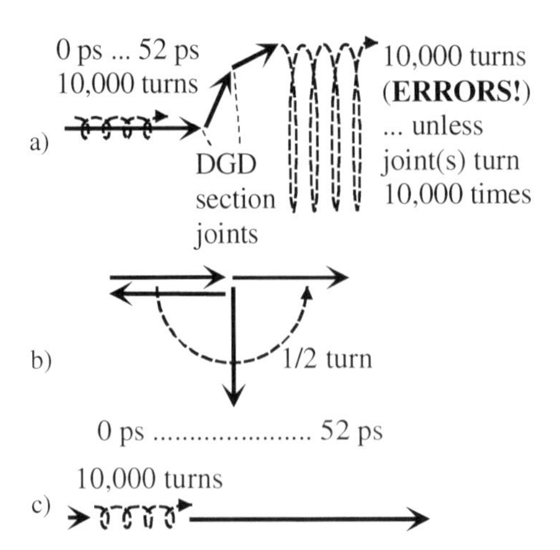

Fig. 2.31 Variable DGD section equalizer has speed problem (a). Transmission span most easily changes PMD vector orientations, and so does a fixed DGD section equalizer (b). Pure DGD vector lengthening is unlikely to happen in a transmission span (c). © 1999 IEEE.

The situation turns out to be different for equalizers with more than one section: What happens if the DGD of the first section (denoted as $-\tilde{\Omega}_N$ in Fig. 2.30) of a two- or even multisection EQ with *variable* DGDs has to slide from 0 to 52 ps? The latter value corresponds to roughly 10,000 periods of a 1550 nm lightwave. The DGD increase corresponds to a lengthening of this vector by a screw motion with a pitch of one lightwave period of DGD change per turn. If the subsequent DGD vectors (denominated as $-\tilde{\Omega}_{N-1}$ ... $-\tilde{\Omega}_1$) are connected to it by fixed joints (= SBAs with frozen parameters) the PMD profile will change shape during each screw turn, which is highly detrimental (Fig. 2.31a). If the joints are of rotary type (= SBAs with variable orientations) the PMD profile may stay the same because all subsequent DGD vectors may revolve in place 10,000 times like axes connected by rotary joints. If the first section is followed by a ball-and-socket joint (= ER with 3 degrees-of-freedom) only this latter has to turn 10,000 times. The issue can also be understood from $\tilde{\Omega}_i = \left(\prod_{j=1}^{i-1} \mathbf{G}_j^T\right)\Omega_i$. Now, if we consider that each of the 10,000 turns of the joint(s) (SBAs or ER) may need between 10 and 100 optimization steps it becomes clear that variable DGD sections are unpractical due to the huge required number of SBA (or ER) adjustment steps, except for the last section which has no subsequent SBA. For comparison, consider two *fixed* 26 ps DGD sections in the compensator. Changing the SBA (or ER) in between these two by a retardation of just π will flip the DGD profile open like a pocketknife from 0 ps to a total DGD of 52 ps (Fig. 2.31b), and this is about 10,000 times faster than the previous case.

But could a fixed DGD equalizer follow if a TS chose to vary its PMD profile by lengthening a DGD vector? Here we explain why substantial DGD vector length changes are unlikely to happen: Consider a TS with two *fixed* 26 ps DGD sections. As already explained, a DGD change of 52 ps requires just a retardation change of π for the SBA (or ER) in between them (Fig. 2.31b). In contrast, a pure 52 ps DGD *growth* (PMD vector lengthening) requires a much higher retardation change of $10{,}000 \cdot 2\pi$ and therefore occurs with a negligibly small probability (Fig. 2.31c).

As a consequence, **an equalizer with fixed DGD sections is a natural PMD compensator for a fiber transmission span** whereas more than one variable DGD section of an equalizer can practically not be used as such. Nevertheless, a single-section variable DGD equalizer is able to compensate 1st-order PMD, better than a single-section fixed DGD equalizer.

According to what we have learnt the DGD profile of the TE-TM converter of Fig. 2.25 is a straight arrow or rod as long as there is no mode conversion. This rod can be bent in any direction by a proper combination of in-phase and quadrature mode conversion. Twisting about the axis will only occur as a function of optical frequency but not electrooptically. The DGD profile of, say, Fig. 2.30 is just an approximation. In reality it is more likely that the DGD profile of the fiber will be smooth, without sharp corners. It is only logical to construct a near-perfect PMD equalizer from a sufficiently large number of such TE-TM mode converters on one chip. The DGD is about 0.26ps/mm. With possible chip sizes of almost 100 mm this is sufficient to compensate one bit duration of DGD at 40 Gbit/s. The DGD profile of the equalizer will be a bend-flexible but torsion-stiff rod, with close to ideal properties. Fig. 2.32 shows a small section of such a distributed PMD equalizer [18]. See also [24, 25] for further discussion on PMD and its compensation.

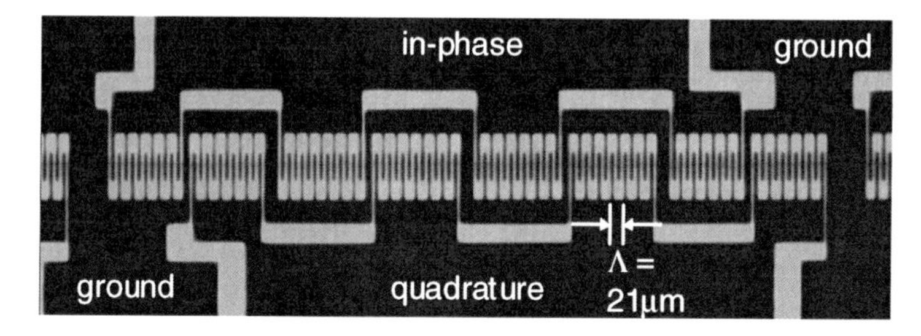

Fig. 2.32 One out of 73 TE-TM mode converters on an X-cut, Y-propagation $LiNbO_3$ chip. The waveguide runs underneath the comb electrodes.

2.6.3 Chromatic Dispersion Compensation

The Jones matrix (2.421) describes a birefringent waveguide with periodic mode coupling. We design it to have a length equal to an integer number m of coupling periods, $\beta_p z = 2\pi m$. The Jones matrix is now

$$\mathbf{J}_{PMC} = e^{-j\bar{\beta}z}(-1)^m \begin{bmatrix} \cos\Psi z - j\frac{\Delta\beta/2}{\Psi}\sin\Psi z & -j\frac{\underline{\kappa}}{\Psi}\sin\Psi z \\ -j\frac{\underline{\kappa}^*}{\Psi}\sin\Psi z & \cos\Psi z + j\frac{\Delta\beta/2}{\Psi}\sin\Psi z \end{bmatrix}. \tag{2.444}$$

We place this device between two circular retarders with retardations of $\pm\pi/2$ and Jones matrices $\frac{1}{\sqrt{2}}\begin{bmatrix} 1 & \mp 1 \\ \pm 1 & 1 \end{bmatrix}$. Also, we choose $\underline{\kappa}$ to be a purely real constant $\mathrm{K} = \underline{\kappa}$. The overall Jones matrix is calculated as

$$\mathbf{J} = \frac{1}{\sqrt{2}}\begin{bmatrix} 1 & 1 \\ -1 & 1 \end{bmatrix}\mathbf{J}_{PMC}\frac{1}{\sqrt{2}}\begin{bmatrix} 1 & -1 \\ 1 & 1 \end{bmatrix}. \tag{2.445}$$

If we feed an input signal into the first mode and select our output signal to be the first mode then its transfer function is the (1,1) element of $\mathbf{J}$, which equals

$$e^{-j\bar{\beta}z}(-1)^m\left(\cos\Psi z - j\frac{\mathrm{K}}{\Psi}\sin\Psi z\right). \tag{2.446}$$

For $|\underline{\kappa}| >> |\Delta\beta/2|$ and real $\underline{\kappa}$ the quotient K/Ψ may be approximated by $\operatorname{sgn}\mathrm{K} = \pm 1$. The resulting optical transfer function is $e^{-j\bar{\beta}z}(-1)^m e^{-j(\operatorname{sgn}\mathrm{K})\Psi z}$. Furthermore,

$$(\operatorname{sgn}\mathrm{K})\Psi z = (\operatorname{sgn}\mathrm{K})\sqrt{(\Delta\beta/2)^2 + |\mathrm{K}|^2}\, z \approx \mathrm{K}z + \Delta\beta^2 z/(8\mathrm{K}) \tag{2.447}$$

holds in that region. Let us assume that $\Delta\beta = 0$ holds at the optical carrier frequency. Using $\Delta\beta = \beta_{TE} - \beta_{TM} \pm \beta_p = (\omega - \omega_0)\Delta n/c$ the optical transfer function turns out to be proportional to

$$e^{-j((\omega-\omega_0)\Delta n)^2 z/(8\mathrm{K}c^2)} = e^{j\varphi} = e^{-j\beta(\omega)z}. \tag{2.448}$$

The optical phase depends quadratically on the optical frequency. The associated group delay is $\tau_g = -d\varphi/d\omega = (\omega - \omega_0)(\Delta n)^2 z/(4\mathrm{K}c^2)$. Since it is proportional to the optical frequency there is a chromatic dispersion

$$D = \frac{d^2\tau_g}{d\lambda dz} = -\frac{\omega^2}{2\pi c}\frac{d^2\beta}{d\omega^2} \quad \text{with} \quad \frac{d^2\beta}{d\omega^2} = \frac{\Delta n^2}{4\mathrm{K}c^2}. \tag{2.449}$$

At first glance there seems to be no direct use for this equation. However, it is possible to construct a device which approximates the above-given mathematics and has a chromatic dispersion which is adjustable by the coupling variable $1/\mathrm{K}$. Such devices usable as adaptive chromatic dispersion compensators [26–31], especially at 40 Gb/s where data transmission through a standard single-mode optical fiber (D = 17ps/nm/km) is limited to typically less than 5 km. In order to make it polarization-independent, two independent waveguides must be used instead of one birefringent waveguide. Since only the (1,1) element of the transfer matrix is needed, the circular retarders at input and output can be replaced by a Y splitter and a Y combiner, respectively.

The DGD profile of this chromatic dispersion compensator consists of a long DGD section. For $1/\mathrm{K} = 0$ it is coiled up with an infinitely small radius. For a non-zero $1/\mathrm{K}$ the radius is finite. Due to the circular retarders before and behind it the "coil" lies in the $\Omega_2 - \Omega_3$ plane whereas input and output arrows point in Ω_1 direction.

Planar lightwave circuits are attractive for the realization of tunable chromatic dispersion compensators. An integration with wavelength division demultiplexers is possible. A possible hardware implementation using a Mach-Zehnder lattice structure is schematically shown in Fig. 2.33. There are n sections DGD_i (i=1...n) with DGDs τ_i, separated by n–1 tunable couplers C_i. At least n = 2, better 4 or more sections are needed. Each DGD section is directly followed by a differential microheater (hatched) which introduces a differential phase shift φ_i. The input light is being split in a coupler C_{in} with equal powers into both branches of DGD_1. After the last DGD section DGD_n an output coupler $\mathrm{C}_{\mathrm{out}}$ directs most of the light into one output waveguide. Each tunable coupler C_i consists of two cascaded 3dB couplers, separated by an auxiliary DGD section with a DGD $\tau_{\mathrm{C},i}$ and an auxiliary differential phase shift $\varphi_{\mathrm{C},i}$. We assume all $\omega\tau_i$, $\omega\tau_{\mathrm{C},i}$ are odd multiples of π at the carrier frequency ω of a channel in the middle of a WDM band. The various $\varphi_{\mathrm{C},i}$ are controlled externally to adjust the chromatic dispersion. All φ_i need to be adjusted in such a way that the powers in both waveguides after passing C_{i+1} are equal. This is possible if we place two taps at the outputs of the next coupler C_{i+1}. The difference of the detected tap signals is integrated, and the integrator output controls φ_i. This local feedback greatly eases the control requirements because it halves the number of external control variables.

If a true 3dB input coupler C_{in} can not be fabricated reliably it can be made variable by a differential phase shift φ_{in} of about $\pm\pi/2$ between two nominal 3dB couplers, the phase shift φ_{in} being controlled to yield equal power distribution in DGD_1. In this case the following differential phase shift, here φ_1, must be diminished by $\pm\pi/2$. An analogous procedure can be implemented to control light steering into the output waveguide. For this the output coupler $\mathrm{C}_{\mathrm{out}}$ must have three outputs. The signal exits at the center waveguide, and the light power difference between the two lateral monitor outputs is made to become zero by suitable adjustment of the last differential phase shift φ_n. These embodiments of C_{in}, $\mathrm{C}_{\mathrm{out}}$ are depicted in Fig. 2.33 whereas the DGD profile of Figs. 2.34, 2.36 assumes 3dB couplers for C_{in}, $\mathrm{C}_{\mathrm{out}}$.

Problem: What is the effect of C_{in} as drawn in Fig. 2.33?

Solution: Using the terminology (2.434) with Jones matrices we derive $\mathrm{MC}(-\pi/2)\mathrm{PS}(\varphi_{\mathrm{in}})\mathrm{MC}(-\pi/2) = \begin{bmatrix} \cos\varphi_{\mathrm{in}}/2 & -\sin\varphi_{\mathrm{in}}/2 \\ \sin\varphi_{\mathrm{in}}/2 & \cos\varphi_{\mathrm{in}}/2 \end{bmatrix} \begin{bmatrix} 0 & -j \\ -j & 0 \end{bmatrix}$. This is a circular retarder with retardation φ_{in}, preceded by a full mode conversion or an interchange of waveguides. Variation of φ_{in} sets the power splitting ratio of C_{in}, while the subsequent DGD section allows to adjust also the phase shift between uncoupled and coupled waves.

DGD profiles can also describe the propagation of the two equipolarized waves in the two waveguides of the Mach-Zehnder lattice. Each of the many 3-dB couplers is described by a $\mathrm{MC}(\pi/2)$ matrix, i.e., a direction change by $\pi/2$ of the whole subsequent rest of the DGD profile.

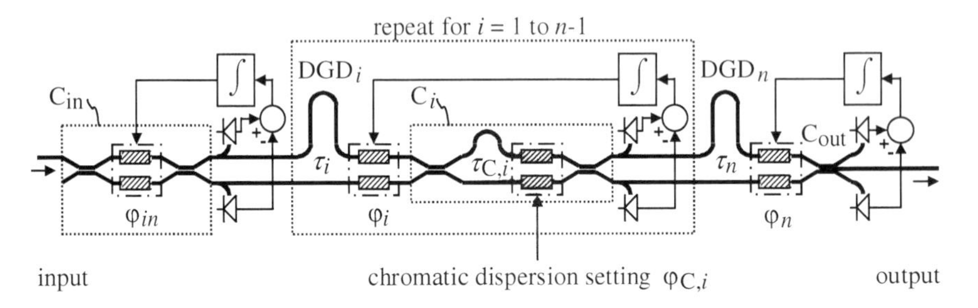

Fig. 2.33 Structure of planar chromatic dispersion compensator based on Mach-Zehnder lattice with tunable couplers

To give an example, we consider n = 12 Mach-Zehnder stages with $\tau_{1,11} = -\tau_{2,12} = 5\mathrm{ps}$, $\tau_{3,5,7,9} = -\tau_{4,6,8,10} = 10\mathrm{ps}$, $\varphi_{1...12} = 0$, and $\tau_{C,1...11} = -25\mathrm{fs}$, $\varphi_{C,i} = 0$ for the tunable couplers. The alternately positive and negative $\tau_{1,12}$ allow for a fairly regular, space-saving meandering waveguide pattern. At the center frequency the DGD profile is depicted in Fig. 2.34. The folded DGD sections correspond to infinite mode coupling, $1/\mathrm{K} = 0$. The input signal which is fed to one, and not the other input arm of the coupler C_{in} is represented by an input arrow in Ω_1 direction. The various rods in the Ω_2-Ω_3 plane represent DGD_i and have lengths τ_i. They must be configured to lie in the Ω_2-Ω_3 plane. This corresponds to equal power splitting between the two arms, which is achieved by the local control loops. Otherwise the dispersion compensation capability could be seriously hampered. Between the DGD_i there are rods along Ω_1 with lengths $\tau_{C,i}$, representing the DGD sections inside coupler C_i. Note that the Ω_1 coordinate in Fig. 2.34 is stretched by a factor of 50 for better viewability. Each rod twists of course about its own axis as a function of ω because the $\omega\tau_i$, $\omega\tau_{C,i}$ are proportional to ω. The angle change between the various rods represents mode coupling. The 3dB couplers C_{in}, C_{out} as well as those inside C_i are represented by concentrated $\pi/2$ turns. Looking at the whole DGD profile, if ζ is the angle between the output arrow and the input arrow (along Ω_1) then $\cos^2\zeta$ is the power fraction transmitted to the output. According to the above DGD vector definition the group delay experienced by the signal is –1/2 times the Ω_1 coordinate of the DGD profile endpoint. The factor 1/2 applies because the DGD equals the difference of the group delays of one mode, which is excited, and the other mode, which could be excited at the other, unused input port of C_{in}.

Now it is easy to graphically understand the function of this CD compensator: A *linear movement of the DGD profile endpoint along the Ω_1 axis as a function of ω* is required, simply because CD is a linear dependence of the group delay on ω! In addition, the endpoint vector must have the same direction as the input vector in order to avoid power loss. If the DGD profile did not lie in the Ω_2-Ω_3 plane at the center frequency not only power would be lost: In that case the DGD profile endpoint would not lie close to the Ω_1 axis off the center frequency, and this would cause an additional, symmetric eye closure like in a system with 1st-order PMD.

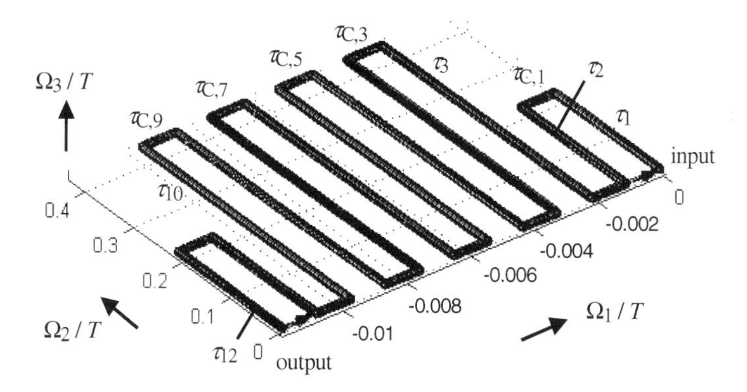

Fig. 2.34 DGD profile without chromatic dispersion at the center frequency. Scaling unit is the bit duration T = 23.3ps.

Simulated 43-Gbit/s NRZ and optically pre-filtered CS-RZ eye diagrams are shown in Fig. 2.35 left. They virtually coincide with the back-to-back eye diagrams. The electrical lowpass filter impulse response is chosen to be a sine halfwave with a length equal to one bit duration.

In the vicinity of the center frequency the DGD profile stays very similar to that of Fig. 2.34 because the frequency-dependent twists of adjacent but oppositely directed sections DGD_i cancel fairly well.

Next we want to compensate for the chromatic dispersion DL (given in ps/nm) of a transmission fiber. By numerical modeling we have found that a fairly good coupler phase shift setting is $\varphi_{C,i} = 0.0098 \cdot DL$. It is valid up to about $|\varphi_{C,i}| = 0.6\pi$. This corresponds to a ~±190ps/nm tuning range. Individual optimization of the $\varphi_{C,i}$ yields a slightly wider tuning range. For $\varphi_{C,i} = -\pi/2$ the DGD profile at the center frequency looks like a "discrete" spiral (right part of Fig. 2.36). All DGD_i rods lie essentially in one plane. 20 GHz higher, the DGD profile spiral is pushed far apart by the twist of the DGD rods about their axes, thereby providing the desired wavelength dependence of the group delay (rest of Fig. 2.36). For negative frequency offsets the spiral extends in the other direction. The larger the $|\varphi_{C,i}|$ are, the larger is the diameter of the "discrete" spiral, the faster moves the DGD profile endpoint as a function of ω and the larger is the amount of positive or negative CD introduced by the compensator.

Obviously, the chosen symmetry of τ_i and $\varphi_{C,i}$ settings with respect to the center of the compensator yields a regular DGD profile and is advantageous.

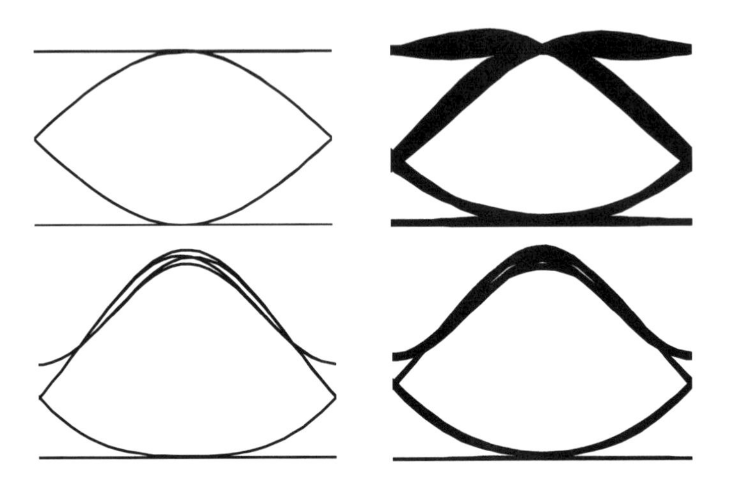

Fig. 2.35 Simulated 43-Gbit/s eye patterns behind compensator, with 0 ps/nm (left) or 160 ps/nm (right) of chromatic dispersion compensated. NRZ (top) and optically pre-filtered CS-RZ (bottom).

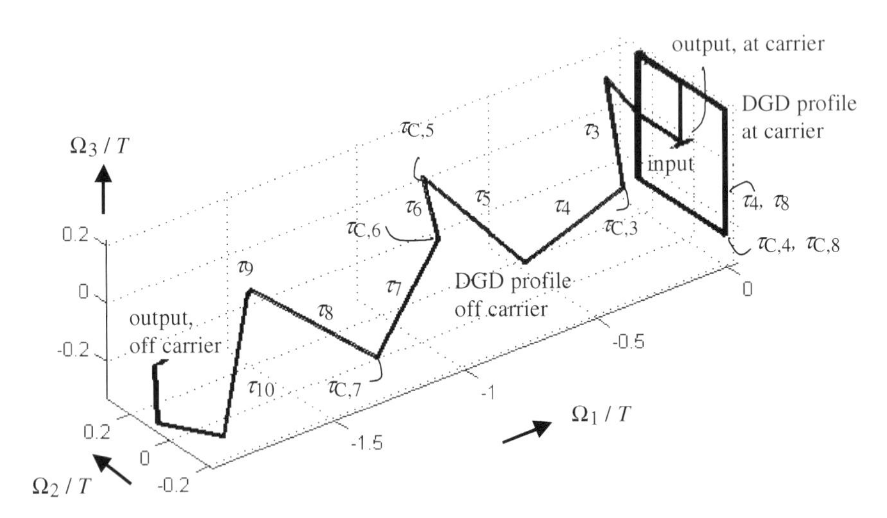

Fig. 2.36 DGD profile at carrier frequency and 20 GHz off carrier when a 160 ps/nm chromatic dispersion is compensated, using $\varphi_{C,i} = -\pi/2$. – When $\varphi_{C,i} = 0$ is used, Fig. 2.34 applies for a centered carrier frequency, and Fig. 2.36 simultaneously applies for a carrier frequency having a –10 THz offset with respect to the centered carrier frequency. A ~2 ps/nm^2 dispersion slope compensation is thereby provided.

Fig. 2.35 right shows the eye diagrams if a 160 ps/nm fiber dispersion is now compensated. If negative CD values are compensated the eye diagrams are identical to those for positive values.

Since all τ_i are integer multiples of 5 ps the DGD profile repeats every 200 GHz, but not exactly because of the non-zero $\tau_{C,i}$. With $\varphi_{C,i} = 0$ the DGD profile at the center frequency is given by Fig. 2.34. A large frequency offset lets the DGD sections inside C_i twist noticeably. Specifically, for a carrier frequency offset of –10 THz, where they twist by 1/4 turn, and for a frequency offset of –9.98 THz, the DGD profiles are again given by Fig. 2.35, and the eye diagram while compensating 160 ps/nm is given by Fig. 2.35 right. This shows that this device compensates a dispersion slope of ~2 ps/nm^2. The slope compensation changes its sign if we set $\varphi_1 = \varphi_{12} = \pi$ and replace the signs of all $\varphi_{C,1...11}$. This has the same effect as a sign change of all $\tau_{C,1...11}$.

Note that the desired dispersion slope must be chosen when compensation starts. Later it can not be changed without temporary loss of compensation. Compensation of much larger (e.g., 10ps/nm^2) or smaller slopes can also be implemented by proportional changes of the $\tau_{C,i}$. More finely selectable or truly variable amounts of slope compensation require more complicated variable coupler structures. These can be found intuitively using DGD profiles.

The local control of the differential phase shifts with the help of the monitor diode taps halves the number of degrees-of-freedom, thereby facilitating chromatic dispersion control. The power splitting control is not compromised by the dramatic, frequency-dependent DGD profile shape changes of Fig. 2.36. They occur in opposite directions for the two sidebands of a WDM channel and cancel each other. This means that at the gravity center of the optical spectrum of a WDM channel the DGD profile will lie in the Ω_2-Ω_3 plane as required, no matter what the chosen CD is. As a consequence, dispersion slope mitigation in a multichannel WDM environment will not cause power splitting control problems either because each WDM channel individually contributes to a correct setting. Moreover, vestigial sideband and similar modulation schemes are also possible because only the gravity center of the spectrum matters.

Another feature of this design example is the apodization of the τ_i vs. i. Here the two first and two last DGD sections are shorter than the middle ones, more than in [26]. The apodization greatly reduces unwanted distortions (eye closure) which are for example due to a non-flat amplitude response of the CD compensator. In our case the apodization with 5 ps and 10 ps DGDs makes it necessary to provide two separate compensators and AWG DEMUXs for even and odd 100GHz-spaced WDM channels. These commensurable τ_i yield periodic dispersion compensation windows. If periodicity is not needed then a τ_i apodization similar to a single sine halfwave (sampled from 0 to π for i = 1 to n) may be more advantageous.

While a purely analytical and/or numerical treatment of such dispersion compensators is of course possible the representation by DGD profiles provides a graphical understanding of Mach-Zehnder lattices and allows for an intuitive design.

2.6.4 Fourier Expansion of Mode Coupling

The definition of 1st-order PMD is undisputed [32]. Poole and Wagner have searched those principal states-of-polarization (PSPs) which will, to first order, not vary as a function of optical frequency at the output of a fiber, and have found that they exhibit a differential group delay (DGD) [8]. Alternatively one can determine the polarization-dependent small-signal intensity transfer function of a fiber (2.306), and will find maximum and minimum delays, hence a DGD, between orthogonal two polarizations, which are the PSPs [10].

Higher-order PMD means that the PMD vector varies with the optical frequency. Regarding higher-order PMD, most work on its definition concerns a truncated **Taylor expansion of the PMD vector** (TEPV),

$$\tilde{\Omega}(\omega)=\tilde{\Omega}(\omega_0)+(\omega-\omega_0)\tilde{\Omega}'(\omega_0)+\frac{(\omega-\omega_0)^2}{2!}\tilde{\Omega}''(\omega_0)+\frac{(\omega-\omega_0)^3}{3!}\tilde{\Omega}'''(\omega_0)+\dots \tag{2.450}$$

The PMD vector $\tilde{\Omega}(\omega_0)$ at a given optical frequency ω_0 defines 1st-order PMD, its first and second derivative with respect to optical (angular) frequency ω define 2nd and 3rd-order PMD, respectively, and so on. It is easy to calculate the PMD vector derivatives from the Jones matrix. The opposite direction is described by Heismann [33]. The advantages of a higher-order PMD definition by the TEPV are the following:

- Easy analytical calculation of higher PMD orders.
- Addition/subtraction of 2nd-order PMD to the fiber chromatic dispersion. The same holds for other orders of PMD and chromatic dispersion.
- Relation of 3rd-order PMD to slope steepness differences of NRZ signals.

Its key disadvantages are:

- No direct relation to physical fiber parameters exists.
- The true frequency-dependent trajectory of the DGD vector in the Stokes space would be described by sums of sinusoids with arguments that depend linearly on frequency. But sinusoids are not well approximated by a Taylor series! Inevitably, an infinite DGD will be predicted far off the optical carrier frequency!

2nd order PMD is usually classified with respect to the direction of 1st order PMD [34]. This is good to discuss statistics. However, and the same holds for 3rd order PMD, its effect on received eye patterns is better described when it is discussed with respect to the signal polarization: Depending on whether 2nd order PMD is parallel or antiparallel to the input polarization it will add and/or subtract to fiber chromatic dispersion.

Most PMD simulation is carried out by assuming a **sequence of cascaded DGD sections (SDGD)**, because it is widely accepted that an infinite number of randomly cascaded sections produces "natural" PMD. A finite SDGD can also be used for PMD description [10]. A restricted 3-section variant for effects up to 2nd order was proposed by Shtaif et al. [35]. Möller [36] has pointed out that the structure of a cascade of DGD sections can be obtained by a layer peeling algorithm [37]. This algorithm has been implemented, and experimental DGD profiles of simple PMD media have been obtained, including a distributed PMD compensator in several configurations [22]. Important advantages of the SDGD method are:

- Easy graphical display of DGD profile
- Building a PMD emulator with a SDGD is much easier than building one for the TEPV with adjustable orders of 1st and higher-order PMD [38].
- It emulates what a real fiber typically does, which is not evident for TEPV-based higher-order PMD emulators.

A disadvantage of the PMD definition by a SDGD is the potentially large number of unknowns. A modification of the scheme might involve choosing a number of DGD sections, for example all with equal, initially unknown lengths, and then finding their length and orientations which describe the PMD medium best. Obviously a single DGD section describes 1st-order PMD alone. Two DGD sections describe this, plus the fairly typical interplay between higher orders of the Taylor-approximated PMD vector. The total DGD is constant, which reminds of [34] where it has been shown that 2nd-order PMD parallel to 1st-order PMD is a lot weaker than perpendicular to it. The typical slope steepness difference, caused by 3rd-order PMD, can easily be observed even in two DGD sections. More accurate PMD modeling is possible with three or more DGD sections.

While all this makes sense we discuss in the following a related technique to describe higher-order PMD [32], a **F̲ourier e̲xpansion of m̲ode c̲oupling (FEMC)**. The direction change between adjacent DGD sections is the retardation of a mode converter. Seen by an observer who looks in the direction of the preceding DGD section this direction change occurs up/down or right/left. This amounts to in-phase or quadrature mode coupling between local PSPs. If the number of DGD sections approaches infinity, mode coupling becomes continuous. We point out that it is possible to describe a PMD medium by FEMC as follows:

- A frequency-independent mode conversion at the fiber input. This is described by 2 parameters, for example retardation and orientation.
- A total DGD.
- A frequency-independent mode conversion at the fiber output. In the general case a mode conversion (2 parameters, as at the input) and a differential phase shift (one more parameter) are needed. In total this means that there is a frequency-independent elliptical retarder at the output.
- Complex Fourier coefficients F_k (2.451), (2.452) of mode coupling along the birefringent medium, which exhibits the above-mentioned total DGD only in the absence of mode conversion.

These are a number of spatial Fourier coefficients of retardation or mode coupling as control variables. For a sequence of SBAs, $i = 1...n$ with individual retardations φ_i and orientation angles ψ_i (2.435) one can write

$$F_k = \sum_{i=1}^{n} \varphi_i e^{j\psi_i} e^{-j2\pi ik/n} . \qquad (2.451)$$

For a fully distributed mode coupling this becomes

$$F_k = \int_0^L \left(\frac{d\varphi(z)}{dz} e^{j\psi(z)} \right) e^{-j2\pi kz/L} dz . \qquad (2.452)$$

Among the four bullet items, the first three simply describe 1st-order PMD. Only the 4th item makes it a FEMC. If there is mode coupling the DGD profile will bend. Bends at discrete positions would correspond to the SDGD model. Fourier coefficients describe DGD profile bending in a continuous manner (Fig. 2.37).

The zero-order coefficient F_0 coils the DGD profile. Whether coiling occurs up/down or right/left or in a mix of these cases depends on the phase angle of F_0. The coiling radius is inversely proportional to the magnitude of this coefficient. Other F_k will wind a spiral when they occur alone. F_k combined with F_{-k} can result in a forth-and-back bending of the DGD profile.

Table 2.4 Order and number of real parameters in higher-order PMD definition methods. Number of real parameters needed for different orders of a chosen method are given, excluding 3 extra parameters for frequency-independent output polarization transformation. One example of the case in boldface is given in Figs. 2.37, 2.38. © 2004 IEEE.

Method (below) and its order (right)	1	2	3	4
Taylor expansion of PMD vector (TEPV)	3	6	9	12
Sequence of DGD sections (SDGD)	3	5	7	9
Fourier expansion of mode coupling (FEMC)	3	5	**9**	13

The number of real parameters needed to describe PMD by the three mentioned methods is listed in Table 2.4. In all cases 3 extra parameters must generally be added to specify a frequency-independent elliptical retarder at the fiber output. Only the order 1, corresponding to 1st-order PMD, is identical for all methods.

The TEPV needs 3 vector components in the Stokes space for each PMD order. Maximum PMD order covered and method order are equal only here.

The SDGD needs the 2 parameters of an SBA, plus one total DGD, for the 1st order. Each additional order is defined by 2 parameters of an SBA. The method order here means how many DGD sections there are.

In the FEMC, no mode coupling occurs in the 1st order PMD case. F_0 adds 2 real parameters. Each higher order of the method adds two Fourier coefficients $F_{\pm k}$, which amounts to 4 more real parameters.

In the following an FEMC example is given for method order 3 ($|k| \leq 1$). It needs 9 real parameters, like 3rd order TEPV. A random PMD medium has been taken as a reference. It is composed of 16 DGD sections with equal lengths. The length of one DGD section defines the normalized unit length in Fig. 2.37. The 1st-order PMD was 5.1 units, the (1st-order) PMD vector was $[-4.98, -1.24, -0.42]^T$. The reference is cascaded with a smoother DGD profile that is an inversion of the FEMC structure. It follows the jagged reference profile with gentle bends and more or less cuts through the "messy" left part of the reference profile. For convenience the FEMC structure was also represented by 16 DGD sections (instead of an infinite number) but this coincidence has no importance because their total DGD is only about half as high as that of the 16 reference DGD sections.

The FEMC coefficients can be determined as follows: A Gaussian input pulse was assumed, with a width equal to the total DGD of the DGD profile used in the FEMC (assuming no mode conversion). This is not the only possible FEMC pulse shape and duration. But it makes sense to choose the total DGD rather than the 1st-order DGD because the former is related to the overall complexity of the PMD situation while the latter may even disappear. Pulse width and the identical total DGD were of course varied during the optimization. The PMD medium (reference) and the inverse of the structure defined by the FEMC were concatenated. The various parameters were adjusted so that the output signal was – as far as possible – in only one (co-)polarization mode, and that the impulse in the other (cross-)polarization had its residual maximum amplitude near the time origin – not elsewhere like in the case of 1st-order PMD. Fig. 2.38 shows the magnitudes of the electric fields in co- and cross-polarized output pulses. The unwanted polarization is ≥37.2 dB down. The Gaussian input pulse is also shown.

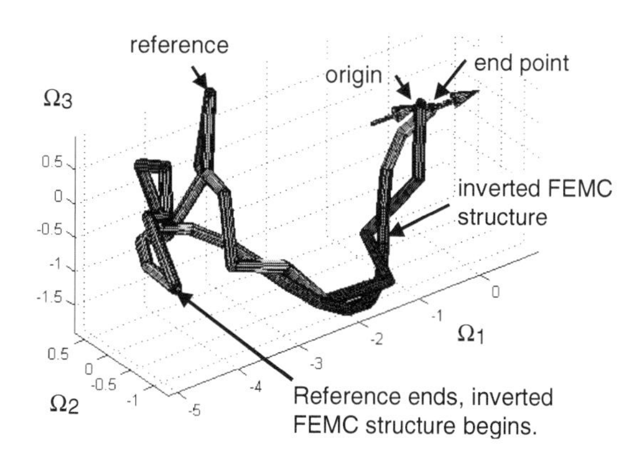

Fig. 2.37 DGD profile of reference (exemplary PMD medium) cascaded with inverted FEMC structure (which thereby forms a PMD equalizer). Scaling unit is 1 DGD section length of the reference structure. © 2004 IEEE.

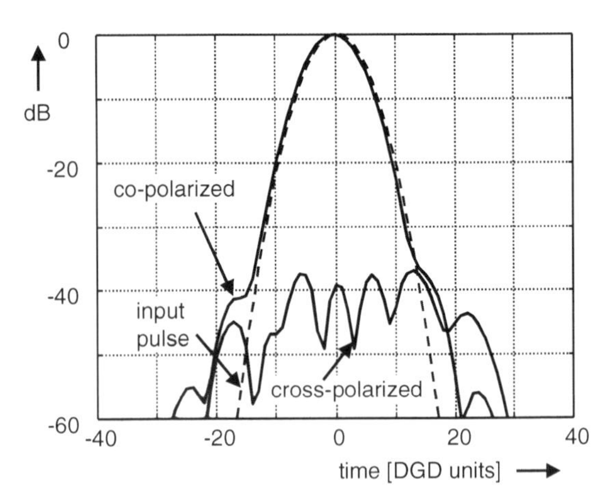

Fig. 2.38 Magnitudes of Gaussian input pulse, and of output pulses resulting from cascaded reference and inverted FEMC structure. Time scale is the same as in Fig. 2.37. Input pulse width is 8.07 DGD units. © 2004 IEEE.

As an alternative to the FEMC, the SDGD method could describe the reference profile exactly, but only if it comprised 16 sections equal to those of the reference.

For comparison, the TEPV method was also tested. The TEPV was calculated up to 3rd order from the Jones matrix of the reference (PMD medium). Then the Jones matrix corresponding to this truncated TEPV was built as described in [33]. The inverse of that matrix was cascaded with the reference. As a third candidate, the EMTY method, an exponential expansion of the Jones matrix by Eyal, Marshall, Tur and Yariv [39, 40], was tested the same way. For all methods the input pulse width was chosen identical to that after convergence of the FEMC for a particular PMD example. Table 2.5 shows orthogonal polarization suppression vs. method and its order, averaged over 75 randomly generated PMD examples. FEMC holds an advantage over the TEPV and the EMTY method. The extinction improvement in dB after addition of higher-order terms is more than twice as large for FEMC compared to TEPV and EMTY. The difference becomes more decisive if one takes into account that part of the extinction improvement of high method orders is due to the broader input pulses alone. All this is not surprising, because the FEMC and SDGD models are closely related to natural PMD, unlike TEPV or EMTY.

Table 2.5 Suppression of cross polarization by equalizers defined by three higher-order PMD definition methods. All methods are equal for order 1. Results are averaged over 75 randomly generated PMD examples, including the one in Fig. 2.37. © 2004 IEEE.

Method order	1	2	3
Chosen input pulse width [units]	5.6	7.2	9.5
Taylor expansion of PMD vector (TEPV)	10.3 dB	14.8 dB	19.9 dB
Exponential Jones matrix expansion (EMTY)	10.3 dB	12.6 dB	16.1 dB
Fourier expansion of mode coupling (FEMC)	10.3 dB	21.6 dB	35.5 dB

These results strongly suggest that a PMDC that could, for example, compensate the traditional PMD orders 1 to 3, is not the most efficient equalizer. Rather, a distributed PMDC could be preferable, with as sharp as possible a polarization transformation at its input, and one more at the output if a defined output polarization is needed, and controlled amounts of mode coupling along its length, for example defined by Fourier coefficients.

In its realization the distributed PMDC has a fixed total DGD. So its DGD profile must be flexible enough to fold, in order eliminate any excess DGD.

Note that for the calculations of Table 2.5, 3 more parameters than shown in Table 2.4 were chosen because the "equalizer" needed to be aligned to the reference to separate the polarizations as shown in Fig. 2.38.

Several variations of the FEMC are conceivable. For example, instead of the Fourier coefficients F_k with orders $k = 0,\pm1,\pm2,...$ one could use the orders $k = \pm1/2, \pm3/2, \pm5/2, ...$. This would mean that the numbers in the last line of Table 2.4 would need to be replaced by 3, 7, 11, 15. Or, the input and output polarization transformers could be made part of the mode conversion process. This is also practically the case if a distributed PMDC is used for PMD compensation plus output polarization control.

Determining the F_k and other FEMC coefficients presently involves the described numerical minimization process. It may indeed be considered as the most important drawback of this PMD description method that an analytical solution is not known. Finding an easy solution might help in the control of distributed PMDCs – this is a question of not getting trapped in local optima during the PMD control process.

The SDGD method is conceptually similar to FEMC and can be an alternative to FEMC, maybe by keeping the DGD section lengths constant and varying their number.

2.7 Nonlinearities in Optical Fibers

A silica optical fiber is strictly speaking not a linear transmission medium. At optical powers beyond 1 mW nonlinearities become more and more noticeable. We investigate several effects and their influence on optical data transmission systems. If a nonlinearity is to be utilized it is beneficial to choose a small core area since this increases the power density and hence the nonlinear effect. When a nonlinearity threatens to impair transmission – and this is more frequent – one chooses if possible a fiber with large core area.

Stimulated Brillouin scattering (SBS) in fibers is mediated by phonons (acoustical quanta). A wave which surpasses in a frequency range of a few 10 MHz a power of a few mW is backscattered in the fiber, and the backscattered wave possesses an optical frequency that is about 11 GHz lower than that of the incident wave. At high powers almost the total incident power is reflected. SBS is a very narrowband effect, and the reaction is slow due to propagation and backreflection. To suppress SBS it is usually sufficient to slightly modulate the pump current of the transmitter laser in the upper kHz regime. In addition to a weak amplitude modulation this also generates a frequency modulation which reduces the power spectral density of the optical carrier. Direct laser modulation or modulation schemes with suppressed carrier amplitude (CS-RZ, PSK) also reduce SBS.

Stimulated Raman scattering (SRS) in fibers is mediated by molecule vibrations. The amplification band is very broad and exhibits a maximum about 50 to 100 nm below the pump frequency. In forward direction SRS causes high-frequency WDM channels to be attenuated more strongly than low-frequency ones. In backward direction SRS can avoid the signal power reduction before reaching an amplifier station so much that larger amplifier spans are allowed. Usually several (in order to achieve a flat amplification profile) longitudinally multimoded (in order to suppress SBS) high-power laser are used as pumps. One needs a few 100mW of optical pump power. Polarization beam splitters and wavelength division multiplexers allow to add the powers of several pump lasers with low losses.

A very fast effect is caused by the nonlinear electronic third-order susceptibility in fibers, also called Kerr effect. In an unmodulated signal it manifests itself in self phase modulation (SPM). SPM provides a chirp to a modulated signal. Self phase modulation can be observed on an unmodulated signal which causes pulse compression in the presence of chromatic dispersion. Special pulses, so-called solitons, can even propagate undistorted within certain limits. Two or more optical signals influence each other by cross phase modulation (XPM). Likewise caused by third-order nonlinear susceptibility is four-wave mixing (FWM), a third-order intermodulation effect which generates new optical frequencies. FWM needs phase matching between the different waves, which means it can easily be suppressed by a non-zero local chromatic dispersion.

2.7.1 *Self Phase Modulation*

Self phase modulation (SPM) is proportional to the optical power, to be precise, to the instantaneous power that is variable within a period of the optical wave. As usual we define a real electric field to be the real part of a time- and position-dependent complex field,

$$\mathbf{E} = \mathrm{Re}\left(\underline{\mathbf{E}}\right) \qquad \mathbf{E} = \frac{1}{2}\left(\underline{\mathbf{E}} + \underline{\mathbf{E}}^*\right) \qquad \underline{\mathbf{E}} = \hat{\underline{\mathbf{E}}} e^{j(\omega t - \beta z)}. \tag{2.453}$$

For the dielectric displacement in the nonlinear optical fiber

$$\mathbf{D} = \varepsilon_0\left(\left(\mathbf{1} + \chi^{(1)}\right)\mathbf{E} + \chi^{(3)}\left(\mathbf{E}\mathbf{E}^T\mathbf{E}\right)\right) \tag{2.454}$$

holds. Nonlinear terms exist because the displacement of electrons in the SiO_2 molecules is no longer proportional to the field at arbitrarily large fields. A second-order term does not exist because the SiO_2 molecules are oriented randomly in the fiber, thereby averaging out the second-order effect. The instantaneous optical power is proportional to $|\mathbf{E}|^2 = \mathbf{E}^T\mathbf{E}$ $= \frac{1}{4}\left(\underline{\mathbf{E}}^T\underline{\mathbf{E}} + \underline{\mathbf{E}}^T\underline{\mathbf{E}}^* + \underline{\mathbf{E}}^+\underline{\mathbf{E}} + \underline{\mathbf{E}}^+\underline{\mathbf{E}}^*\right)$. It follows

$$\mathbf{E}\mathbf{E}^T\mathbf{E} = \frac{1}{8}\begin{pmatrix} \underline{\mathbf{E}}\underline{\mathbf{E}}^T\underline{\mathbf{E}} + \underline{\mathbf{E}}\underline{\mathbf{E}}^T\underline{\mathbf{E}}^* + \underline{\mathbf{E}}\underline{\mathbf{E}}^+\underline{\mathbf{E}} + \underline{\mathbf{E}}\underline{\mathbf{E}}^+\underline{\mathbf{E}}^* \\ + \underline{\mathbf{E}}^*\underline{\mathbf{E}}^T\underline{\mathbf{E}} + \underline{\mathbf{E}}^*\underline{\mathbf{E}}^T\underline{\mathbf{E}}^* + \underline{\mathbf{E}}^*\underline{\mathbf{E}}^+\underline{\mathbf{E}} + \underline{\mathbf{E}}^*\underline{\mathbf{E}}^+\underline{\mathbf{E}}^* \end{pmatrix}. \tag{2.455}$$

In order to continue with complex quantities in the form

$$\underline{\mathbf{D}} = \varepsilon_0\left(\left(\mathbf{1} + \underline{\chi}^{(1)}\right)\underline{\mathbf{E}} + \chi^{(3)}\underline{\left(\mathbf{E}\mathbf{E}^T\mathbf{E}\right)}\right) \qquad \left(\mathbf{E}\mathbf{E}^T\mathbf{E} = \mathrm{Re}\,\underline{\left(\mathbf{E}\mathbf{E}^T\mathbf{E}\right)}\right) \tag{2.456}$$

only the terms at positives frequencies are kept, but are multiplied by 2. This results in the complex quantity

$$\underline{\left(\mathbf{E}\mathbf{E}^T\mathbf{E}\right)} = \frac{1}{4}\left(\underline{\mathbf{E}}\underline{\mathbf{E}}^T\underline{\mathbf{E}} + \underline{\mathbf{E}}\underline{\mathbf{E}}^T\underline{\mathbf{E}}^* + \underline{\mathbf{E}}\underline{\mathbf{E}}^+\underline{\mathbf{E}} + \underline{\mathbf{E}}^*\underline{\mathbf{E}}^T\underline{\mathbf{E}}\right). \tag{2.457}$$

The first term at the right side possesses a tripled frequency and can be neglected. That wave is not strongly excited because there is no phase matching due to fiber dispersion. Second and third term can be combined,

$$\underline{\left(\mathbf{E}\mathbf{E}^T\mathbf{E}\right)} = \frac{1}{4}\left(2\hat{\underline{\mathbf{E}}}\left|\hat{\underline{\mathbf{E}}}\right|^2 + \hat{\underline{\mathbf{E}}}^*\left(\hat{\underline{\mathbf{E}}}^T\hat{\underline{\mathbf{E}}}\right)\right)e^{j(\omega t - \beta z)}. \tag{2.458}$$

In order to avoid ambiguity the unit $\sqrt{W}$ is added to the normalized fields in (2.346). The summation must extend only over the two orthogonal polarization modes ($s = x, y$) which travel with the same propagation constant. If one neglects the longitudinal components and divides by $n_{eff}^{1/2}$

$$\begin{aligned}&\frac{d\underline{E}_x^{\sqrt{W}}}{dz}\underline{\mathbf{E}}_{t,x}^{(1/m)}+\frac{d\underline{E}_y^{\sqrt{W}}}{dz}\underline{\mathbf{E}}_{t,y}^{(1/m)}\\&=-\frac{j\beta_0}{2\varepsilon_0 n_{eff}}\left(\underline{E}_x^{\sqrt{W}}\underline{\boldsymbol{\varepsilon}}_{p,t}\underline{\mathbf{E}}_{t,x}^{(1/m)}+\underline{E}_y^{\sqrt{W}}\underline{\boldsymbol{\varepsilon}}_{p,t}\underline{\mathbf{E}}_{t,y}^{(1/m)}\right)\end{aligned}\tag{2.459}$$

results. In addition, the modes are almost linearly polarized, with field amplitudes of nearly Gaussian shapes in the transversal plane. Therefore it is possible to substitute $\underline{\mathbf{E}}_{t,i}^{(1/m)}=E^{(1/m)}(x,y)\mathbf{e}_i$ with $\iint\left|E^{(1/m)}(x,y)\right|^2dxdy=1$, and a Jones vector $\underline{\mathbf{E}}=\underline{E}_x\mathbf{e}_x+\underline{E}_y\mathbf{e}_y=\left[\underline{E}_x\quad\underline{E}_y\right]^T$ can be introduced. This leads to

$$\frac{d\underline{\mathbf{E}}^{\sqrt{W}}}{dz}E^{(1/m)}=-\frac{j\beta_0}{2\varepsilon_0 n_{eff}}\underline{\boldsymbol{\varepsilon}}_{p,t}\underline{\mathbf{E}}^{\sqrt{W}}E^{(1/m)}.\tag{2.460}$$

From the derivation of (2.360) follows

$$\begin{aligned}&\varepsilon_0\left(\mathbf{1}+\underline{\chi}_t^{(1)}\right)\\&=\varepsilon_0\mathbf{1}n_{eff}^2-\begin{bmatrix}\Delta\varepsilon_1(\mathbf{r})/2 & \Delta\varepsilon_2(\mathbf{r})/2+j\Delta\varepsilon_3(\mathbf{r})/2\\ \Delta\varepsilon_2(\mathbf{r})/2-j\Delta\varepsilon_3(\mathbf{r})/2 & -\Delta\varepsilon_1(\mathbf{r})/2\end{bmatrix}\end{aligned}\tag{2.461}$$

with $\Delta\varepsilon_{1...3}(\mathbf{r})$ ($\mathbf{r}=[x\quad y\quad z]^T$) which for the time being are not only longitudinally but also transversally variable. Here $\varepsilon_0\mathbf{1}n_{eff}^2$ is the transversal part of the unperturbed dielectricity tensor, and the matrix at the right side is the linear part of the perturbation. Now this, (2.456) and $\underline{\hat{\mathbf{E}}}=\underline{\mathbf{E}}^{\sqrt{W}}E^{(1/m)}$ are inserted in (2.460). The result is multiplied by $E^{(1/m)^*}$ and integrated over the transversal plane,

$$\begin{aligned}&\frac{d\underline{\mathbf{E}}^{\sqrt{W}}}{dz}\iint\left|E^{(1/m)}\right|^2dxdy=-\frac{j\beta_0}{2\varepsilon_0 n_{eff}}\\&\left(\begin{aligned}&-\iint\left(\left|E^{(1/m)}(x,y)\right|^2\begin{bmatrix}\Delta\varepsilon_1(\mathbf{r})/2 & \Delta\varepsilon_2(\mathbf{r})/2+j\Delta\varepsilon_3(\mathbf{r})/2\\ \Delta\varepsilon_2(\mathbf{r})/2-j\Delta\varepsilon_3(\mathbf{r})/2 & -\Delta\varepsilon_1(\mathbf{r})/2\end{bmatrix}\right)dxdy\,\underline{\mathbf{E}}^{\sqrt{W}}\\&+\varepsilon_0\chi^{(3)}\iint\left|E^{(1/m)}(x,y)\right|^4dxdy\frac{1}{4}\left(\begin{aligned}&2\underline{\mathbf{E}}^{\sqrt{W}}\left|\underline{\mathbf{E}}^{\sqrt{W}}\right|^2\\&+\underline{\mathbf{E}}^{\sqrt{W}^*}\left(\underline{\mathbf{E}}^{\sqrt{W}^T}\underline{\mathbf{E}}^{\sqrt{W}}\right)\end{aligned}\right)\end{aligned}\right).\end{aligned}\tag{2.462}$$

The integral at the left hand side equals 1. As the result of the integral over the linear perturbation we set a matrix with $\Delta\varepsilon_{1...3}(z)$ (proportional to but not identical with $\Delta\varepsilon_{1...3}(\mathbf{r})$) which depends on z only. The integral $\iint \left|E^{(1/m)}(x,y)\right|^4 dxdy = A_{eff}^{-1}$ has the unit m^{-2} and defines the effective mode field area A_{eff}. Also, $\underline{\mathbf{E}}^{\sqrt{W}}$ can be replaced by a dimensionless and normalized Jones vector $\underline{\mathbf{E}}$ with $\left|\underline{\mathbf{E}}\right|^2 = 1$ (not to be confused with $\underline{\mathbf{E}}$ in (2.456), where the dimension is V/m), if in agreement with (2.333) the optical power $P = \frac{1}{2}\left|\underline{\mathbf{E}}^{\sqrt{W}}\right|^2$ (usually z-dependent due to attenuation) is inserted. The differential equation, to be solved as a function of z, is now

$$\frac{d\underline{\mathbf{E}}}{dz} = -\frac{j\beta_0}{2n_{eff}} \left(\begin{array}{l} -\frac{1}{\varepsilon_0} \begin{bmatrix} \Delta\varepsilon_1/2 & \Delta\varepsilon_2/2 + j\,\Delta\varepsilon_3/2 \\ \Delta\varepsilon_2/2 - j\,\Delta\varepsilon_3/2 & -\Delta\varepsilon_1/2 \end{bmatrix} \underline{\mathbf{E}} \\ + \frac{\chi^{(3)}P}{2A_{eff}} \left(2\underline{\mathbf{E}} + \underline{\mathbf{E}}^*\left(\underline{\mathbf{E}}^T\underline{\mathbf{E}}\right)\right) \end{array} \right). \tag{2.463}$$

The nonlinear perturbation is proportional to the effective power density P/A_{eff}. For the time being let the linear perturbation be zero, $\Delta\varepsilon_{1...3} = 0$. Using the nonlinearity constant

$$\gamma = \frac{3\beta_0\chi^{(3)}}{4n_{eff}A_{eff}}, \tag{2.464}$$

$$\frac{d\underline{\mathbf{E}}}{dz} = -j\frac{\gamma P}{3}\left(2\underline{\mathbf{E}} + \underline{\mathbf{E}}^*\left(\underline{\mathbf{E}}^T\underline{\mathbf{E}}\right)\right) \tag{2.465}$$

is obtained. The summand $2\underline{\mathbf{E}}$ at the right side causes pure SPM while preserving the state of polarization, i.e. multiplication by a phasor $e^{j\psi(z)}$. The term $\underline{\mathbf{E}}^*\left(\underline{\mathbf{E}}^T\underline{\mathbf{E}}\right)$ generally causes SPM and at the same time polarization changes, unless it disappears or is identical with $\underline{\mathbf{E}}$. It is identical if we set linear polarization $\underline{\mathbf{E}} = \begin{bmatrix} \cos\vartheta \\ \sin\vartheta \end{bmatrix} e^{j\psi(z)}$. The resulting differential equation $\frac{d\underline{\mathbf{E}}}{dz} = -j\gamma P\underline{\mathbf{E}}$ can be reduced to a scalar differential equation $\frac{de^{j\psi(z)}}{dz} = -j\gamma P e^{j\psi(z)}$ with the solution $\psi(z) = \psi(0) - \gamma\int_0^z P(\zeta)d\zeta$. For a transmitted power $P(0)$ and an amplitude attenuation constant α,

$$\int_0^z P(\zeta)d\zeta = P(0)\int_0^z e^{-2\alpha\zeta}d\zeta = P(0)\frac{1-e^{-2\alpha z}}{2\alpha} \tag{2.466}$$

holds. In the case of elliptical polarization with elevation angle ϑ and ellipticity angle ε

$$\underline{\mathbf{E}} = \begin{bmatrix} \cos\vartheta\cos\varepsilon + j\sin\vartheta\sin\varepsilon \\ \sin\vartheta\cos\varepsilon - j\cos\vartheta\sin\varepsilon \end{bmatrix} e^{j\psi} = \underline{\mathbf{E}}_{\|} \tag{2.467}$$

holds. This means $\left(\underline{\mathbf{E}}^T\underline{\mathbf{E}}\right) = \cos 2\varepsilon\, e^{j2\psi}$. Eqn. (2.465) can no longer be reduced to a scalar differential equation. Rather, $\vartheta = \vartheta(z), \varepsilon = \varepsilon(z), \psi = \psi(z)$ holds. The derivative with respect to z reads

$$\begin{aligned} \frac{\partial\underline{\mathbf{E}}}{\partial z} &= \frac{\partial\underline{\mathbf{E}}}{\partial\vartheta}\frac{\partial\vartheta}{\partial z} + \frac{\partial\underline{\mathbf{E}}}{\partial\varepsilon}\frac{\partial\varepsilon}{\partial z} + \frac{\partial\underline{\mathbf{E}}}{\partial\psi}\frac{\partial\psi}{\partial z} = \frac{\partial\underline{\mathbf{E}}}{\partial\vartheta}\vartheta' + \frac{\partial\underline{\mathbf{E}}}{\partial\varepsilon}\varepsilon' + \frac{\partial\underline{\mathbf{E}}}{\partial\psi}\psi' \\ &= e^{j\psi}\left(\begin{aligned} &\begin{bmatrix} -\sin\vartheta\cos\varepsilon + j\cos\vartheta\sin\varepsilon \\ \cos\vartheta\cos\varepsilon + j\sin\vartheta\sin\varepsilon \end{bmatrix}\vartheta' \\ &+ \begin{bmatrix} -\cos\vartheta\sin\varepsilon + j\sin\vartheta\cos\varepsilon \\ -\sin\vartheta\sin\varepsilon - j\cos\vartheta\cos\varepsilon \end{bmatrix}\varepsilon' \\ &+ j\begin{bmatrix} \cos\vartheta\cos\varepsilon + j\sin\vartheta\sin\varepsilon \\ \sin\vartheta\cos\varepsilon - j\cos\vartheta\sin\varepsilon \end{bmatrix}\psi' \end{aligned} \right) \end{aligned} \tag{2.468}$$

Both left and right hand side contain various states of polarization. Both sides are now split into mutually orthogonal components $\underline{\mathbf{E}}_{\|}, \underline{\mathbf{E}}_{\perp}$ with $\underline{\mathbf{E}}_{\perp}{}^{+}\underline{\mathbf{E}}_{\|} = 0$. With respect to (2.467),

$$\underline{\mathbf{E}}_{\perp} = \begin{bmatrix} -\sin\vartheta\cos\varepsilon - j\cos\vartheta\sin\varepsilon \\ \cos\vartheta\cos\varepsilon - j\sin\vartheta\sin\varepsilon \end{bmatrix} e^{j\psi} \tag{2.469}$$

holds. If (2.465) is multiplied from the left side by $\underline{\mathbf{E}}_{\|}{}^{+} = \underline{\mathbf{E}}^{+}$ one obtains

$$j\left(\sin 2\varepsilon\,\vartheta' + \psi'\right) = -j\gamma P\left(1 - (1/3)\sin^2 2\varepsilon\right). \tag{2.470}$$

If (2.465) is multiplied from the left side by $\underline{\mathbf{E}}_{\perp}{}^{+}$ one obtains

$$\cos 2\varepsilon\,\vartheta' - j\varepsilon' = \frac{\gamma P}{3}\sin 2\varepsilon\cos 2\varepsilon\,. \tag{2.471}$$

Separated into real and imaginary parts these are three equations $\varepsilon' = 0$, $\vartheta' = \frac{\gamma P}{3}\sin 2\varepsilon$, $\psi' = -\gamma P\left(1 - (1/3)\sin^2 2\varepsilon\right) - \sin 2\varepsilon\,\vartheta'$ with the solution

$$\varepsilon(z)=\varepsilon(0) \qquad \vartheta(z)=\vartheta(0)+\frac{\gamma\sin 2\varepsilon}{3}\int_0^z P(\zeta)d\zeta \qquad (2.472)$$
$$\psi(z)=\psi(0)-\gamma\int_0^z P(\zeta)d\zeta$$

In the case of elliptical polarization the polarization ellipse rotates in addition to the pure SPM effect! For circular polarization ($\varepsilon=\pm\pi/4, \sin 2\varepsilon=\pm 1$) (2.467) results in

$$\underline{\mathbf{E}}=\frac{1}{\sqrt{2}}\begin{bmatrix}1\\ \mp j\end{bmatrix}e^{j(\psi\pm\vartheta)}=\frac{1}{\sqrt{2}}\begin{bmatrix}1\\ \mp j\end{bmatrix}e^{j\left(\psi(0)\pm\vartheta(0)-(2/3)\gamma\int_0^z P(\zeta)d\zeta\right)}. \qquad (2.473)$$

It is seen that for circular polarization a change of elevation angle corresponds to a phase change of the wave. On the other hand the bracketed term in $\underline{\mathbf{E}}^*\left(\underline{\mathbf{E}}^T\underline{\mathbf{E}}\right)$ disappears in the circular case. Therefore $\frac{d\underline{\mathbf{E}}}{dz}=-j\frac{2\gamma P}{3}\underline{\mathbf{E}}$ holds, from which, after scalar calculation, an unaltered state-of-polarization and the same result as in (2.473) follows: For circular polarization self phase modulation is only 2/3 as strong as for linear polarization. Reason for this is that for circular polarization the instantaneous power constantly equals the mean power. But for linear polarization it fluctuates between 0 and twice the mean power, and the power maxima coincide with the extrema of the fundamental wave so that the phase of the linearly polarized wave is influenced particularly strongly.

Now we return to the general differential equation (2.463). The state-of-polarization in the fiber changes due to the linear perturbation. In a long fiber in which at least two of the perturbations $\Delta\varepsilon_{1...3}$ exist in a statistically independent manner the state-of-polarization is equidistributed on the surface of the Poincaré sphere. For this to be the case it is for example sufficient if a slightly elliptical core cross section exists with principal axes of the ellipse being randomly variable as a function of position. Expressed in spherical coordinates the equidistribution means that the doubled elevation angle 2ϑ is equidistributed and the doubled ellipticity angle 2ε is distributed like a cosine,

$$p_{2\vartheta}(2\vartheta)=\frac{1}{2\pi} \quad (|2\vartheta|\le\pi) \qquad p_{2\varepsilon}(2\varepsilon)=(1/2)\cos(2\varepsilon) \quad (|2\varepsilon|\le\pi/2). \qquad (2.474)$$

Let

$$\underline{\mathbf{E}}_{\|}(z)=\underline{\mathbf{J}}_l(z)\underline{\mathbf{E}}_{nl,\|}(z), \qquad (2.475)$$

where the matrix equation $\underline{\mathbf{E}}_{\|}(z)=\underline{\mathbf{J}}_l(z)\underline{\mathbf{E}}_{nl,\|}(0)$ is the solution of the linear differential equation and $\underline{\mathbf{E}}_{\|}(0)=\underline{\mathbf{E}}_{nl,\|}(0)$ is the transmitted signal. Partially inserted into (2.463) this results in

$$\left(\frac{d\underline{\mathbf{J}}_l}{dz}\underline{\mathbf{E}}_{nl,\parallel}+\underline{\mathbf{J}}_l\frac{d\underline{\mathbf{E}}_{nl,\parallel}}{dz}\right)$$
$$=-\frac{j\beta_0}{2n_{eff}}\left(\begin{array}{l}-\frac{1}{\varepsilon_0}\begin{bmatrix}\Delta\varepsilon_1/2 & \Delta\varepsilon_2/2+j\Delta\varepsilon_3/2\\ \Delta\varepsilon_2/2-j\Delta\varepsilon_3/2 & -\Delta\varepsilon_1/2\end{bmatrix}\underline{\mathbf{J}}_l\underline{\mathbf{E}}_{nl,\parallel}\\ +\frac{\chi^{(3)}P}{2A_{eff}}\left(2\underline{\mathbf{E}}_\parallel+\underline{\mathbf{E}}_\parallel{}^*\underline{\mathbf{E}}_\parallel{}^T\underline{\mathbf{E}}_\parallel\right)\end{array}\right). \tag{2.476}$$

On both sides, the first terms in the brackets describe linear propagation and can be left out. From the simplified system of equations

$$\underline{\mathbf{J}}_l\frac{d\underline{\mathbf{E}}_{nl,\parallel}}{dz}=-j\frac{\gamma P}{3}\left(2\underline{\mathbf{E}}_\parallel+\underline{\mathbf{E}}_\parallel{}^*\underline{\mathbf{E}}_\parallel{}^T\underline{\mathbf{E}}_\parallel\right) \tag{2.477}$$

a multiplication from left by $\underline{\mathbf{E}}_\parallel{}^+=\left(\underline{\mathbf{J}}_l\underline{\mathbf{E}}_{nl,\parallel}\right)^+$ allows to obtain

$$\underline{\mathbf{E}}_{nl,\parallel}{}^+\frac{d\underline{\mathbf{E}}_{nl,\parallel}}{dz}=-j\frac{\gamma P}{3}\left(2+\left|\underline{\mathbf{E}}_\parallel{}^T\underline{\mathbf{E}}_\parallel\right|^2\right). \tag{2.478}$$

$\underline{\mathbf{E}}_{nl,\parallel}$ is defined just as in (2.467) but with the subscript *nl*. At the right side $\underline{\mathbf{E}}_\parallel=\underline{\mathbf{J}}_l\underline{\mathbf{E}}_{nl,\parallel}$ is defined according to (2.467). Due to $\left(\underline{\mathbf{E}}_\parallel{}^T\underline{\mathbf{E}}_\parallel\right)=\cos 2\varepsilon\, e^{j2\psi}$ one obtains

$$j\left(\sin 2\varepsilon_{nl}\,\vartheta'_{nl}+\psi'_{nl}\right)=-j\gamma P\left(1-(1/3)\sin^2 2\varepsilon\right). \tag{2.479}$$

By multiplication of (2.477) by the orthogonal polarization $\underline{\mathbf{E}}_\perp{}^+=\left(\underline{\mathbf{J}}_l\underline{\mathbf{E}}_{nl,\perp}\right)^+$, where $\underline{\mathbf{E}}_{nl,\perp}$ is defined according to (2.469) with subscript *nl*,

$$\underline{\mathbf{E}}_{nl,\perp}{}^+\frac{d\underline{\mathbf{E}}_{nl,\parallel}}{dz}=-j\frac{\gamma P}{3}\underline{\mathbf{E}}_\perp{}^+\underline{\mathbf{E}}_\parallel{}^*\underline{\mathbf{E}}_\parallel{}^T\underline{\mathbf{E}}_\parallel \tag{2.480}$$

and finally

$$\cos 2\varepsilon_{nl}\,\vartheta'_{nl}-j\varepsilon'_{nl}=\frac{\gamma P}{3}\sin 2\varepsilon\cos 2\varepsilon \tag{2.481}$$

is found. It immediately follows $\varepsilon'_{nl}=0$, $\varepsilon_{nl}=\text{const.}$. Without loss of generality we assume $\cos 2\varepsilon_{nl}\neq 0$. The change of polarization or phase on one hand and of power on the other hand can be considered as statistically independent. Integration over dz on both sides therefore results in

$$\begin{aligned}\cos 2\varepsilon_{nl}\left(\vartheta_{nl}(z)-\vartheta_{nl}(0)\right)&=\int_0^z\frac{\gamma P}{3}\sin 2\varepsilon\cos 2\varepsilon\, d\zeta\\ &=\frac{\gamma}{3}\left(\int_0^z P(\zeta)d\zeta\right)\left(\int_{-\pi/2}^{\pi/2}\sin 2\varepsilon\cos 2\varepsilon\, p_{2\varepsilon}(2\varepsilon)d(2\varepsilon)\right)=0\end{aligned} \tag{2.482}$$

or $\vartheta_{nl} = \text{const.}$, $\vartheta'_{nl} = 0$. If this result is used when integrating (2.479) over dz, one obtains

$$\begin{aligned}\psi_{nl}(z) - \psi_{nl}(0) &= -\int_0^z \gamma P\left(1 - (1/3)\sin^2 2\varepsilon\right) d\zeta \\ &= \gamma\left(\int_0^z P(\zeta) d\zeta\right)\left(\int_{-\pi/2}^{\pi/2}\left(1 - (1/3)\sin^2 2\varepsilon\right) p_{2\varepsilon}(2\varepsilon) d(2\varepsilon)\right). \\ &= \bar{\gamma}\int_0^z P(\zeta) d\zeta \qquad \bar{\gamma} = \frac{8\gamma}{9}\end{aligned} \tag{2.483}$$

In long fibers therefore ideally no SPM-related polarization change will occur but only pure SPM that is 8/9 times as strong as for linear polarization. In (2.465) therefore $\underline{\mathbf{E}}^*\left(\underline{\mathbf{E}}^T \underline{\mathbf{E}}\right)$ can be replaced by $(2/3)\underline{\mathbf{E}}$ and it holds

$$\frac{d\underline{\mathbf{E}}}{dz} = -j\bar{\gamma}P\underline{\mathbf{E}} \text{ in the perturbed fiber,} \tag{2.484}$$

which of course can be reduced to a scalar equation. However, if the polarization state is not equidistributed on the Poincaré sphere within lengths which are small compared to the length $1/(2\alpha)$ on which the power is reduced to the $1/e$-fold of its initial value, then SPM-related polarization changes will generally occur. To be precise, equidistribution is not strictly necessary but (2.482), (2.483) are already fulfilled if for example 2ε alternatively assumes the values $\pm\arcsin\sqrt{2/3}$ over lengths << $1/(2\alpha)$ with equal probabilities. To fulfill (2.482) and suppress polarization changes even $p_{2\varepsilon}(2\varepsilon) = p_{2\varepsilon}(-2\varepsilon)$ is sufficient. This is the case in all birefringent fibers except for more or less hypothetical, circularly polarization-maintaining fibers which might just be produced by a strong twist of a fabricated fiber.

In usual fibers the effective nonlinearity constant is roughly $\bar{\gamma} = 1{,}5\ \mathrm{W}^{-1}\cdot\mathrm{km}^{-1}$ at $\lambda = 1550$ nm.

We will now describe the interplay of fiber nonlinearity and dispersion. The phase constant of a propagating wave can be expanded around the carrier frequency ω_0 by a truncated Taylor series

$$\beta(\omega) = \beta_0 + (\omega - \omega_0)\beta_1 + (\omega - \omega_0)^2\frac{\beta_2}{2!} + (\omega - \omega_0)^3\frac{\beta_3}{3!} \tag{2.485}$$

with $\beta_i = \left(d\beta/d\omega\right)\big|_{\omega=\omega_0}$. We assume now an electrical field amplitude of the form

$$\underline{E}(t,z) = \underline{\hat{E}}(t',z)e^{j(\omega_0 t - \beta_0 z)} \quad \text{with} \quad t' = t - \beta_1 z\,. \tag{2.486}$$

$\underline{\hat{E}}(t',z)$ is the electrical field envelope, in a coordinate system which moves with the group velocity $v_g = \beta_1^{-1}$ along the fiber. $\underline{\hat{E}}$ also has the second argument z

since the envelope can change even if t' is constant, for example due to fiber attenuation. For simplicity we will use the same symbols $\underline{E}$ in time and frequency domains, and distinguish them only by the argument t or t' vs. ω. Using $\underline{E}(\omega, z) = \int \underline{E}(t, z) e^{-j\omega t} dt$, $\hat{\underline{E}}(\omega, z) = \int \hat{\underline{E}}(t', z) e^{-j\omega t'} dt'$, we find

$$\underline{E}(\omega, z) = e^{-j(\beta_0 + (\omega - \omega_0)\beta_1)z} \hat{\underline{E}}(\omega - \omega_0, z) \tag{2.487}$$

The attenuation is assumed to be frequency-independent. In the frequency domain the linear wave propagation has the solution $\underline{E}(\omega, z) = e^{-(\alpha + j\beta(\omega))z} \underline{E}(\omega, 0)$. When (2.487) and (2.485) are inserted, and with a final substitution $\omega' = \omega - \omega_0$, one obtains

$$\hat{\underline{E}}(\omega', z) = e^{-\left(\alpha + j\left(\omega'^2 \frac{\beta_2}{2} + \omega'^3 \frac{\beta_3}{6}\right)\right)z} \hat{\underline{E}}(\omega', 0). \tag{2.488}$$

The partial derivative with respect to z is

$$\frac{\partial \hat{\underline{E}}(\omega', z)}{\partial z} = -\left(\alpha + j\left(\omega'^2 \frac{\beta_2}{2} + \omega'^3 \frac{\beta_3}{6}\right)\right) \hat{\underline{E}}(\omega', z). \tag{2.489}$$

An inverse Fourier transform results in

$$\frac{\partial \hat{\underline{E}}(t', z)}{\partial z} = \left(-\alpha + j\frac{\beta_2}{2}\frac{\partial^2}{\partial t'^2} + \frac{\beta_3}{6}\frac{\partial^3}{\partial t'^3}\right) \hat{\underline{E}}(t', z). \tag{2.490}$$

Note that $dt' = dt$. In scalar form, and in the moving coordinate system, (2.484) becomes $\frac{d\hat{\underline{E}}(t', z)}{dz} = -j\bar{\gamma} P(t', z) \hat{\underline{E}}(t', z)$. $P(t', z)$ is the power of the field envelope in W. This contribution is added to (2.490) in order to take also the nonlinear propagation into account. We thereby obtain a generalized form of the *nonlinear Schrödinger equation*

$$\frac{\partial \hat{\underline{E}}(t', z)}{\partial z} = \left(-\alpha + j\frac{\beta_2}{2}\frac{\partial^2}{\partial t'^2} + \frac{\beta_3}{6}\frac{\partial^3}{\partial t'^3} - j\bar{\gamma} P(t', z)\right) \hat{\underline{E}}(t', z). \tag{2.491}$$

For a given $\hat{\underline{E}}(t', z)$ at a certain position z the field evolution along the fiber can thereby be predicted. The simplified form

$$\frac{\partial \hat{\underline{E}}(t', z)}{\partial z} = j\left(\frac{\beta_2}{2}\frac{\partial^2}{\partial t'^2} - \bar{\gamma} P(t', z)\right) \hat{\underline{E}}(t', z), \tag{2.492}$$

without higher-order dispersion and attenuation, has as one possible solution a

$$\underline{\hat{E}}(t', z) = \sqrt{2P_0}\,\mathrm{sech}(t'/t_0)e^{jz\beta_2/\left(2t_0^2\right)} \qquad \text{1st-order } \textit{soliton}. \quad (2.493)$$

Here we have assumed $\underline{\hat{E}}(t', z)$ to have the unit $\sqrt{W}$ with the definition $P = \frac{1}{2}\left|\underline{\hat{E}}\right|^2$. The *hyperbolic secant* function is defined as $\mathrm{sech}\,x = \cosh^{-1} x = 2/\left(e^x + e^{-x}\right)$. We define the

$$L_D = \frac{t_0^2}{-\beta_2} = \frac{t_0^2}{D}\frac{2\pi c}{\lambda^2} \qquad \textit{dispersion length} \quad (2.494)$$

and the

$$L_{NL} = \frac{1}{\gamma P_0} \qquad \textit{nonlinear length}. \quad (2.495)$$

For a 1st-order soliton to propagate undistorted, the peak power P_0 and the soliton duration t_0 need to obey

$$L_D = L_{nl} \Leftrightarrow P_0 t_0^2 = \frac{-\beta_2}{\overline{\gamma}} = \frac{D}{\overline{\gamma}}\frac{\lambda^2}{2\pi c}. \quad (2.496)$$

Since $\overline{\gamma}$ is positive in silica (SiO_2), solitons can exist for positive fiber dispersion $D > 0$, also called anomalous dispersion due to $\beta_2 < 0$. Negative or normal dispersion ($D < 0$, $\beta_2 > 0$) does not allow soliton propagation; it would require a negative $\overline{\gamma}$. For transmission at telecom wavelengths (λ = 1550 nm) through standard singlemode fiber (D = 17 ps/nm/km, $\overline{\gamma} = 1{,}5\ \mathrm{W}^{-1} \cdot \mathrm{km}^{-1}$) it holds

$$P_0 t_0^2 \approx 14\ \mathrm{W} \cdot \mathrm{ps}^2. \quad (2.497)$$

A soliton with a duration t_0 = 10 ps needs a peak power of 140 mW. In non-zero dispersion-shifted fiber with a D = 3 ps/nm/km, solitons with duration t_0 = 100 ps need a peak power of just 0.25 mW.

Solitons of like polarity repel each other, solitons of opposite polarity attract each other. Solitons having different carrier frequencies travel of course at different group velocities, due to dispersion. When a faster soliton catches up with a slower one the resulting wave envelope is distorted, but as propagation continues they will again separate. The envelope distortion is subsequently restored, and after some time it looks as if the fast soliton had passed the slow one without distortion of either. Solitons tend to travel solitarily, which has determined their name.

Fiber attenuation is neglected in (2.492). Solitons therefore will not retain their shapes in real fiber. However, if (2.496) holds on average in a transmission system with a number of fiber spans between optical amplifiers, then the soliton still can be recovered at the receive end.

To understand the dynamical SPM effect qualitatively it is best to consider a single, rounded pulse, for example a soliton. The pulse center where there is maximum power is subject to the largest phase delay due to SPM. This means that the carrier frequency is reduced at the leading pulse edge and increased at the trailing pulse edge. In fibers with $D > 0$, the group delay increases with wavelength and decreases with frequency. The leading pulse edge with the reduced frequency is therefore delayed, and the trailing edge is accelerated. This compresses the pulse. The effect of chromatic dispersion is thereby reduced or, for solitons, completely compensated.

The phase of a soliton varies parabolically with time, corresponding to a negative chirp factor $\alpha_H < 0$ for $D > 0$. A Gaussian pulse with duration t_0 and chirp $\alpha_H = \Gamma^{-1}\left(1-\sqrt{1-\Gamma^2}\right)$ ($|\Gamma| \leq 1$, $\Gamma = \frac{4\beta_2 L}{t_0^2} = -\frac{2\lambda^2}{\pi c t_0^2} DL$) appears with the same duration and shape at the end of a fiber with length L, but the chirp has changed its sign. In contrast, SPM continuously restores the chirp, thereby allowing an ideal soliton to travel undistorted over arbitrary distance.

Let us assess the usefulness of solitons for optical fiber transmission:

- SPM broadens the optical spectrum, which can limit the WDM channel number.
- The generation of true solitons is difficult.
- Higher order chromatic dispersion impairs soliton propagation.
- Fiber attenuation perturbs the equilibrium of SPM and chromatic dispersion. As has been mentioned, it is sufficient to maintain this equilibrium on average in the whole fiber while the power is locally increased in the optical amplifiers and reduced in between. But also in that case each WDM channel must have a certain mean power in order to produce undistorted pulses at the fiber end instead of strongly broadened ones.

Practically it is often preferable to transmit rounded RZ pulses which can be generated fairly easily, in order to enjoy the positive properties of solitons at least partly. Even normal NRZ pulses are somewhat more tolerant to chromatic dispersion at moderately increased power levels than in the linear case.

2.7.2 Cross Phase Modulation

Now we consider the interaction of two signals with different frequencies and the total field

$$\underline{\mathbf{E}} = \underline{\mathbf{E}}_1 e^{j(\omega_1 t - \beta_1 z)} + \underline{\mathbf{E}}_2 e^{j(\omega_2 t - \beta_2 z)}. \tag{2.498}$$

Similarly to (2.453) to (2.458), leaving out the terms with negative frequencies and with frequencies $3\omega_1$, $2\omega_1 + \omega_2$, $\omega_1 + 2\omega_2$ and $3\omega_2$ results in

$$
\underline{\left(\mathbf{E}\mathbf{E}^T\mathbf{E}\right)} = \frac{1}{4}\left(\begin{array}{ll}
\left(2\underline{\mathbf{E}}_1\left|\underline{\mathbf{E}}_1\right|^2 + \underline{\mathbf{E}}_1^*\left(\underline{\mathbf{E}}_1^T\underline{\mathbf{E}}_1\right)\right)e^{j(\omega_1 t-\beta_1 z)} & \text{(SPM)} \\
+\left(2\underline{\mathbf{E}}_1\left|\underline{\mathbf{E}}_2\right|^2 + 2\underline{\mathbf{E}}_2\left(\underline{\mathbf{E}}_2^+\underline{\mathbf{E}}_1\right) + 2\underline{\mathbf{E}}_2^*\left(\underline{\mathbf{E}}_2^T\underline{\mathbf{E}}_1\right)\right)e^{j(\omega_1 t-\beta_1 z)} & \text{(XPM)} \\
+\left(2\underline{\mathbf{E}}_2\left|\underline{\mathbf{E}}_2\right|^2 + \underline{\mathbf{E}}_2^*\left(\underline{\mathbf{E}}_2^T\underline{\mathbf{E}}_2\right)\right)e^{j(\omega_2 t-\beta_2 z)} & \text{(SPM)} \\
+\left(2\underline{\mathbf{E}}_2\left|\underline{\mathbf{E}}_1\right|^2 + 2\underline{\mathbf{E}}_1\left(\underline{\mathbf{E}}_1^+\underline{\mathbf{E}}_2\right) + 2\underline{\mathbf{E}}_1^*\left(\underline{\mathbf{E}}_1^T\underline{\mathbf{E}}_2\right)\right)e^{j(\omega_2 t-\beta_2 z)} & \text{(XPM)} \\
+\left(2\underline{\mathbf{E}}_1\left(\underline{\mathbf{E}}_2^+\underline{\mathbf{E}}_1\right) + \underline{\mathbf{E}}_2^*\left(\underline{\mathbf{E}}_1^T\underline{\mathbf{E}}_1\right)\right)e^{j((2\omega_1-\omega_2)t-(2\beta_1-\beta_2))z} & \text{(FWM)} \\
+\left(2\underline{\mathbf{E}}_2\left(\underline{\mathbf{E}}_1^+\underline{\mathbf{E}}_2\right) + \underline{\mathbf{E}}_1^*\left(\underline{\mathbf{E}}_2^T\underline{\mathbf{E}}_2\right)\right)e^{j((-\omega_1+2\omega_2)t-(-\beta_1+2\beta_2)z)} & \text{(FWM)}
\end{array}\right). \tag{2.499}
$$

In addition to SPM of signals 1 and 2 there is cross phase modulation (XPM) of a signal by the power of the other signal, and four-wave mixing (FWM) that generates new optical frequencies. With regard to XPM we first consider two simple cases: If both polarizations are linear and identical the three nonlinear terms can be combined to become $(3/2)\underline{\mathbf{E}}_i\left|\underline{\mathbf{E}}_k\right|^2$ ($i = 1,2$, $k = 3-i$). If they are linear but orthogonal to each other only the first term remains, $(1/2)\underline{\mathbf{E}}_i\left|\underline{\mathbf{E}}_k\right|^2$. XPM varies by a factor of 3 depending on polarization. If the signals are linearly and identically polarized XPM alone is twice as strong as SPM. If there is PMD the linear perturbations of the two signals may differ.

We restrict ourselves to the case with linear perturbation but without PMD, where both polarizations are equidistributed on the Poincaré sphere. To simplify the calculation we abandon (2.499) and consider the sum (2.498) to be a single signal. Its power is

$$
P = \frac{1}{2}\left|\underline{\mathbf{E}}\right|^2 = \frac{1}{2}\left(\left|\underline{\mathbf{E}}_1\right|^2 + 2\,\mathrm{Re}\left(\underline{\mathbf{E}}_2^+\underline{\mathbf{E}}_1 e^{j((\omega_1-\omega_2)t-(\beta_1-\beta_2)z)}\right) + \left|\underline{\mathbf{E}}_2\right|^2\right). \tag{2.500}
$$

Inserted into (2.484) it results

$$
\begin{aligned}
\frac{d\underline{\mathbf{E}}}{dz} = -j\frac{\overline{\gamma}}{2}&\left(\left|\underline{\mathbf{E}}_1\right|^2 + 2\,\mathrm{Re}\left(\underline{\mathbf{E}}_2^+\underline{\mathbf{E}}_1 e^{j((\omega_1-\omega_2)t-(\beta_1-\beta_2)z)}\right) + \left|\underline{\mathbf{E}}_2\right|^2\right) \\
&\cdot\left(\underline{\mathbf{E}}_1 e^{j(\omega_1 t-\beta_1 z)} + \underline{\mathbf{E}}_2 e^{j(\omega_2 t-\beta_2 z)}\right)
\end{aligned}. \tag{2.501}
$$

The mixed term in the first bracket on the right hand side causes an upward or downward frequency shift. Also on the left hand side there are terms with several frequencies. On both sides we consider only the terms of one frequency ω_i and divide on both sides by $e^{j(\omega_i t-\beta_i z)}$. This way FWM is eliminated and just SPM and XPM remain,

$$
\begin{aligned}
\frac{d\underline{\mathbf{E}}_1}{dz} &= -j\frac{\overline{\gamma}}{2}\left(\left(\left|\underline{\mathbf{E}}_1\right|^2 + \left|\underline{\mathbf{E}}_2\right|^2\right)\underline{\mathbf{E}}_1 + \left(\underline{\mathbf{E}}_2^+\underline{\mathbf{E}}_1\right)\underline{\mathbf{E}}_2\right) \\
\frac{d\underline{\mathbf{E}}_2}{dz} &= -j\frac{\overline{\gamma}}{2}\left(\left(\underline{\mathbf{E}}_1^+\underline{\mathbf{E}}_2\right)\underline{\mathbf{E}}_1 + \left(\left|\underline{\mathbf{E}}_1\right|^2 + \left|\underline{\mathbf{E}}_2\right|^2\right)\underline{\mathbf{E}}_2\right)
\end{aligned}. \tag{2.502}
$$

We generate coupled differential equations of the Stokes vectors from (2.502). Eqn.

$$\frac{d}{dz}\left(\left|\underline{E}_{i1}\right|^2 \pm \left|\underline{E}_{i2}\right|^2\right) = 2\mathrm{Re}\left(\frac{d\underline{E}_{i1}}{dz}\underline{E}_{i1}{}^* \pm \frac{d\underline{E}_{i2}}{dz}\underline{E}_{i2}{}^*\right)$$
$$= \frac{\overline{\gamma}}{2} 2\,\mathrm{Im}\left(\underline{E}_{k2}{}^* \underline{E}_{i2}\underline{E}_{k1}\underline{E}_{i1}{}^* \pm \underline{E}_{k1}{}^* \underline{E}_{i1}\underline{E}_{k2}\underline{E}_{i2}{}^*\right). \tag{2.503}$$

holds. If one chooses the upper sign the expression at the left side equals dS_{i0}/dz and one obtains $dS_{i0}/dz = 0$. This means there is lossless mode conversion between the two signals. The lower sign results in

$$\frac{dS_{i1}}{dz} = \frac{\overline{\gamma}}{2}\left(S_{i2}S_{k3} - S_{i3}S_{k2}\right) \tag{2.504}$$

Similarly, one obtains

$$\frac{dS_{i2}}{dz} + j\frac{dS_{i3}}{dz} = \frac{d}{dz}\left(2\underline{E}_{i1}\underline{E}_{i2}{}^*\right) = 2\frac{d\underline{E}_{i1}}{dz}\underline{E}_{i2}{}^* + 2\frac{d\underline{E}_{i2}{}^*}{dz}\underline{E}_{i1}$$
$$= \frac{\overline{\gamma}}{2}\left(S_{i3}S_{k1} - S_{i1}S_{k3}\right) + j\frac{\overline{\gamma}}{2}\left(S_{i1}S_{k2} - S_{k1}S_{i2}\right). \tag{2.505}$$

The findings can be combined into

$$\frac{d\mathbf{S}_i}{dz} = -\frac{d\mathbf{S}_k}{dz} = \frac{\overline{\gamma}}{2}\left(\mathbf{S}_i \times \mathbf{S}_k\right). \tag{2.506}$$

Here the Stokes vectors $\mathbf{S}_i = \left[S_{i1} \quad S_{i2} \quad S_{i3}\right]^T$ (and $\mathbf{S}_k$) are composed of true, not normalized Stokes parameters 1...3. These Stokes vectors gyrate on two circles with equal radii $\left|\mathbf{S}_i \times \left(\mathbf{S}_i + \mathbf{S}_k\right)\right| / \left|\mathbf{S}_i + \mathbf{S}_k\right| = \left|\mathbf{S}_k \times \left(\mathbf{S}_i + \mathbf{S}_k\right)\right| / \left|\mathbf{S}_i + \mathbf{S}_k\right|$, through the centers of which the vector $\mathbf{S}_1 + \mathbf{S}_2$ passes, which is perpendicular to them.

XPM disturbs the signal transmission in wavelength-division multiplex (WDM) systems. In the simplest case, XPM in the interplay with chromatic dispersion just changes the impulse shape. The pure, phase-modulating XPM effect is particularly disturbing in the case of phase modulation. Polarization conversion impairs transmission especially when polarization division multiplex or polarization modulation are used.

XPM can be used in interferometers in order to switch optical signals all-optically. Even though this method has a fs resolution in fibers the polarization changes are a significant practical impediment against the usage of XPM. In semiconductor materials the effect is much stronger, but also much slower.

2.7.3 Four Wave Mixing

Four-wave mixing (FWM) is a distributed third-order intermodulation in the nonlinear fiber. In the simplest case which is just to be mentioned briefly, three waves with frequencies ω_i, ω_m, ω_k generate a fourth with frequency ω_l. Since there is an interaction of electromagnetic radiation with matter FWM must also be considered quantum-mechanically: Out of two photons with frequencies ω_i, ω_m one each of frequencies ω_k, ω_l is generated. Energy $\hbar\omega_i + \hbar\omega_m = \hbar\omega_l + \hbar\omega_k$ and momentum $\beta_i + \beta_m = \beta_l + \beta_k$ are conserved.

We now restrict ourselves to the case $m = i$, where two waves generate a third. In (2.499) the FWM terms

$$\frac{1}{4}\left(2\underline{\mathbf{E}}_i\left(\underline{\mathbf{E}}_k^{+}\underline{\mathbf{E}}_i\right)+\underline{\mathbf{E}}_k^{*}\left(\underline{\mathbf{E}}_i^{T}\underline{\mathbf{E}}_i\right)\right)e^{j((2\omega_i-\omega_k)t-(2\beta_i-\beta_k))z} \tag{2.507}$$

disappear if and only if the two signals are linearly and mutually orthogonally polarized. Just as for SPM and XPM, mainly the case with randomly birefringent fiber but without PMD is of interest. We therefore augment (2.501) at the right side within the bracket, and at the left side, by the term $\underline{\mathbf{E}}_l e^{j(\omega_l t-\beta_l z)}$ having the frequency $\omega_l = 2\omega_i - \omega_k$ and its derivative, respectively,

$$\begin{aligned}
&\frac{d\underline{\mathbf{E}}_i}{dz}e^{j(\omega_i t-\beta_i z)}+\frac{d\underline{\mathbf{E}}_k}{dz}e^{j(\omega_k t-\beta_k z)}+\frac{d\underline{\mathbf{E}}_l}{dz}e^{j(\omega_l t-\beta_l z)}= \\
&-j\overline{\gamma}\left(\begin{array}{l}\operatorname{Re}\left(\underline{\mathbf{E}}_k^{+}\underline{\mathbf{E}}_i e^{j((\omega_i-\omega_k)t-(\beta_i-\beta_k)z)}\right)+\operatorname{Re}\left(\underline{\mathbf{E}}_l^{+}\underline{\mathbf{E}}_i e^{j((\omega_i-\omega_l)t-(\beta_i-\beta_l)z)}\right) \\ +\operatorname{Re}\left(\underline{\mathbf{E}}_l^{+}\underline{\mathbf{E}}_k e^{j((\omega_k-\omega_l)t-(\beta_k-\beta_l)z)}\right)\end{array}\right) \\
&\cdot\left(\underline{\mathbf{E}}_i e^{j(\omega_i t-\beta_i z)}+\underline{\mathbf{E}}_k e^{j(\omega_k t-\beta_k z)}+\underline{\mathbf{E}}_l e^{j(\omega_l t-\beta_l z)}\right)
\end{aligned} \tag{2.508}$$

Here in the first bracket on the left side, which contains the instantaneous power P, the terms $\frac{1}{2}\left|\underline{\mathbf{E}}_n\right|^2$ ($n = i,k,l$) have been left out since they have already been discussed in the context of SPM and XPM and can be taken into account in the context of FWM by marginally changed phase constants.

The resulting terms at frequencies ω_i, ω_k, ω_l are always considered separately because the integral over z averages out to zero unless the integration is conducted with the right phases. After multiplying by $e^{-j(\omega_n t-\beta_n z)}$ three coupled differential equations

$$\begin{aligned}
\frac{d\underline{\mathbf{E}}_i}{dz}&=-j\frac{\overline{\gamma}}{2}\left(\begin{array}{l}\left(\underline{\mathbf{E}}_k^{+}\underline{\mathbf{E}}_i\right)\underline{\mathbf{E}}_k+\left(\underline{\mathbf{E}}_l^{+}\underline{\mathbf{E}}_i\right)\underline{\mathbf{E}}_l \\ +\left(\left(\underline{\mathbf{E}}_i^{+}\underline{\mathbf{E}}_k\right)\underline{\mathbf{E}}_l+\left(\underline{\mathbf{E}}_i^{+}\underline{\mathbf{E}}_l\right)\underline{\mathbf{E}}_k\right)e^{j(2\beta_i-\beta_k-\beta_l)z}\end{array}\right) \\
\frac{d\underline{\mathbf{E}}_k}{dz}&=-j\frac{\overline{\gamma}}{2}\left(\left(\underline{\mathbf{E}}_i^{+}\underline{\mathbf{E}}_k\right)\underline{\mathbf{E}}_i+\left(\underline{\mathbf{E}}_l^{+}\underline{\mathbf{E}}_k\right)\underline{\mathbf{E}}_l+\left(\underline{\mathbf{E}}_l^{+}\underline{\mathbf{E}}_i\right)\underline{\mathbf{E}}_i e^{-j(2\beta_i-\beta_k-\beta_l)z}\right) \\
\frac{d\underline{\mathbf{E}}_l}{dz}&=-j\frac{\overline{\gamma}}{2}\left(\left(\underline{\mathbf{E}}_i^{+}\underline{\mathbf{E}}_l\right)\underline{\mathbf{E}}_i+\left(\underline{\mathbf{E}}_k^{+}\underline{\mathbf{E}}_l\right)\underline{\mathbf{E}}_k+\left(\underline{\mathbf{E}}_k^{+}\underline{\mathbf{E}}_i\right)\underline{\mathbf{E}}_i e^{-j(2\beta_i-\beta_k-\beta_l)z}\right)
\end{aligned} \tag{2.509}$$

result. We consider the case where at the fiber input there are only the waves i, k, i.e., $\left|\underline{\mathbf{E}}_{i0}\right|>0$, $\left|\underline{\mathbf{E}}_{k0}\right|>0$, $\left|\underline{\mathbf{E}}_{l0}\right|=0$. Let initially the polarizations i, k be identical. Now also the remaining XPM terms can be left out because they can be taken into account in the context of FWM by adaptation of the phase constants. Also, amplitudes can be substituted for the vectors. One thereby obtains

$$\begin{aligned} \frac{d\underline{E}_i}{dz} &= -j\frac{\bar{\gamma}}{2}\left(\underline{E}_i^+\underline{E}_k\underline{E}_l+\underline{E}_i^+\underline{E}_l\underline{E}_k\right)e^{j(2\beta_i-\beta_k-\beta_l)z} \\ \frac{d\underline{E}_k}{dz} &= -j\frac{\bar{\gamma}}{2}\underline{E}_l^+\underline{E}_i\underline{E}_ie^{-j(2\beta_i-\beta_k-\beta_l)z} \\ \frac{d\underline{E}_l}{dz} &= -j\frac{\bar{\gamma}}{2}\underline{E}_k^+\underline{E}_i\underline{E}_ie^{-j(2\beta_i-\beta_k-\beta_l)z} \end{aligned} \quad . \tag{2.510}$$

Due to $\left|\underline{E}_{l0}\right|=0$ the $\underline{E}_{i,k}$ can initially be considered as constant. Integration of the third equation yields

$$\begin{aligned} \underline{E}_l(z) &= -j\frac{\bar{\gamma}}{2}\underline{E}_k^+\underline{E}_i\underline{E}_ie^{-j(2\beta_i-\beta_k-\beta_l)z/2}\frac{2\sin((2\beta_i-\beta_k-\beta_l)z/2)}{2\beta_i-\beta_k-\beta_l} \\ &= -j\frac{\bar{\gamma}}{2}\underline{E}_k^+\underline{E}_i\underline{E}_ize^{-j(2\beta_i-\beta_k-\beta_l)z/2}z\,\mathrm{sinc}((2\beta_i-\beta_k-\beta_l)z/2). \end{aligned} \tag{2.511}$$

For successful FWM, phase matching is needed, ideally $\left|(2\beta_i-\beta_k-\beta_l)z/2\right|<<1$. For this to be the case, chromatic dispersion must be small. For large $(2\beta_i-\beta_k-\beta_l)/2$ the field $\underline{E}_l(z)$ oscillates as a function of z. $\underline{E}_l$ is proportional to $P_i=\frac{1}{2}\left|\underline{E}_i\right|^2$.

In principle, FWM can enormously impair WDM transmission because the FWM signal $\underline{E}_l$ must stay very small, $\left|\underline{\mathbf{E}}_l\right|<<\left|\underline{\mathbf{E}}_{i,k}\right|$, if it coexists with another wanted signal of frequency ω_l. This is aggravated by the fact that in general there are many, many possibilities to generate a carrier at ω_l, in particular also with different ω_i, ω_m, ω_k. However, this is valid only if there is phase matching. If for example positive dispersion is chosen on the first, and negative dispersion on the second half of a fiber, or if one transmits over standard single-mode fiber and subsequently compensates chromatic dispersion by DCF, then chromatic dispersion and FWM are very well suppressed globally and locally, respectively. To estimate the necessary local dispersion we have to consider fiber loss: FWM will no longer increase even for zero dispersion if the signal is strongly attenuated. The effective length is then $P^{-1}(0)\int_0^\infty P(\zeta)d\zeta=(2\alpha)^{-1}$, i.e., 22 km at λ = 1550 nm. On the other hand

$$2\beta_i - \beta_k - \beta_l \approx -(\omega_k - \omega_i)^2 \beta_T'' = (\omega_k - \omega_i)^2 \frac{\lambda^2}{2\pi c} D \tag{2.512}$$

holds. With D = 17 ps/nm/km and a WDM channel separation $(\omega_k - \omega_i)/2\pi$ = 100 GHz one obtains $(2\beta_i - \beta_k - \beta_l)/2$ = 1/(234 m). Due to $|z\,\mathrm{sinc}((2\beta_i - \beta_k - \beta_l)z/2)| \le 2/|2\beta_i - \beta_k - \beta_l|$ an effective length of no more than 234 m results. As long as there is $|\underline{\mathbf{E}}_l| << |\underline{\mathbf{E}}_{i,k}|$ (which, however, would not be the case for $D = 0$ and powers in the mW range), FWM is therefore suppressed by dispersion by something like 40 dB. FWM is therefore generally of subordinate importance in well-designed practical WDM systems. In contrast, chromatic dispersion, SPM and XPM are accumulated over the whole fiber length.

$\underline{E}_l$ is also proportional to $\underline{E}_k^+$. With respect to $\underline{E}_k$ this means that there is a phase conjugation, i.e. the phase and the frequency offset with respect to the carrier change signs. There is a spectral inversion: The modulation sidebands are flipped over, exchanged. Of course the carrier frequency is also changed, $\omega_l \ne \omega_k$. If approximately in the middle of a fiber the signal $\underline{E}_k$ is converted by FWM into the signal $\underline{E}_l$ then distortions due to chromatic dispersion and SPM can be removed. Pump wave $\underline{E}_i$ and unconverted signal wave $\underline{E}_k$ must be eliminated by filtering before the second half of the fiber, and $\underline{E}_l$ must appropriately amplified. However, the expenses to realize this scheme are so high that it is not being used in practice so far.

Out of interest we calculate from (2.510) the derivatives of all powers $n = i, k, l$,

$$\frac{dP_n}{dz} = \frac{1}{2}\frac{d|\underline{E}_n|^2}{dz} = \mathrm{Re}\left(\underline{E}_n^* \frac{d\underline{E}_n}{dz}\right), \tag{2.513}$$

with the result

$$\frac{dP_k}{dz} = \frac{dP_l}{dz} = -\frac{1}{2}\frac{dP_i}{dz} = \frac{\overline{\gamma}}{2}\mathrm{Im}\left(\underline{E}_l^+ \underline{E}_k^+ \underline{E}_i \underline{E}_i e^{-j(2\beta_i - \beta_k - \beta_l)z}\right). \tag{2.514}$$

While conserving the total energy (due to $dP_k/dz + dP_l/dz + dP_i/dz = 0$) the signals k, l are amplified by equal power increments at the expense of signal i.

Finally we consider the case of orthogonal polarizations, $\underline{\mathbf{E}}_k^+ \underline{\mathbf{E}}_i = 0$. From (2.509) it then follows that the wave l is not generated or amplified. It is just cross-phase modulated – if it exists at all. If in a WDM system with constant channel separation $\omega_k - \omega_i$ and negligible PMD, neighbor channels are alternatively transmitted with orthogonal polarizations then FWM and the polarization-converting part of XPM between even and odd channels are suppressed. XPM therefore will not be polarization-converting at all. FWM then takes place only (a) between all even and (b) between all odd channels, and the doubled frequency separation further reduces FWM.

The most important reason for transmitting WDM channels with alternating orthogonal polarizations is the possibility to suppress linear interference. Behind a filter which is designed to pass only $\underline{\mathbf{E}}_1$, the neighbor signal $\underline{\mathbf{E}}_2$ may happen not to be sufficiently well attenuated yet. In the total power $P = \frac{1}{2}\left(\left| \underline{\mathbf{E}}_1 \right|^2 + 2\,\mathrm{Re}\left(\underline{\mathbf{E}}_2^+ \underline{\mathbf{E}}_1 e^{j((\omega_1 - \omega_2)t - (\beta_1 - \beta_2)z)} \right) + \left| \underline{\mathbf{E}}_2 \right|^2 \right)$ the first term is wanted, the third is negligible. The second term is most disturbing but is suppressed by orthogonal neighbor channel polarizations ($\underline{\mathbf{E}}_2^+ \underline{\mathbf{E}}_1 = 0$) (unless PMD in the fiber causes $\underline{\mathbf{E}}_2^+ \underline{\mathbf{E}}_1 \neq 0$).

Chapter 3
Optical Fiber Communication Systems

3.1 Standard Systems with Direct Optical Detection

Optical data transmission systems differ considerably from their counterparts in the electrical domain, e.g., by extremely high carrier frequencies, extremely high data rates, no direct measurability of the transmitted fields, squaring elements at the receiver input, quantum effects, multimoded reciprocal amplifiers, dispersive and somewhat nonlinear transmission medium. Starting from the simple, important techniques for the design of optical data transmission systems shall be presented in this chapter.

3.1.1 Signal Generation, Transmission, and Detection

For serial data transmission different signal formats are being used, depending on requirements and technical boundary conditions. A symbol alphabet with M data symbols b possesses a *mean entropy* of $H=-\sum_{m=0}^{M-1} P_m \log_2 P_m$ bit. Here P_m is the probability that symbol m is being transmitted. For statistical equidistributed symbols with $P_m = 1/M$ the entropy assumes a maximum $H = \log_2 M$. This is assumed in the following. The information rate or bit rate is $R_B = H/T = (\log_2 M)/T$ bit/s. T is the constant time difference between two subsequent symbols. The symbol rate $1/T$ is also called the *clock frequency*. Binary optical signals (M = 2 and in general $P_1 = P_0 = 1/2$) are generated most easily. Here the bit rate equals the symbol rate. In a simplifying manner a binary data symbol is often called a bit. The data symbols b are coded as transmitted symbols $\underline{c}$. For trivial coding in the binary case, $\underline{c} = b$ holds with $b \in \{0;1\}$.

We generate a data signal

$$\underline{a}(t)=\sum_{i=-\infty}^{\infty}\underline{d}_i(t-iT) \qquad \underline{d}_i(t)=\underline{c}_i\underline{s}(t), \tag{3.1}$$

which may be regarded as dimensionless for the time being. In spite of possible signal distortions the data signal can often be approximated by a linear superposition of data pulses $\underline{d}_i(t)$ which are temporally separated by integer multiples of the symbol duration T. Each data pulse is the product of a transmitted symbol $\underline{c}_i$ and a single pulse $\underline{s}(t)$. In order to allow for description of more complicated modulation schemes and/or modulation means with complex impulse responses we admit complex transmitted symbols and single pulses. In an ideal amplitude modulator without chirp, the output field of which is under consideration, $\underline{s}(t)$ is real and positive.

The photocurrent in the optical receiver is proportional to the optical power, which may be proportional to the pump current of a directly modulated laser. It therefore may be useful to exclusively consider electrical signals (pump current, photo current) or the optical power which is proportional to them. A "power impulse response" of the fiber does not exist. Rather, wave propagation must be described with respect to optical field strengths. Also, transmitter chirp and dispersion of the medium must be represented using fields. The transmitted field must first be determined from the electrical modulation signal, and the photocurrent must also be calculated from the received field.

A practical difficulty exists because the optical fields are not directly measurable in general. However, the optical power spectral density (PSD) can be measured in an optical spectrum analyzer, consisting of a tunable monochromator with attached photodetector. Of course electrical power spectral densities can also be measured, for example that of the photocurrent in the receiver.

The autocorrelation function

$$\underline{l}(\tau) = \left\langle \underline{a}(t)\underline{a}^*(t-\tau) \right\rangle = \lim_{\tilde{T}\to\infty} \frac{1}{\tilde{T}} \int_{-\tilde{T}/2}^{\tilde{T}/2} \underline{a}(t)\underline{a}^*(t-\tau)dt \tag{3.2}$$

of the data signal is here given at the left side in the usual definition based on an ensemble average and at the right side based on a temporal average. If the latter is correct (as usually is the case) the signal is called ergodic. After insertion of (3.1) into the temporal average with $\tilde{T} = (2N+1)T$ the summation over i and the integration can be exchanged. Under the condition that the single pulses decay within a finite time the summation boundaries of i rather than the integration boundaries may be chosen finite,

$$\begin{aligned} \underline{l}(\tau) &= \lim_{N\to\infty} \frac{1}{(2N+1)T} \int_{-(N+1/2)T}^{(N+1/2)T} \sum_{i=-\infty}^{\infty} \underline{c}_i \underline{s}(t_1 - iT) \sum_{k=-\infty}^{\infty} \underline{c}_k^* \underline{s}^*(t_1 - \tau - kT)dt_1 \\ &= \lim_{N\to\infty} \frac{1}{(2N+1)T} \sum_{i=-N}^{N} \int_{-\infty}^{\infty} \underline{c}_i \underline{s}(t_1) \sum_{m=-\infty}^{\infty} \underline{c}_{i+m}^* \underline{s}^*(t_1 - \tau - mT)dt_1 \end{aligned} \tag{3.3}$$

$t_1 - iT$ has also been replaced by t_1, and it has been substituted $m = k - i$. The time shift by mT can be taken into account by the convolution with respect to τ of a time-shifted Dirac impulse,

$$\underline{l}(\tau)=\lim_{N\to\infty}\frac{1}{(2N+1)T}\sum_{i=-N}^{N}\int_{-\infty}^{\infty}\int_{-\infty}^{\infty}\underline{s}(t_1)\sum_{m=-\infty}^{\infty}\underline{c}_i\underline{c}_{i+m}^{*}\underline{s}^{*}(t_1-t_2)\delta(t_2-\tau-mT)dt_2dt_1$$

$$=\lim_{N\to\infty}\frac{1}{(2N+1)T}\sum_{i=N}^{N}\int_{-\infty}^{\infty}\left(\int_{-\infty}^{\infty}\underline{s}(t_1)\underline{s}^{*}(t_1-t_2)dt_1\right)\sum_{m=-\infty}^{\infty}\underline{c}_i\underline{c}_{i+m}^{*}\delta(t_2-\tau-mT)dt_2$$

$$=\underbrace{\frac{1}{T}\int_{-\infty}^{\infty}\underline{s}(t_1)\underline{s}^{*}(t_1-\tau)dt_1}_{x(\tau)}*\underbrace{\lim_{N\to\infty}\frac{1}{2N+1}\sum_{i=-N}^{N}\sum_{m=-\infty}^{\infty}\underline{c}_i\underline{c}_{i+m}^{*}\delta(-\tau-mT)}_{y(\tau)}. \tag{3.4}$$

The third line follows indeed from the second, as can be seen from $x(\tau)*y(\tau)=\int_{-\infty}^{\infty}x(t_2)y(\tau-t_2)dt_2$. The summation over $2N+1$ symbols and the division of the sum by $2N+1$ mean that the ensemble average is taken. We may write symbolically

$$\underline{l}(\tau)=\frac{1}{T}\int_{-\infty}^{\infty}\underline{s}(t)\underline{s}^{*}(t-\tau)dt*\sum_{m=-\infty}^{\infty}\left\langle\underline{c}_i\underline{c}_{i+m}^{*}\right\rangle\delta(\tau+mT). \tag{3.5}$$

The ensemble average $\left\langle\underline{c}_i\underline{c}_{i+m}^{*}\right\rangle$ depends on the displacement m. Considered as a time-domain function it needs to be defined only at times $\tau=mT$. The larger $|m|$, the smaller is the correlation in general. We put this as the sum of a non-periodic sequence $\underline{v}_m$ that vanishes for $m\to\pm\infty$ and a function $\underline{w}(\tau)$ which has the period NT and is itself the sum of N harmonic oscillations with amplitudes $\underline{W}_n$,

$$\underline{l}(\tau)=\frac{1}{T}\int_{-\infty}^{\infty}\underline{s}(t)\underline{s}^{*}(t-\tau)dt$$
$$*\left(\sum_{m=-\infty}^{\infty}\underline{v}_m\delta(\tau+mT)+\underline{w}(\tau)\sum_{m=-\infty}^{\infty}\delta(\tau+mT)\right) \tag{3.6}$$
$$\left(\lim_{m\to\pm\infty}\left|\underline{v}_m\right|=0,\qquad \underline{w}(\tau)=\sum_{n=0}^{N-1}\underline{W}_n e^{j2\pi\tau n/(NT)}\right)$$

The PSD is the Fourier transform (with respect to τ) of the autocorrelation function, $\underline{l}(\tau)$ ○—● $\underline{L}(f)$. The convolution is replaced by a multiplication of the corresponding spectra. For the single pulse $\underline{s}(t)$ ○—● $\underline{S}(f)$ holds. The comb spectrum in the time domain corresponds to a comb spectrum also in the frequency domain. Since it is multiplied by $\underline{v}(\tau)$ in the time domain it must be convolved with respect to f with the Fourier transform thereof,

$$\underline{L}(f)=\frac{1}{T}\left|\underline{S}(f)\right|^2\left(\begin{array}{c}\sum_{m=-\infty}^{\infty}\underline{v}_m e^{j2\pi fmT}+\frac{1}{T}\left(\sum_{m=-\infty}^{\infty}\delta(f-m/T)\right)\\ *\left(\sum_{n=0}^{N-1}\underline{W}_n\delta(f-n/(NT))\right)\end{array}\right). \tag{3.7}$$

In general the transmitted symbols are statistically independent. For binary trivial coding $\underline{c}=b$ with $b\in\{0;1\}$ the result $\left\langle \underline{c}_i\underline{c}_{i+m}^*\right\rangle=\begin{cases}1/2 & \text{for } m=0\\ 1/4 & \text{for } m\neq 0\end{cases}$ holds. This results in $\underline{v}_m=\begin{cases}1/4 & \text{for } m=0\\ 0 & \text{for } m\neq 0\end{cases}$, $N=1$, $\underline{W}_0=1/4$. $\underline{W}_0\neq 0$ means that there is a DC component. In this case the PSD is

$$\underline{L}(f)=\frac{1}{4T}\left|\underline{S}(f)\right|^2\left(1+\frac{1}{T}\sum_{m=-\infty}^{\infty}\delta(f-m/T)\right). \tag{3.8}$$

In order to avoid intersymbol interference (ISI) the single pulses are often chosen such that they are as long as or shorter than a symbol duration T. They may be slightly longer than T only if they are decaying. For the non-return-to-zero-signal format (NRZ; Fig. 3.1 top) it holds $\sum_i s(t-iT)=\text{const.}$, so that if $\underline{c}_i=1$ for all i, i.e. for consecutive ones in the trivially coded binary case, a constant transmitted signal results. For an idealized single NRZ pulse with vertical impulse edges it holds

$$\underline{s}(t)=\begin{cases}1 & |t|\leq T/2\\ 0 & |t|>T/2\end{cases} \quad \circ\!\!-\!\!\bullet \quad \underline{S}(f)=\frac{\sin \pi fT}{\pi f}=T\,\mathrm{sinc}\,\pi fT\,. \tag{3.9}$$

Hence the PSD of an idealized NRZ signal with DC component is

$$\underline{L}_{NRZ}(f)=\frac{1}{4}\left(T\,\mathrm{sinc}^2\,\pi fT+\delta(f)\right). \tag{3.10}$$

There is only one Dirac line corresponding to the DC component because $\underline{S}(m/T)=0$ holds for integer $m\neq 0$. Practically there may be slope asymmetries between 01 and 10 slopes while the data signal is constant for continuous ones or zeros. These nonlinear distortions can not be taken into account by our calculus. They result in non-vanishing Dirac lines $\delta(f\pm 1/T)$ at the symbol clock frequency. A directly modulated laser is generally operated above threshold to keep relaxation oscillations short. Therefore $|\underline{c}|$ does not vanish for $b=0$. In addition there is significant chirp. The 01 (fast, generally with overshoot) and 10 (slower) transitions differ significantly so that directly modulated lasers are usually modeled accurately by integration of the rate equations in the time domain.

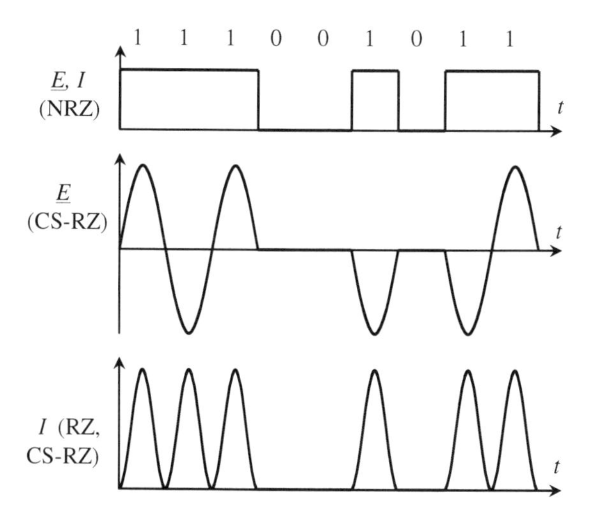

Fig. 3.1 Idealized optical NRZ signal (top), optical CS-RZ signal (middle) and corresponding normalized photocurrents (NRZ: top; CS-RZ: bottom) as a function of time

If the data signal $\underline{a}(t)$ is an electrical base band signal the PSD is centered around $f = 0$. In an electrical spectrum analyzer we observe for $f > 0$ the single-sided PSD $2\underline{L}(f)$. If the data signal $\underline{a}(t)$ is a carrier frequency signal, in particular an optical signal having the carrier frequency f_0, an optical spectrum analyzer allows to display $|\underline{L}(f - f_0)|$, i.e. a frequency-shifted PSD for $f > 0$.

Light emitting diodes (LEDs) as signal sources are always modulated through the current in their power. For laser diodes this is true up to about 3.3 Gbit/s. For data rates from 10 to 13 Gbit/s an electroabsorption modulator (EAM) is often employed, which can be integrated together with the laser on the same chip, but direct laser modulation is also possible.

If we use normalized field vectors with the dimension $\sqrt{W}$ the relations $P = \frac{1}{2}|\underline{E}|^2$, $|\underline{E}| = \sqrt{2P}$ hold. If the chirp factor α_H of the EAM is known, the missing phase can be calculated: From

$$P = \frac{1}{2}|\underline{E}|^2 \qquad \alpha_H = \frac{dn/dU}{dn_i/dU} \qquad \underline{n} = n - jn_i \qquad \underline{E} = E_0 e^{-jk_0\underline{n}l} \tag{3.11}$$

it follows

$$\begin{aligned} &|\underline{E}| = \sqrt{2P} = E_0 e^{-k_0 n_i l} \qquad n = \alpha_H n_i + \varphi_0/(k_0 l) \\ &\underline{E} = \sqrt{2P}e^{-jk_0 nl} = \sqrt{2P}e^{j(-\alpha_H n_i k_0 l - \varphi_0)} = \sqrt{2P}e^{j\left(\alpha_H\left(\ln\sqrt{2P} - \ln E_0\right) - \varphi_0\right)} \end{aligned} \tag{3.12}$$

with an arbitrary phase offset φ_0. For directly modulated lasers matters are more complicated because in the stationary case n does not modulate the phase but the optical frequency.

In long distance transmission with wavelength-division multiplex (WDM) usually the tolerance with respect to chromatic dispersion is to be maximized. In that case one has to avoid EAM or, worse, direct laser modulation and uses instead a Mach-Zehnder modulator in $LiNbO_3$, see discussion on p. 84. Table 3.1 summarizes trends in optical intensity modulation.

Table 3.1 Methods for optical intensity modulation

	Data rate up to	Remarks
Directly modulated laser	3.3 Gbit/s	10.7 Gbit/s is also in reach.
Electroabsorption modulator (EAM)	10.7 Gbit/s	Integrated with DFB laser. 43 Gbit/s stand-alone EAMs under development.
Mach-Zehnder modulator	43 Gbit/s	Lithium niobate ($LiNbO_3$). Will be challenged by III/V materials.

The transmitted polarization results from transmitter geometry and layout. Fibers with small lengths L can be modeled by a (nearly frequency-independent) attenuation. Usually one must also take into account at least chromatic dispersion. To that purpose the spectrum of the transmitted field (scalar or Jones vector) is multiplied by the complex transfer function $e^{-(\alpha + j\beta(\omega))L}$ in the frequency domain. If polarization mode dispersion (PMD) is to be included the transfer function consists of frequency-dependent Jones matrices. However, PMD can also be modeled in the time domain. To that purpose the time-dependent Jones vector is convolved from the right side with a Jones matrix which has been transformed into the time domain. Nonlinear fiber behavior must be modeled in the time domain.

The received optical power is $P = \frac{1}{2}|\underline{E}|^2 = \frac{1}{2}|\underline{\mathbf{E}}|^2$ for our field definition. At the receiver side PIN photodiodes are often used, or sometimes avalanche photodiodes (APDs). For PIN photodiodes the photo current is $I = RP$, where R is the responsivity. So all requirements are fulfilled to regenerate the binary signal.

The NRZ signal format is not the only possible. For return-to-zero signals (RZ) each one consists of an isolated pulse which is separate from neighbor ones. Carrier-suppressed RZ signals (CS-RZ) are obtained from NRZ signals if they are modulated in another Mach-Zehnder modulator with a sinusoidal signal having half the symbol rate as frequency. This generates a ternary optical signal (Fig. 3.1 middle). Due to $\sin x \approx x$ for $|x| << 1$ the additional modulator with field transfer function $\sin((c_2 - c_3)U/2)$ can be regarded as linear up to about $|(c_2 - c_3)U/2| \leq \pi/4$. In that case the achievable peak power equals after all 1/2 of the input value. If attenuation is neglected and after suitable normalization the electric field as a data signal $\underline{a}(t)$ (idealized binary NRZ signal) is multiplied by $\sqrt{2}\cos\pi t/T$. The resulting normalized photocurrent I under the (physically meaningless) assumption $R = 2$ is shown in Fig. 3.1 bottom for CS-RZ. For NRZ it is shown in Fig. 3.1 top.

In order to judge the quality of an electric data signal, e.g., the photocurrent I, it is useful to display it oscillographically. If the oscilloscope is triggered with the symbol clock the *eye diagram* or eye pattern is generated (Fig. 3.2). Rounding of the signals is caused by lowpass properties of various devices but is not detrimental here.

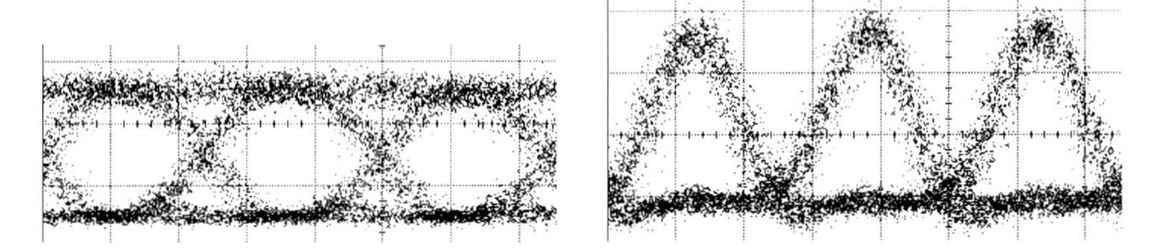

Fig. 3.2 Measured binary electrical 40Gbit/s eye diagrams for NRZ (left) and CS-RZ (right) measured in an optical receiver. Horizontal scale: 10 ps/division.

Problem: Calculate the PSD of an optical CS-RZ signal (optical field $\underline{E}$) and of the resulting normalized photocurrent I! Repeat the calculation for an RZ signal with $\underline{s}(t) = \begin{cases} \sqrt{2}\cos\pi t/T & |t| \le T/2 \\ 0 & |t| > T/2 \end{cases}$! Compare also with an idealized binary NRZ signal!

Solution: The CS-RZ modulation (Fig. 3.1 middle) can be taken into account by two steps. Firstly, the single optical pulses are now

$$\underline{s}(t) = \begin{cases} \sqrt{2}\cos\pi t/T & |t| \le T/2 \\ 0 & |t| > T/2 \end{cases} \circ\!\!-\!\!\bullet \quad \underline{S}(f) = \sqrt{2}\frac{2T}{\pi}\frac{\cos\pi fT}{1-4f^2T^2}. \tag{3.13}$$

Secondly, the transmitted symbols change polarity from step to step. Here $\underline{v}_m = \begin{cases} 1/4 & \text{für } m = 0 \\ 0 & \text{für } m \ne 0 \end{cases}$, $N = 2$, $\underline{W}_0 = 0$, $\underline{W}_1 = 1/4$ holds. The comb spectrum is translated by half a period. For $f = (m+1/2)/T$ the function $\cos\pi fT$ is zero. At $f = \pm 1/(2T)$ we need to take $\lim_{f \to \pm 1/(2T)} \frac{\cos\pi fT}{1-4f^2T^2} = \frac{\pi}{4}$. So there are only two spectral lines while the original carrier is suppressed. From (3.7) the PSD of the optical signal is obtained,

$$\underline{L}(f) = \frac{2T}{\pi^2}\left(\frac{\cos\pi fT}{1-4f^2T^2}\right)^2 + \frac{1}{2\pi}\left(\delta(f-1/2/T) + \delta(f+1/2/T)\right). \tag{3.14}$$

The normalized photocurrent $I = |\underline{E}|^2$ (Fig. 3.1 bottom) consists of RZ impulses

$$\underline{s}(t) = \begin{cases} 2\cos^2 2\pi t/T & |t| \le T/2 \\ 0 & |t| > T/2 \end{cases} \quad \circ\!\!-\!\!\bullet \quad \underline{S}(f) = \frac{\sin \pi fT}{\pi f\left(1 - f^2T^2\right)}. \tag{3.15}$$

They are shorter than idealized NRZ impulses. Practically this reduces intersymbol interference and allows it to choose a shorter impulse response of the base band filter. All this increases the receiver sensitivity. The transmitted symbols always have equal polarity, $\underline{v}_m = \begin{cases} 1/4 & \text{für } m = 0 \\ 0 & \text{für } m \ne 0 \end{cases}$, $N = 1$, $\underline{W}_0 = 1/4$. The PSD of the normalized photocurrent is therefore

$$\underline{L}(f) = \frac{T}{4}\left(\frac{\operatorname{sinc} \pi fT}{1 - f^2T^2}\right)^2 + \frac{1}{4}\delta(f) + \frac{1}{8}\delta(f - 1/T) + \frac{1}{8}\delta(f + 1/T). \tag{3.16}$$

In contrast, for the defined RZ signal it holds $N = 1$, $\underline{W}_0 = 1/4$. The PSD of the optical signal is

$$\underline{L}(f) = \frac{2}{\pi^2}\left(T\left(\frac{\cos \pi fT}{1 - 4f^2T^2}\right)^2 + \sum_{m=-\infty}^{\infty}\left(\frac{1}{1 - 4m^2}\right)^2 \delta(f - m/T)\right), \tag{3.17}$$

that of the normalized photocurrent is again (3.16). Yet the RZ signal needs more optical bandwidth than the CS-RZ signal and tolerates less chromatic dispersion.

As an exception, for an idealized NRZ signal (Fig. 3.1 top) $I = \underline{E}$ holds because $\underline{E}$ assumes only the values 0 and 1. The PSD of the normalized photocurrent is again (3.10).

If the receiver sensitivity is to be maximized RZ and CS-RZ are preferable over NRZ. The chromatic dispersion tolerance is slightly reduced for CS-RZ, and strongly for RZ. This is due to the broader main lobe of the spectrum. Rounded signal edges (also possible for NRZ) facilitates neighbor channel suppression because the spectra decay faster at high freuqencies. The tolerance with respect to self phase modulation (chapter 3.7) is larger for RZ and CS-RZ than for NRZ.

Problem: Duobinary modulation is similar to binary modulation. But strings of one or more ones ($b_i = 1$) are transmitted either as positive ($\underline{c}_i = 1$) or as negative ($\underline{c}_i = -1$) symbols. The polarity of two subsequent ones differs if there is an odd number of zeros (and nothing else) between them: For $b_{i-l} = 1$, $l \ge 1$, $b_k = 0$ for all k with $i - l < k < i$, $b_i = 1$ it holds $\underline{c}_i = (-1)^{l-1}\underline{c}_{i-l}$. Find a recursive encoding relation between the data symbols b_i with $b \in \{0;1\}$ and the transmitted symbols $\underline{c}_i$! Calculate the PSD of an optical duobinary NRZ signal (optical field $\underline{E}$) and of the resulting normalized photocurrent I!

Solution: For convenience we define a bipolar data symbol $d = 2b - 1 \in \{-1;1\}$. It is differentially encoded, $f_i = d_i f_{i-1} \in \{-1;1\}$. (An equivalent alternative is $g_i = b_i \oplus g_{i-1} \oplus 1 \in \{0;1\}$, $f_i = 2g_i - 1$.) The transmitted symbols are obtained by averaging of two neighbor symbols,

$\underline{c}_i = (1/2)(f_i + f_{i-1}) \in \{-1;0;1\}$. Insertion of the previous equations results in $\underline{c}_i = (1/2)(d_i f_{i-1} + f_{i-1}) = (1/2)(d_i + 1) f_{i-1} = b_i f_{i-1}$.

The autocorrelation function of the transmitted symbols is

$$\begin{aligned} \left\langle \underline{c}_i \underline{c}_{i+m}^* \right\rangle &= \left\langle b_i f_{i-1} b_{i+m} f_{i-1+m} \right\rangle = \left\langle f_{i-1}^2 b_i b_{i+m} \prod_{k=i}^{i-1+m} d_k \right\rangle \\ &= \left\langle f_{i-1}^2 (b_i d_i) b_{i+m} \prod_{k=i+1}^{i-1+m} d_k \right\rangle = \left\langle b_i b_{i+m} \left(\prod_{k=i+1}^{i-1+m} d_k \right) \right\rangle \end{aligned} . \tag{3.18}$$

We can immediately calculate $\left\langle \underline{c}_i \underline{c}_i^* \right\rangle = \left\langle b_i b_i \right\rangle = 1/2$, $\left\langle \underline{c}_i \underline{c}_{i+1}^* \right\rangle = \left\langle b_i b_{i+1} \right\rangle = \left\langle b_i \right\rangle \left\langle b_{i+1} \right\rangle = 1/4$. For $m > 1$ all three multiplicands are statistically independent and we evaluate (3.18), $\left\langle \underline{c}_i \underline{c}_{i+m}^* \right\rangle = \left\langle b_i \right\rangle \left\langle b_{i+m} \right\rangle \left\langle \prod_{k=i+1}^{i-1+m} d_k \right\rangle = (1/2)(1/2) \cdot 0 = 0$. Also, $\left\langle \underline{c}_i \underline{c}_{i+m}^* \right\rangle = \left\langle \underline{c}_i \underline{c}_{i-m}^* \right\rangle$ holds.

This results in $\underline{v}_m = \begin{cases} 1/2 & \text{for } m = 0 \\ 1/4 & \text{for } |m| = 1 \\ 0 & \text{else} \end{cases}$, $N = 1$, $\underline{W}_0 = 0$. The PSD of duobinary NRZ signals is

$$\underline{L}(f) = T \frac{\sin^2 \pi f T}{(\pi f T)^2} \left(\frac{1}{2} + \frac{1}{4} e^{j2\pi f T} + \frac{1}{4} e^{-j2\pi f T} \right) = T \frac{\sin^2 \pi f T}{(\pi f T)^2} \cos^2 \pi f T = T \frac{\sin^2 2\pi f T}{(2\pi f T)^2} . \tag{3.19}$$

The doubinary NRZ spectrum is identical in shape but half as broad as the standard binary NRZ spectrum. Nevertheless this implementation offers only a moderate advantage over binary modulation. The normalized photocurrent $I = |\underline{E}|^2$ is insensitive to the sign of the transmitted ones, and therefore is identical to that of binary NRZ transmission. Its PSD is $T \operatorname{sinc}^2 \pi f T$.

If $\underline{c}$ is generated from f not by averaging of neighbor symbols but by a Bessel lowpass filter with a cutoff frequency of about 0.3/*T*, then the chromatic dispersion tolerance becomes more than twice that of binary NRZ, whereas the receiver sensitivity is reduced by about 1...2 dB. In that implementation duobinary modulation is a good candidate for upgrading WDM links to 40 Gbit/s which were initially designed for 10 Gbit/s and have a 50 GHz channel spacing.

The distortion which can be inflicted by chromatic dispersion on (chirp-free) NRZ signals is illustrated in Fig. 3.3. It is valid at $\lambda = 1.55$ μm for 10 Gb/s data rate and 100 km of SSMF, or 40 Gb/s and 6.25 km of SSMF. Larger distances usually require the insertion of dispersion-compensating fiber.

Fig. 3.3 Reference eye pattern (left) and eye pattern under chromatic fiber dispersion (right), valid for 1.55 μm wavelength, 40 Gb/s data rate and 6.25 km of standard single-mode fiber with a dispersion of 17 ps/nm/km

In optical receivers noise is generated. *Thermal noise* is observed even in the absence of a photocurrent. A thermally noisy impedance $\underline{Z}$ delivers to a complex conjugated load $\underline{Z}^*$ the two-sided noise PSD $kT/2$. This corresponds to PSDs of $kT\,\mathrm{Re}\,\underline{Z}/2$, $kT/(2\,\mathrm{Re}\,\underline{Z})$ for voltage and current, respectively. The likewise noisy load $\underline{Z}^*$ delivers the same noise PSD to $\underline{Z}$. PSDs of open circuit voltage at $\underline{Z}$ and short-circuit current of $\underline{Z}$ are $2kT\,\mathrm{Re}\,\underline{Z}$ and $2kT/\mathrm{Re}\,\underline{Z}$, respectively. In data sheets usually physical, one-sided PSDs are given. The dimension is W/Hz, V^2/Hz or A^2/Hz. Often the square root of the PSD of a voltage or a current is given, i.e. a rms amplitude density spectrum with the dimension $\mathrm{V}/\sqrt{\mathrm{Hz}}$ or $\mathrm{A}/\sqrt{\mathrm{Hz}}$, respectively.

If there is a direct current then also *shot noise* is generated, since current flow is caused by elementary charges. While electronic amplifier noise contains shot noise generated by DC, the shot noise of a photodiode must normally be taken into account separately. Through the photodiode flows the sum of photocurrent (caused by photon absorption) and dark current (leakage current which exists even without illumination; for PIN diodes made from InP in the order of 10^{-9} A).

The time-dependent current

$$i(t) = e\sum\nolimits_k \delta(t - t_k) \tag{3.20}$$

contains an infinite sum of elementary charge impules which occur at randomly distributed times t_k. In a short time interval Δt the number of impules ΔN is Poisson distributed with the expectation value

$$\langle \Delta N \rangle = I\Delta t/e\;. \tag{3.21}$$

We subdivide the time into equal time steps and replace the current by an equivalent function

$$\tilde{i}(t)=\frac{e\Delta N_k}{\Delta t} \quad \text{for} \quad |t-k\Delta t|<\frac{\Delta t}{2} \qquad \text{with } \langle \Delta N_k \rangle = \langle \Delta N \rangle, \ \lim_{\Delta t \to 0} \tilde{i}(t)=i(t). \tag{3.22}$$

For small Δt it holds $0 \le \langle \Delta N \rangle << 1$. We find $\langle \Delta N_k \Delta N_k \rangle = \langle \Delta N \rangle$, because $0 \cdot 0 = 0$, $1 \cdot 1 = 1$, and the occurrence of larger impulse numbers $\Delta N_k \ge 2$ in the k-th time interval is negligible. Impulse numbers in different time intervals are statistically independent. Hence $\langle \Delta N_k \Delta N_l \rangle = \langle \Delta N_k \rangle \langle \Delta N_l \rangle = \langle \Delta N \rangle^2 << \langle \Delta N \rangle$ is valid for $k \ne l$. The autocorrelation function of the equivalent function becomes

$$l_{\tilde{i}}(\tau)=\frac{e^2}{\Delta t^2}\begin{cases} \langle \Delta N \rangle^2 |\tau|/\Delta t + \langle \Delta N \rangle (\Delta t - |\tau|)/\Delta t & \text{for} |\tau| < \Delta t \\ \langle \Delta N \rangle^2 & \text{else} \end{cases}. \tag{3.23}$$

Due to $\lim_{\Delta t \to 0} \langle \Delta N \rangle = 0$ it is a triangle of total width $2\Delta t$ embedded in a vanishing baseline.

The autocorrelation function of the current $i(t)$ and the corresponding two-sided power spectral density $L_I(f)$ (calculated by Fourier transform) are

$$l_I(\tau)= \lim_{\Delta t \to 0} l_{\tilde{i}}(\tau) = eI\delta(\tau) \quad \circ\!\!-\!\!\bullet \quad L_I(f)=eI\,. \tag{3.24}$$

The one-sided PSD of shot noise is $2eI$. The photocurrent shot noise can be the dominating noise source in optical receivers for analog signals (cable TV) and in coherent optical receivers. In direct-detection optical receivers for intensity the photocurrent and hence also the shot noise depend in a time-variable manner on the transmitted information. This should be taken into account.

In the case of an APD the shot noise equation must be modified. The higher the avalanche multiplication factor, the stronger is the deviation from the usual shot noise. One sets

$$L_{I_M}(f)=eI_pM^{2+x}=e\frac{\eta e}{hf}PM^{2+x} \tag{3.25}$$

for the two-sided PSD of the current I_M that can be measured at the APD terminals. According to $I_M = MI_p$, if avalanche multiplication were noise-free the diode would be an ideal amplifier for the primary photocurrent I_p. The PSD would be M^2 times as strong as that of a normal photodiode, and the *excess noise exponent* x of avalanche multiplication would be $x=0$. In real avalanche photodiodes it is, dependent on the semiconductor material, about 0.3 (Si), 0.7 (InP) or 1 (Ge). So Si APDs allow to construct very sensitive optical receivers because the noise is not much stronger than that of an idealized APD. For $\lambda > 1\mu\text{m}$ Si does not absorb. Ge has a high excess noise factor and a photodetection efficiency which decreases for $\lambda > 1{,}5\mu\text{m}$. So InP is the material of choice in the long-wavelength region. The empirical fitting of measured noise

PSDs by characteristic excess noise exponents x is exact only in a limited range. Therefore M^x is sometimes replaced by an excess noise factor $F(M)$. The higher M^x or $F(M)$ are the less the assumption of a Gaussian noise distribution is justified.

In special cases like analog receivers the relative intensity noise (RIN) with a one-sided PSD $\mathrm{RIN}(f)$ can not be neglected. Noise signals which are generated by different sources are statistically independent. Therefore the corresponding variances can be added. Using one-sided noise PSDs the photocurrent variance is therefore

$$\sigma_I^2 = \int_0^\infty \Big(\underbrace{RIN(f) I^2}_{\text{relative inensity noise}} + \underbrace{2eI}_{\text{shot noise}} + \underbrace{\overline{di_{th}^2}/df}_{\text{thermal noise}} \Big) df \,. \tag{3.26}$$

Here $\overline{di_{th}^2}/df$ is the one-sided thermal noise PSD of the electronic circuit attached to the photodiode. Note that the three terms are constant, proportional to I and its square, respectively. Since the three terms all represent Gaussian noise the photocurrent also has a Gaussian noise distribution.

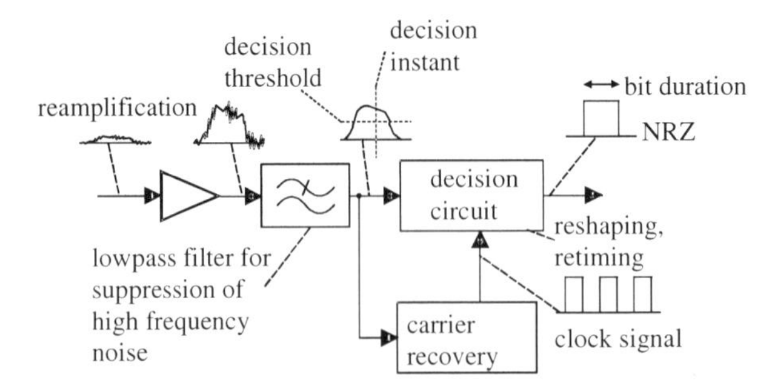

Fig. 3.4 Schematical drawing of an electrical regenerator for binary signals

In Fig. 3.4 a regenerator for binary signals is shown. It operates in the baseband where the data signal is quasi a DC signal, not an optical carrier frequency signal. The regenerator has three tasks: reamplification, reshaping, retiming of received pulses (3R). If a sufficiently strong signal occurs at the regenerator input the decision circuit will, after preamplification and filtering, fully regenerate the transmitted impulses, with exception of a very small number of usually single bits which are negated. The decision circuit acts as an edge-triggered D flip-flop with high preamplification and hence small hysteresis. Always at a specific time, approximately in the middle of a bit, it is checked whether the input signal ranges above or below a threshold. Correspondingly the decision circuit outputs a zero or a one. The clock pulses for the decision circuit are produced by a clock recovery circuit for the symbol clock. If the bit error ratio (BER), i.e. the proportion of falsified bits, is for example 10^{-12} on average, so even a cascading of 10 data transmission lines with regenerators will yield a very small BER of about 10^{-11}.

The complete regenerator for optical signals consists of an opto-electric converter (photodiode, O/E), the electrical regenerator (3R) and an electro-optic converter (laser and maybe modulator, E/O). In long transmission links optical amplifiers are used on the line which can amplify many wavelength division multiplex (WDM) channels simultaneously (Fig. 3.5). Though fiber attenuation between optical amplifiers must be smaller than between regenerators the construction of optical amplifiers is much simpler than that of regenerators. This means optical amplification (1R: reamplification) is always helpful. The standard choice are erbium-doped fiber amplifiers (EDFAs). At the end of the link complete regeneration is necessary of course. The signals with different optical frequencies ω_i are combined and separated with low loss through optical multiplexers (MUX) and demultiplexers (DEMUX), respectively. The electrical signals are brought to the bit rate $1/T$ at the transmitter and are demultiplexed again in the receivers to lower data rates.

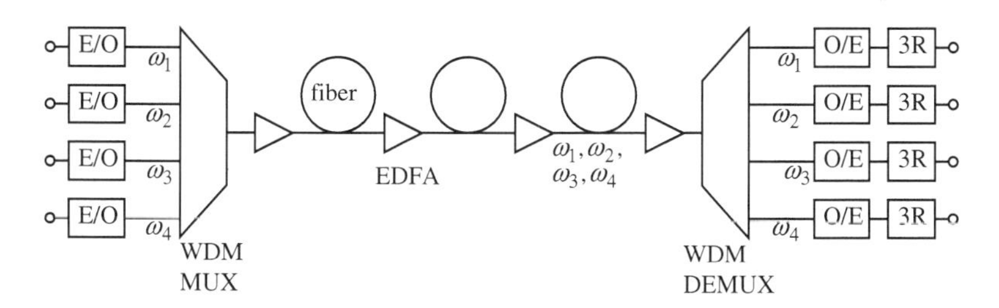

Fig. 3.5 Schematic drawing of an optical data transmission link with wavelength division multiplex

3.1.2 Regeneration of Binary Signals

A communication problem is the detection of digital signals with added zero-mean white Gaussian noise (AWGN) $n(t)$ with a correlation function proportional to a Dirac impulse and with constant PSD,

$$l_n(\tau) = \delta(\tau) \qquad \circ\!\!-\!\!\bullet \qquad L_n(f) = 1. \tag{3.27}$$

This holds also for regenerators of optical signals but often only with significant restrictions. The autocorrelation function of white noise is a Dirac function and the rms value $l_n(0)$ of the time-domain signal $n(t)$ is therefore infinity for a non-zero noise PSD. Since this is physically impossible noise is always band-limited, where a finite noise PSD results in a finite rms value. Gaussian signals occur, due to the central limit theorem of statistics, if a large number of signals such as thermal movements of electrons are linearly superimposed and possibly transmitted through a linear system. In contrast, squaring or other nonlinear processing transforms the Gaussian into another distribution.

If Gaussian noise with the time-domain function $n(t)$ is transmitted through a linear system then the output noise $n_{out}(t) = n(t) * h(t) = \int n(t-t')h(t')dt'$ is obtained by a convolution with the impulse response $h(t)$. Evaluated at a time t the convolution integral can be understood as an infinite number of random variables (RVs) $n_{out}(t-t')$ with probability density functions (PDFs) which, due to the multiplication of the RV by $h(t')$ (= generation of a new RV by a linear function) are broadened $h(t')$-fold and enlarged $1/|h(t')|$-fold. (The area under the PDF remains unchanged.) If $n(t)$ is a white Gaussian noise process so $n_{out}(t)$ is also Gaussian according to the central limit theorem. The statement remains true also for colored Gaussian noise $n(t)$, i.e. $l_n(\tau) \neq \delta(\tau)$: Any Gaussian noise can be generated from white Gaussian noise by a convolution with an additional impulse response $g(t)$. This means that $n_{out}(t)$ is generated by the convolution of white Gaussian noise with the combined impulse response $(h(t) * g(t))$. $n_{out}(t)$ is therefore Gaussian. The property of linear systems, to always generate Gaussian output signals for Gaussian input signals is important for the subsequent calculation of BERs. In general, linear systems will not generate Gaussian output signals for non-Gaussian input signals. However, the output signal PDF is generally more similar to a Gaussian PDF than the input signal PDF.

In the receiver the data signal $\underline{a}(t)$ plus noise $\underline{n}(t)$ are filtered by a linear system with an impulse response $\underline{h}(t)$,

$$\begin{gathered}\underline{x}(t) = \underline{h}(t) * (\underline{a}(t) + \underline{n}(t)) = \sum_{i=-\infty}^{\infty} \underline{c}_i \underline{s}_{out}(t - iT) + n_{out}(t) \\ \underline{s}_{out}(t) = \underline{h}(t) * \underline{s}(t) \qquad \underline{n}_{out}(t) = \underline{h}(t) * \underline{n}(t)\end{gathered} \quad . \tag{3.28}$$

Since the noise is Gaussian already before the filter it is sufficient for the minimization of the BER to maximize the signal-to-noise ratio (SNR) by a suitable choice of the impulse response $h(t)$. For a constant input noise PSD L_n, the noise variance $\sigma_{n,out}^2$, decision sample at time t_0 for a transmitted single pulse $s(t)$ and resulting SNR are

$$\sigma_{n,out}^2 = \int_{-\infty}^{\infty} L_{n,out}(f)df = \int_{-\infty}^{\infty} |H(f)|^2 L_n(f)df = L_n \int_{-\infty}^{\infty} |H(f)|^2 df \, , \tag{3.29}$$

$$\underline{s}_{out}(t_0) = \int_{-\infty}^{\infty} \underline{S}_{out}(f) e^{j2\pi f t_0} df = \int_{-\infty}^{\infty} \underline{H}(f)\underline{S}(f) e^{j2\pi f t_0} df \, , \tag{3.30}$$

$$\mathrm{SNR} = \frac{\left|\underline{s}(t_0)\right|^2}{\sigma_{n,out}^2} = \frac{\left|\int_{-\infty}^{\infty} \underline{H}(f)\underline{S}(f)e^{j2\pi f t_0}df\right|^2}{L_n \int_{-\infty}^{\infty} \left|\underline{H}(f)\right|^2 df}$$

$$= \frac{\int_{-\infty}^{\infty} \left|\underline{S}(f)\right|^2 df}{L_n} \cdot \frac{\left|\int_{-\infty}^{\infty} \underline{H}(f)\underline{S}(f)e^{j2\pi f t_0}df\right|^2}{\int_{-\infty}^{\infty} \left|\underline{H}(f)\right|^2 df \int_{-\infty}^{\infty} \left|\underline{S}(f)e^{j2\pi f t_0}\right|^2 df} \le \frac{\int_{-\infty}^{\infty} \left|\underline{S}(f)\right|^2 df}{L_n}, \quad (3.31)$$

respectively. In the last step the right fraction was bounded by the Cauchy-Schwarz inequality. The equal sign for a SNR maximization is valid only for

$$\underline{H}(f) = \underline{C}\underline{S}^*(f)e^{-j2\pi f t_0} \quad \bullet\!\!-\!\!\circ$$

$$\underline{h}(t) = \underline{C}\int_{-\infty}^{\infty} \underline{S}^*(f)e^{-j2\pi f (t_0 - t)}df = \underline{C}\underline{s}^*(t_0 - t). \quad (3.32)$$

The delay t_0 of the decision instant must be chosen large enough to let the impulse response $h(t)$ be causal. This impulse response is obtained from the transmission impulse by a mirroring and a translation of the time axis so that $\underline{h}(t) = 0$ holds for $t < 0$, and complex conjugation. $\underline{C}$ is a constant, and $L_{n,out}(f)$, L_n are two-sided PSDs. A filter according to (3.32) is called a *matched filter*. In the baseband part of the regenerator there are real-valued electrical signals and one obtains $h(t) = Cs(t_0 - t)$.

According to (3.28) the output signal of a linear receiver with lowpass characteristic can be written as

$$\underline{x}(t) = \underline{c}_i \underline{s}_{out}(t - iT) + \sum_{k \ne i} \underline{c}_k \underline{s}_{out}(t - kT) + n_{out}(t). \quad (3.33)$$

At the decision instant the first term is the wanted signal and the third is the unavoidable noise signal. The second term, responsible for intersymbol interference (ISI), is a random variable. Its expectation value depends for statistically independent transmitted symbols not on i oder $\underline{c}_i$, and is therefore constant. Consequently, ISI does not change the <u>mean</u> distance between the voltage values for received 0 and 1. But it does change the PDF of this distance and those of the distances from the threshold y_0. Since the ISI term varies bit pattern dependent, the mean BER is obtained by ensemble or time averaging of those BERs which occur for all possible bit sequences. A reduction of the distance from the threshold is always worse than an increase by the same amount. The (mean) BER is therefore increased by ISI. The most important aim of data transmission is to guarantee very low ISI at an acceptable baseband filter width.

We limit ourselves to real binary signals $\underline{c}_i = b_i \in \{0;1\}$ and consider the isolated detection of a data symbol b_i. Compared to (3.33), the time t is increased by iT so that we refer to a constant time in each symbol. and sample index k are referred to that of the symbol to be detected. For a suitable decision instant $t = t_0 + iT$ the expectation value of the output signal is

$$\langle x(t_0 + iT) \rangle = b_i \underline{s}_{out}(t_0) + \sum_{k \neq i} b_k s_{out}(t_0 + (i-k)T) := \mu_{0,b_i} . \tag{3.34}$$

Quantity t_0 is a fixed delay which is the same for all data symbols and is determined by the clock recovery.

The PDF of the decision value with AWGN depends on the data symbol b (index i is omitted),

$$p_x(x) = p_x(x|b) = \frac{1}{\sqrt{2\pi}\sigma_{n,b}} e^{-(x-\mu_{0,b})^2 / \left(2\sigma_{n,b}^2\right)} \qquad (b \in \{0;1\}) . \tag{3.35}$$

Here an instationary noise signal (having time-dependent moments) was admitted, which can have a different standard deviation σ_{n1} for a transmitted 1 than for a transmitted 0, σ_{n0}. For example, shot noise increases the standard deviation of received ones. Even larger differences are observed if an APD is used. In each case it must be checked of course whether the assumption of Gaussian signals really holds. APD noise is not really Gaussian, and even pure shot noise becomes Poissonian if we consider very short time intervals and small currents.

Decision works as follows: The hypothesis $b_i = 1$ is adopted for $x > x_s$, and $b_i = 0$ is assumed for $x \leq x_s$. The value x_{th} is the *decision threshold.* The conditional probability of a wrong 1 decision although a 0 has been transmitted is

$$P(1|0) = \int_{x_{th}}^{\infty} p_x(x|0) dx = \frac{1}{2} \operatorname{erfc}\left(\frac{Q_0}{\sqrt{2}}\right) \qquad Q_0 = \frac{x_{th} - \mu_{0,0}}{\sigma_{n0}} . \tag{3.36}$$

where $\operatorname{erfc}(z) = \frac{2}{\sqrt{\pi}} \int_z^{\infty} e^{-\xi^2} d\xi$ has been used. The conditional probability of a wrong 0 decision for a transmitted 1 is

$$P(0|1) = \int_{-\infty}^{x_{th}} p_x(x|1) dx = \frac{1}{2} \operatorname{erfc}\left(\frac{Q_1}{\sqrt{2}}\right) \qquad Q_1 = \frac{\mu_{0,1} - x_{th}}{\sigma_{n1}} . \tag{3.37}$$

The *bit error ratio* (BER) is the sum of these conditional probabilities, each multiplied by the probability P_0, P_1 of the respective data symbol,

$$\text{BER} = P(0|1)P_1 + P(1|0)P_0 = \frac{P_1}{2} \operatorname{erfc}\left(\frac{Q_1}{\sqrt{2}}\right) + \frac{P_0}{2} \operatorname{erfc}\left(\frac{Q_0}{\sqrt{2}}\right) . \tag{3.38}$$

The middle expression is general, the one on the right holds only for AWGN. Staying within a specified BER limit is a central requirement for digital communication systems. SNR and BER deteriorate for decreasing received power. During long times the receiver sensitivity was defined by that received power which is necessary for a $\mathrm{BER}=10^{-9}$. As data rates rose to 10 Gbit/s and above, $\mathrm{BER}=10^{-12}$ was usually required. Recently there is a trend towards error-correcting codes (*forward error correction, FEC*). Here a BER of about 10^{-4}, 10^{-3} or even higher is used to define the receiver sensitivity.

For a given amplitude deviation $\mu_{0,1}-\mu_{0.0}$ of the output signal the BER can be minimized by an optimum choice of the decision threshold. By setting the derivative

$$\frac{d\,\mathrm{BER}}{dx_{th}}=P_1\frac{dP(0|1)}{dx_{th}}+P_0\frac{dP(1|0)}{dx_{th}}=P_1 p_x(x_{th}|1)-P_0 p_x(x_{th}|0) \tag{3.39}$$

equal to zero we obtain the general result

$$P_1 p_x(x_{th,opt}|1)=P_0 p_x(x_{th,opt}|0). \tag{3.40}$$

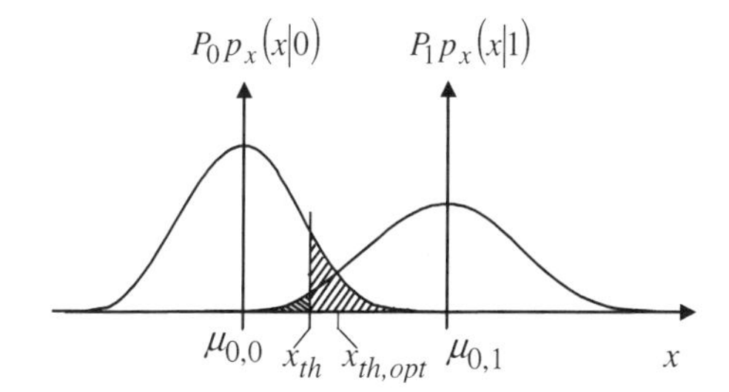

Fig. 3.6 Conditional PDFs for transmitted 0 and 1, each of them weighted with the respective symbol probability. The hatched areas define the probabilities of a falsification of a 0 into a 1 or vice versa. The sum of the hatched areas, the BER, is minimal if the threshold is placed at the intersection of the two curves.

In Fig. 3.6 the hatched area marks the BER, and this is indeed minimized by (3.40). In the case of Gaussian PDFs one obtains a quadratic equation for the optimum decision threshold, with the solution

$$x_{th,opt}=\mu_{0,0}+\frac{1}{1-\sigma_{n1}^2/\sigma_{n0}^2}\left(\mu_{0,1}-\mu_{0,0}-\frac{\sigma_{n1}}{\sigma_{n0}}\sqrt{(\mu_{0,1}-\mu_{0,0})^2+2\left(\sigma_{n0}^2-\sigma_{n1}^2\right)\ln\frac{P_1\sigma_{n0}}{P_0\sigma_{n1}}}\right). \tag{3.41}$$

The other, positive sign in front of the square root can be excluded because in that case $\sigma_{n1} \to \sigma_{n0}$ would result in $x_0 \to \pm\infty$. The second term under the square root can often be neglected because the symbol probabilities and the symbol-dependent noise amplitudes generally do not vary too much. This results in the simpler expression

$$x_{th,opt} = \frac{\mu_{0,1}\sigma_{n0} + \mu_{0,0}\sigma_{n1}}{\sigma_{n0} + \sigma_{n1}}. \tag{3.42}$$

If we insert this into (3.38) we obtain (Fig. 3.7)

$$\text{BER} = \frac{1}{2}\text{erfc}\left(Q/\sqrt{2}\right) \qquad Q = \frac{\mu_{0,1} - \mu_{0,0}}{\sigma_{n1} + \sigma_{n0}}. \tag{3.43}$$

The so-called Q factor is the distance between the expectation values of ones and zeros, divided by the sum of the respective noise standard deviations. For $P_0 = P_1 = 1/2$, $\sigma_{n0} = \sigma_{n1} = \sigma_n$ (stationary noise) the optimum decision threshold $x_{th,opt}$ lies in the middle between $\mu_{0,0}$ and $\mu_{0,1}$, and $Q = \left(\mu_{0,1} - \mu_{0,0}\right)/\left(2\sigma_n\right)$ holds.

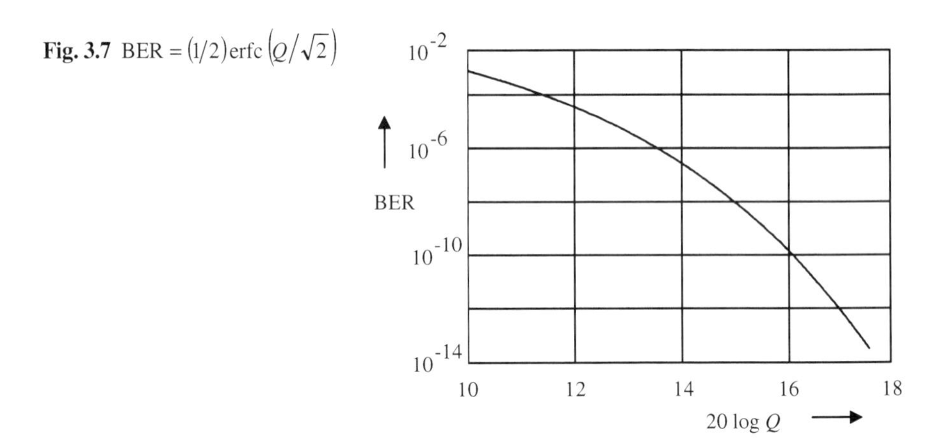

Fig. 3.7 $\text{BER} = (1/2)\text{erfc}\left(Q/\sqrt{2}\right)$

Very low BERs can not be minimized by (3.40) this would take too much measurement time. In those cases it is useful to return to (3.38) [41]. We assume $P_0 = P_1 = 1/2$. The threshold is varied, and BER values are measured wherever they are not too low to be accurate. For too low threshold $x_{th} < x_{th,opt}$ it holds $Q_1 > Q_0$ and $\text{BER} \approx \frac{1}{4}\text{erfc}\left(Q_0/\sqrt{2}\right)$. For $x_{th} > x_{th,opt}$ we find $Q_1 < Q_0$ and

$\mathrm{BER} \approx \frac{1}{4}\operatorname{erfc}\left(Q_1/\sqrt{2}\right)$. Separate for each case, we calculate values of Q_1, Q_0 corresponding to measured BER values. This is possible by inversion of the function $Y = \frac{1}{2}\operatorname{erfc}\left(Q/\sqrt{2}\right)$. A very accurate expression, valid for $Q \in [2, 8]$, is

$$\begin{aligned} Q = 0.12094 - 0.60831 \cdot y - 0.034563 \cdot y^2 - 1.4748 \cdot 10^{-3}\, y^3 \\ - 3.3408 \cdot 10^{-5}\, y^4 - 3.0133 \cdot 10^{-7}\, y^5 \qquad \text{with} \qquad y = \ln Y \end{aligned} . \tag{3.44}$$

Due to $\mathrm{BER} \approx \frac{1}{4}\operatorname{erfc}\left(Q_{0,1}/\sqrt{2}\right)$ we must set $Y = 2 \cdot \mathrm{BER}$. One obtains two sets Q_1, Q_0 as a function of x_{th}. Due to the ideally linear relationship between x_{th} and Q, they are approximated by straight lines. Optimum threshold $x_{th,opt}$ and Q factor Q_{opt} lie where these lines intercept. It is now possible to adjust x_{th} and to estimate the optimum $\mathrm{BER}_{opt} = \frac{1}{2}\operatorname{erfc}\left(Q_{opt}/\sqrt{2}\right)$. The procedure is illustrated in Fig. 3.8. It is well applicable also for non-Gaussian noise.

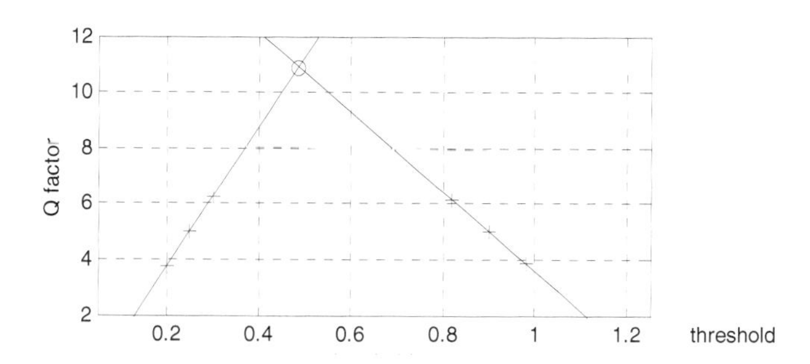

Fig. 3.8 Extrapolation of Q factor for estimation of optimum threshold and BER. Those BER which are expected for $P_0 = P_1 = 1/2$, $\mu_{0,0} = 0.05$, $\sigma_{n0} = 0.04$, $\mu_{0,1} = 1.25$, $\sigma_{n1} = 0.07$ were calculated for $x_{th} \in \{0.2, 0.25, 0.3, 0.82, 0.9, 0.98\}$, thereby substituting real measurements. The lowest BER is about 10^{-10}, which is still well measurable. From the BER the corresponding Q factors were calculated, using (3.44). The two straight lines are linear fits through the data points. They intercept at $x_{th,opt} = 0.486$, $Q_{opt} = 10.91$, $\mathrm{BER}_{opt} = 5 \cdot 10^{-28}$. Even though the BER is unmeasurably small, the BER estimation differs from the true value by only 5%.

Until now we have regarded $\mu_{0,0}$, $\mu_{0,1}$ as constants but in reality they vary according to (3.34) from symbol to symbol. The actual BER is obtained by taking the expectation value over symbols occuring before and after b_k,

$$\mathrm{BER} = \left\langle P(0|1)P_1 + P(1|0)P_0 \right\rangle . \tag{3.45}$$

In the case of Gaussian noise (3.36), (3.37), (3.34) must again be used. Usually a subsequent symbol ($k \in \{i+1; i+2; ...\}$) acts by precursors, and at least one precedent symbol ($k \in \{...; i-2; i-1\}$) acts by postcursors of the filtered and sampled single pulses $s_a(t_0 + (i-k)T)$ on $\mu_{0,0}$ and $\mu_{0,1}$. If $s_a(t_0 - kT) \neq 0$ holds for n different $k \neq i$, averaging must be performed over all 2^n possible different sequences of n bits. We recognize that the BER can be optimized by proper choice of the threshold and the sampling instant, usually by experiment. If $s_a(t_0 + (i-k)T) = 0$ holds for all $k \neq i$, ISI will vanish.

ISI is usually the most difficult problem in data transmission, especially at high data rates. In principle an infinite number of unwanted contributions is added to the wanted sample from neighbor symbols. They can act beneficially or, more often, detrimentally on the BER obtained in the detection of a specific symbol. In total ISI may have very negative consequences, as can be understood from Fig. 3.7: If for beneficial ISI the BER is reduced by a factor of 100, but enlarged by about the same factor for detrimental ISI, then the BER which is an arithmetic average is strongly increased. Which bit sequences are least favorable depends on the transmitted impulse shape, the link dispersion, possibly on nonlinear distortions, and on the baseband filtering in the receiver. A single symbol in the vicinity of many opposite symbols, such as the 1 in the sequence ...0001000... and the 0 in the sequence ...1110111... is often most difficult to detect. Since the leading edges of the impulse response are usually shorter than the falling edges the leading bits influence the sampled value usually more than the trailing bits. Generally the influence of neighbor bits decays strongly as the temporal distance increases, due to causality of filters and due a usually exponential decay of the impulse response. Therefore only a finite number n of samples must practically be taken into account.

Even if the BER is the ultimate quality criterion of a data transmission system one usually looks also the eye diagram (Fig. 3.2) because it often indicates the reason for a too high BER (too much noise or too much ISI?). An eye pattern is fine if there is a sampling instant t_0 at which all traces of ones assume one voltage, and all traces of zeros assume another voltage. These are the sampling values $b_i s_a(t_0)$ of a signal that is free of ISI. For a linear system with stationary Gaussian noise the optimum decision threshold lies in the middle of the vertical eye opening.

If the oscilloscope is triggered with the sequence clock of a pseudo random bit sequence (PRBS) one recognizes singular distorted bit sequences and may eliminate noise by averaging.

The BER is much affected if the frequency response of the lowpass filter falls off sharply because this means the duration of the impulse response is long. This is known for example from $\mathrm{rect}(fT) \circ\!\!-\!\!\bullet\; T^{-1}\,\mathrm{sinc}(\pi t/T)$. In order to realize a matched filter for NRZ impulses the opposite case $\mathrm{sinc}(\pi fT) \bullet\!\!-\!\!\circ\; T^{-1}\,\mathrm{rect}(t/T)$ must be fulfilled. If the impulse response is longer than T, ISI rises sharply. If it is shorter or not rectangular the wanted signal is reduced more strongly than the noise. Practically a compromise must be adopted to minimize the BER.

If there is neither an APD nor an optical amplifier then thermal noise usually dominates over shot noise, and the noise PSD can therefore be considered as approximately constant. The SNR in a matched-filter direct-detection optical receiver is in that case, due to the proportionality of received optical power and photocurrent

$$\mathrm{SNR} = \frac{1}{L_n} \int_{-\infty}^{\infty} |S(f)|^2 df = \frac{1}{L_n} \int_{-\infty}^{\infty} |s(t)|^2 dt \sim \frac{1}{L_n} \int_{-\infty}^{\infty} P_{sig,1}^2(t) dt \,. \quad (3.46)$$

Here $P_{sig,1}(t) \sim s(t)$ be the time-domain signal of the optical power for a single transmitted 1.

Erbium-doped fiber amplifiers (EDFAs) have a large excited state lifetime (some 100 µs ... 1 ms). For data rates in the Gbit/s range therefore even the unlikely case that the number of ones exceeds the number of zeros by 1000 in a short time would reduce the inversion only marginally compared to the equilibrium that is found for equiprobable symbols. Therefore all single pulses (NRZ, RZ, CS-RZ) are also transmitted without distortion. If the EDFA is used in or near saturation the output power will not or not much increase if the input power is increased because the inversion is reduced by the large number of photons generated. However, this is true only for signals longer than the mentioned 100 µs ... 1 ms. For shorter signals the EDFA is linear even where it is (quasistatically) in saturation. This property is called pseudolinearity.

EDFAs have an essentially unlimited output power P_{sig}. Only the mean output power is limited: $P_1 T^{-1} \int P_{sig,1}(t) dt = \text{const.}$ (P_1 = probability of transmitted ones). If EDFA noise can be neglected the optical power should be chosen large which is possible if the impulse duration is shortened (RZ pulses). This is true under the conditions of signal-independent noise, matched filter and limited energy of the transmitted pulse. Practical limits are imposed by signal-dependent noise (from EDFA and shot noise). In addition RZ signals are particularly strongly affected by fiber dispersion. Practically, with EDFAs RZ pulses can yield a sensitivity gain of 1…2 dB.

Semiconductor optical amplifiers (SOAs) have short carrier lifetimes. Operation in saturation would therefore strongly distort the signal. In such systems (without subsequent EDFAs) essentially the transmitted, and the received optical power P_{sig} can be considered as limited. This holds also for the electrical signal amplitude $s(t)$. A similar limitation exists in most transmission systems in electrical communication. In order to maximize the SNR the transmitted impulse should have maximum amplitude during the full bit duration. In amplitude- or peak-power-limited transmission systems NRZ signals are therefore ideal to achieve smallest BER and best receiver sensitivity.

3.1.3 Circuits and Clock Recovery

The subsequently described circuits and others such as receiver preamplifier and modulator driver should preferably be designed balanced, fully symmetrical. This reduces unwanted interference and radiation. Suitable semiconductor technologies are CMOS, SiGe, GaAs, InP, with fabrication cost and upper frequency limit increasing in this order. InP has the advantage of allowing the integration of photodiodes and, in principle, also lasers and modulators, with electronics on one substrate. But hybrid mounting technologies such as flip-chip can perform just as well.

For transmission of high data rates in general a large number of information streams with smaller signals are transmitted using time division multiplex (TDM, Fig. 3.9). The multiplexer (MUX) is built in a power-of-two hierarchy. An n-stage multiplexer combines 2^n data streams with bit rates $2^{-n}/T$ into a single with bit rate $1/T$. It modulates the transmitted optical signal. At the receive end an n-stage demultiplexer (DEMUX) distributes the data signal over 2^n data streams with bit rates $2^{-n}/T$. The clock signals are generated by voltage-controlled oscillators (VCOs) and divider chains. Here the highest needed clock frequency is only half the symbol rate, $2^{-1}/T$.

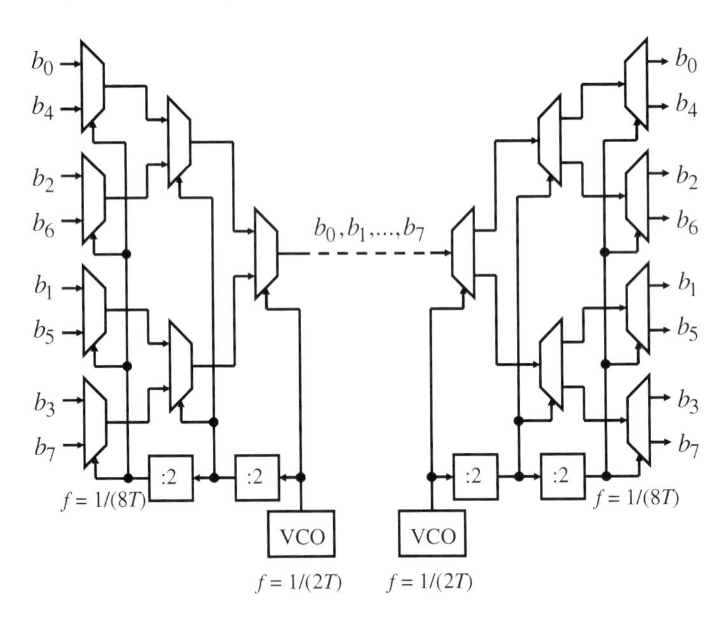

Fig. 3.9 Time division multiplex and demultiplex

A single 2:1 multiplexer cell (Fig. 3.10 left) consists of a selector with D flip-flops at its inputs. In order to provide the selector input signals at the right time one of them must be delayed by half a period of the selector clock. This occurs in an additional D flip-flop with opposed clock polarity. After transmission a D flip-flop which is triggered by the bit clock with frequency $1/T$ can act as a regenerator

(REG). In order to be able to work with a lower maximum frequency, REG is often left out altogether and the signal is regenerated in the subsequent 1:2 demultiplexer (right). It features edge-triggered D flip-flops which are clocked at half the bit rate with opposite clock polarities. Among all flip-flops shown in principle only that of the regenerator must be edge-triggered – or those two at the input of the 1:2 demultiplexer. Another D flip-flop synchronizes the two demultiplexer branches with respect to each other. The other (de-)multiplexer cells of Fig. 3.9 are built very similar.

The receiver VCO must be locked to the symbol clock in a phase-locked loop (PLL). To this purpose one needs first of all a clock phase detector. It can be combined with the decision circuit. (Fig. 3.11 left). DFF2 (D flip-flop) regenerates the data signal. DFF4 – clocked in antiphase – delays the data signal by another half bit duration so that it is synchronous with the output signals of DFF1, DFF3 which are also clocked in antiphase. Let the sampling instant which is ideal for DFF2 be $t_0 = 0$. We restrict ourselves to the case $|t_0| << T/2$. In DFF2 (almost) no wrong decision occurs; let the corresponding data symbol be b_i. The decision instant of DFF1 is $t_0 - T/2$. Roughly at $t_0 = 0$, half of the bits will result in a decision for $b_i \in \{0;1\}$ and the other half will result in a decision for the subsequent bit b_{i+1}.

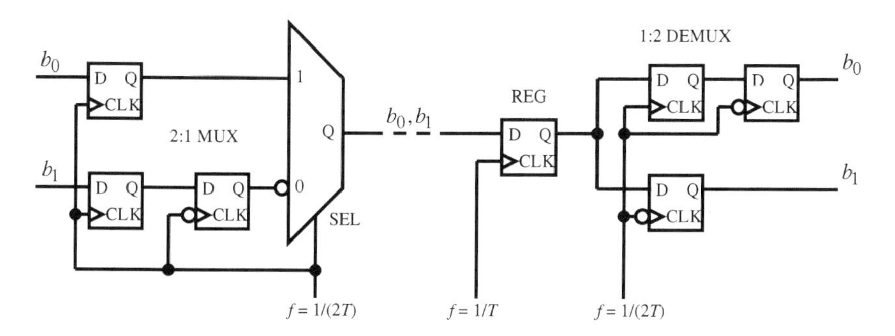

Fig. 3.10 Multiplexer cell, decision circuit (dispensable) and demultiplexer cell

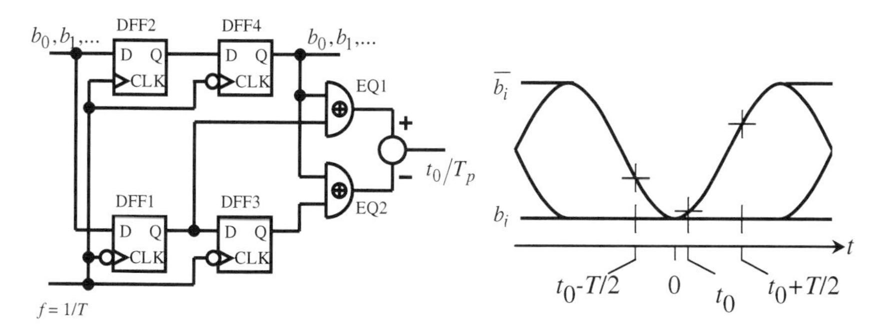

Fig. 3.11 Decision circuit chain with clock phase detector (left), corresponding input signal (right)

In the general case, in particular in the presence of noise the probability for a decision for b_{i+1} (rather than for b_i, which is the other possible value) can be approximated linearly by $1/2+t_0/T_p$. Here $T_p>0$ depends e.g. on the mean noise amplitude. Generally $T_p<T$ holds. The expectation value of the output voltage of the modulo-2 adder EQ1 is therefore $\langle b_i \oplus b_{i+1}\rangle(1/2+t_0/T_p)$ $=(1/2)(1/2+t_0/T_p)$. DFF3 is fed with the data from DFF1, delayed by *T*. With a probability of $1/2+t_0/T_p$ its output signal is b_i, else b_{i-1}. The second modulo-2 adder EQ2 generates an output voltage with the expectation value $\langle b_{i-1} \oplus b_i\rangle(1/2-t_0/T_p)=(1/2)(1/2-t_0/T_p)$. A subtractor generates from this a phase detection signal with the expectation value t_0/T_p. Since the probabilities range between 0 and 1, it ranges between $\pm 1/2$ times the voltage deviation of the binary signal. A noiseless, lowpass-filtered data input signal is shown in Fig. 3.11 right. The priniciple of the clock phase detector can also be transferred to decision circuits which are clocked at half the bit rate.

For completeness we mention further clock phase detection methods which are rather suitable for lower data rates. RZ signals possess a spectral component at the clock frequency. In contrast, an ideal NRZ signal has no clock frequency component (Fig. 3.12). But it has spectral components at half the clock frequency which are multiplied by ±1. If one removes the DC component, applies lowpass filtering and finally squares this signal these polarity jumps will disappear. One obtains a kind of inverted RZ signal with alterated bit pattern which contains the wanted clock frequency component.

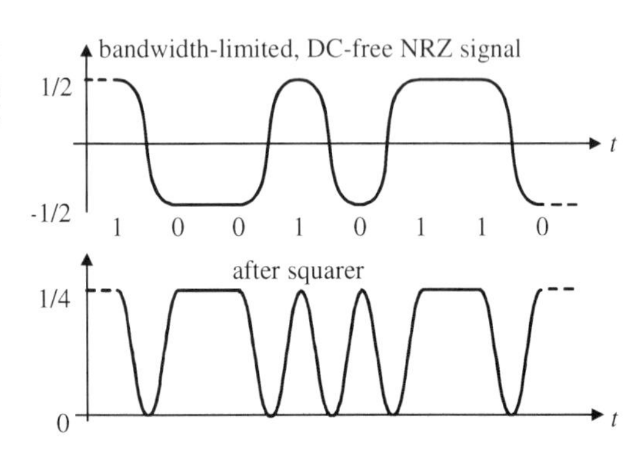

Fig. 3.12 Generation of a clock frequency component by squaring of a band-limited, DC-free NRZ signal

It would be possible to select the clock frequency component by a filter and to use it as recovered clock. However, this would require a very narrowbanded filter with a correspondingly large group delay. Unavoidable temperature variations could therefore cause unacceptable decision instant variations. The same would hold if a synchronized oscillator were used. It is therefore better to use a phase-locked loop (PLL) with a multiplier as a clock phase detector, or most usually with the clock phase detector of Fig. 3.11, which may all the same be described in an abstract manner by the following equations. A PLL (Fig. 3.13) is almost always part of the clock recovery.

Here is how a PLL works: Let sinusoidal signals

$$u_1(t) = u_{10} \sin(\omega_0 t + \varphi_1), \tag{3.47}$$

$$u_2(t) = u_{20} \cos(\omega_0 t + \varphi_2) \tag{3.48}$$

be fed to the multiplier inputs. These stem from the signal to be regenerated and from the VCO, respectively. Without loss of generality the frequencies have been assumed to be identical; possible frequency differences result in a linear time dependence of the phase difference $\varphi_1 - \varphi_2$. At the multiplier output the $2\omega_0$ spectral component is eliminated by a filter and just the lowpass part $u_d = K_d \sin(\varphi_1 - \varphi_2)$ is used. For an ideal multiplier which generates the product $K u_1(t) u_2(t)$ with $[K] = 1\,\text{V}^{-1}$, $K_d = (1/2) K u_{10} u_{20}$ holds. For $|\varphi_1 - \varphi_2| << \pi/2$ we may set

$$u_d = K_d(\varphi_1 - \varphi_2), \qquad [K_d] = 1\,\text{V/rad}. \tag{3.49}$$

In the clock phase detector according to Fig. 3.11, with a binary voltage shift of u_b, it holds $\varphi_1 - \varphi_2 = 2\pi t_0/T$, $u_d = u_b t_0/T_p \Rightarrow K_d = u_b T/(2\pi T_p)$.

The PLL contains also a loop filter and a VCO (Fig. 3.13). The resonator of the VCO can be a surface acoustic wave filter (at low frequencies) or a dielectric resonator (at high frequencies). Most frequently it is an LC tank inside an integrated circuit.

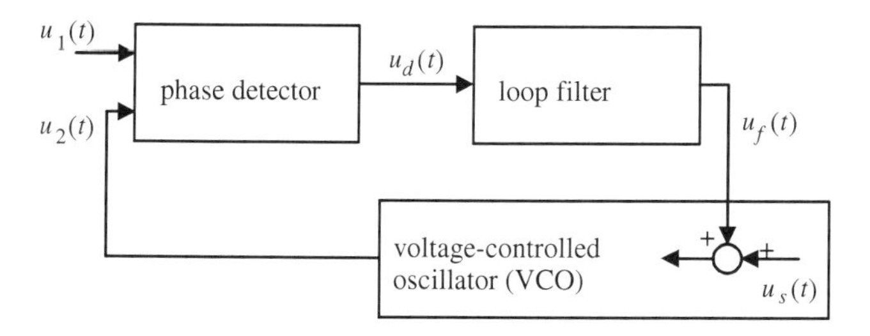

Fig. 3.13 Schematic of a phase-locked loop (PLL)

Input and output signal of the loop filter or controller are linked by its impulse response $h_R(t)$ according to

$$u_f(t) = h_R(t) * u_d(t). \tag{3.50}$$

For the VCO frequency

$$\frac{d\varphi_2(t)}{dt} = K_O\left(u_f(t)+u_s(t)\right) \qquad \left[K_O\right]=\frac{1}{\mathrm{Vs}} \tag{3.51}$$

holds. K_O is the VCO constant. The sum of loop filter output voltage $u_f(t)$ and a possible ficticious voltage $u_s(t)$ acts as a control quantity. The latter voltage is introduced in order to take into account a static offset between the input frequency and the VCO frequency which is generated for $u_f(t)=0$. $u_s(t)$ may drift as a function of time (aging, temperature variations). After Laplace transformation one obtains from (3.49) to (3.51)

$$\begin{aligned} &U_d(p)=K_d\left(\Phi_1(p)-\Phi_2(p)\right) \qquad U_f(p)=H_R(p)U_d(p) \\ &\Phi_2(p)=\frac{1}{p}K_O\left(U_f(p)+U_s(p)\right) \end{aligned} , \tag{3.52}$$

which after insertion of the open-loop transfer function $H_o(p)$

$$\begin{aligned} &\Phi_2(p)=H_o(p)\left(\Phi_1(p)-\Phi_2(p)\right)+\frac{1}{p}K_O U_s(p) \\ &H_o(p)=\frac{1}{p}K_O H_R(p)K_d \end{aligned} \tag{3.53}$$

yields

$$\begin{aligned} &\Phi_2(p)=H(p)\Phi_1(p)+H_s(p)U_s(p) \\ &H(p)=\frac{K_O H_R(p)K_d}{p+K_O H_R(p)K_d} \qquad H_s(p)=\frac{K_O}{p+K_O H_R(p)K_d} \end{aligned} , \tag{3.54}$$

$H(p)$ is the *phase transfer function*, $H_s(p)$ a disturbance transfer function with respect to voltage u_s. For the phase error φ_F and the error transfer function $H_F(p)$ it holds

$$\begin{aligned} &\varphi_F(t)=\varphi_1(t)-\varphi_2(t) \qquad \Phi_F(p)=\Phi_1(p)-\Phi_2(p) \\ &\Phi_F(p)=H_F(p)\Phi_1(p)-H_s(p)U_s(p) \\ &H_F(p)=1-H(p)=\frac{p}{p+K_O H_R(p)K_d} \end{aligned} . \tag{3.55}$$

Sufficient for a stable PLL is that the open-loop transfer function $H_o(j\omega)$ has a magnitude smaller than 1 for all frequencies where its phase equals π. With a pure integral controller, $H_R(p)=K_i/p$, the PLL would be unstable because of $\lim_{\omega\to 0+0} H_o(j\omega)=-\infty$. A proportional controller is suitable to enforce stable control, due to $\lim_{\omega\to 0} H_R(j\omega) = K_p$. A constant frequency offset which occurs at $t = 0$ can be described by $U_s(p) = u_{s0}/p$. Here a proportional controller produces a stationary phase error $\lim_{t\to\infty} \varphi_F(t) = \lim_{p\to 0} p\Phi_F(p) = -u_{s0}/(K_p K_d)$ which usually can not be tolerated. It disappears if the controller has also an integral component. The usual PLL loop filter is therefore a PI controller.

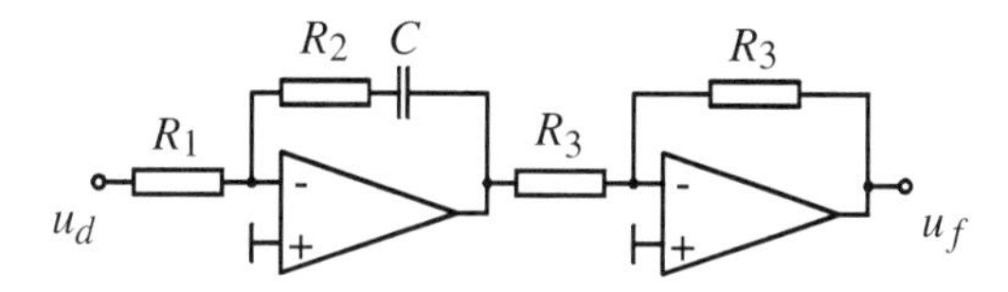

Fig. 3.14 PI controller schematic (active first-order loop filter)

From the PI controller schematic (Fig. 3.14) we find the transfer function

$$H_R(p) = \frac{R_2 + 1/pC}{R_1} = \frac{\tau_2}{\tau_1} + \frac{1}{p\tau_1} = \frac{1 + p\tau_2}{p\tau_1} . \tag{3.56}$$

with $\tau_1 = R_1 C \quad \tau_2 = R_2 C$

After substitution of resonance frequency ω_r and damping constant ξ

$$\omega_r = \sqrt{\frac{K_0 K_d}{\tau_1}} \qquad \xi = \frac{\tau_2}{2}\omega_r = \frac{\tau_2}{2}\sqrt{\frac{K_0 K_d}{\tau_1}} \tag{3.57}$$

we obtain a damped 2nd-order system (Figs. 3.15 and 3.16) with

$$H(p) = \frac{2p\xi\omega_r + \omega_r^2}{p^2 + 2p\xi\omega_r + \omega_r^2} \qquad H_F(p) = \frac{p^2}{p^2 + 2p\xi\omega_r + \omega_r^2} . \tag{3.58}$$

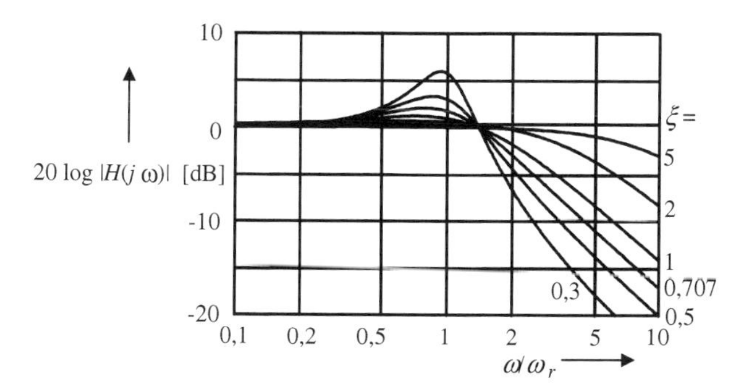

Fig. 3.15 Phase transfer function of a 2nd-order PLL in linear approximation

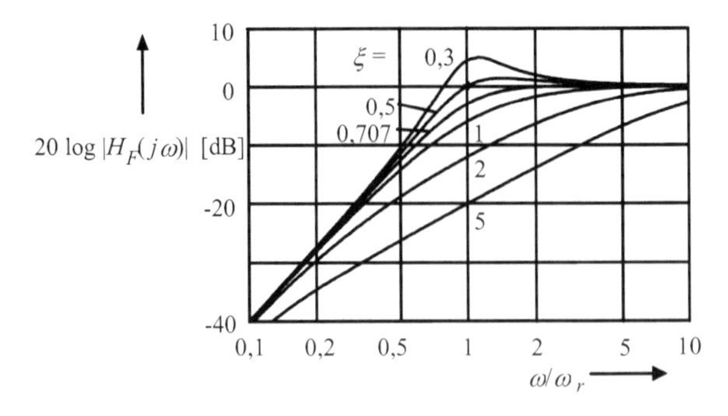

Fig. 3.16 Error transfer function of a 2nd-order PLL in linear approximation

For high frequencies $H(j\omega)$ decays with 20 dB per decade. The error transfer function $H_F(j\omega)$, which is complementary, rises by 40 dB per decade for low frequencies.

If the PLL is in a stable state in which the phase difference approaches zero (or, if a proportional controller is used, a fixed value) then it is *locked*. The frequency range in which ω_0 may change at least slowly without loss of lock is the hold range. If a controller with integral contribution (usually a PI controller) is used the hold range is just as large as the tuning range of the VCO, and the stationary phase error equals zero (see above) since the integrator just compensates the disturbing voltage u_s. In the following we therefore neglect u_s.

In addition to the clock frequency components the phase detector input voltage generally exhibits other frequency components, and there is also noise. Therefore even for constant φ_1 phase differences between input and output signal may exist. Without loss of generality let $\varphi_1(t)=0$. In order to recover a clean clock signal it may be helpful to first reduce the input frequency range by a bandpass filter with a passband width that is small compared to ω_0. The input voltage (3.47) with added narrowband noise is

$$u_1(t)=u_{10}\sin(\omega_0 t)+u_p(t)\sin(\omega_0 t)+u_q(t)\cos(\omega_0 t). \tag{3.59}$$

The noise components in phase $u_p(t)$ and in quadrature $u_q(t)$ are Gaussian, have zero means, and change slowly compared to the clock oscillation. In addition we assume $\overline{u_p^2(t)}=\overline{u_q^2(t)}<<u_{10}^2$. In a first-order approximation one obtains similarly to (3.49)

$$u_d(t)=K_d\left(\frac{u_q(t)}{u_{10}}-\varphi_2(t)\right) \tag{3.60}$$

The quadrature noise component acts like a time-variable phase $\varphi_1(t)=u_q(t)/u_{10}$. For the power spectral density of the phase angle one finds

$$L_{\varphi_2}(f)=|H(j2\pi f)|^2 L_{\varphi_1}(f). \tag{3.61}$$

The variance of φ_2 equals the value of the autocorrelation function at $\tau = 0$. It is obtained by inverse Fourier transformation at $\tau = 0$,

$$\sigma_{\varphi_2}^2 = l_{\varphi_2}(0) = \int_{-\infty}^{\infty} L_{\varphi_2}(f)df = \int_{-\infty}^{\infty} \left|H(j2\pi f)\right|^2 L_{\varphi_1}(f)df \,, \tag{3.62}$$

with $L_{\varphi_1}(f) = u_{10}^{-2} L_{u_q}(f)$. The *noise bandwidth* is

$$B_r = \left|H(0)\right|^{-2} \int_0^{\infty} \left|H(j2\pi f)\right|^2 df \,. \tag{3.63}$$

If $L_{\varphi_1}(f)$ can be considered constant over a sufficiently wide range, $L_{\varphi_1}(f) = L_{\varphi_1}$, one may write due to $\left|H(0)\right| = 1$

$$\sigma_{\varphi_2}^2 = 2B_r L_{\varphi_1} \,. \tag{3.64}$$

The PLL noise bandwidth B_r is usually chosen several orders of magnitude lower than the bit rate. This allows to recover a clock signal with a very small phase error in spite of noise and disturbances.

The clock recovery PLL is a standard regenerator component. In a well-designed PLL the VCO input signal is roughly speaking the temporal derivative of the clock phase. The clock phase is proportional to the signal arrival time. It is therefore possible to detect arrival time or group delay variations of the transmitted signal by evaluating the VCO input signal or the clock phase detector output signal. A condition is that the modulation frequency of the arrival time is well beyond the linewidth of the free-running VCO. Using this scheme, polarization mode dispersion has been detected with a sensitivity ranging from 680 fs [42] to 84 fs [43], and chromatic dispersion with an accuracy of down to about 50 fs/nm [44].

3.2 Optical Amplifiers in Systems with Direct Detection

In optical communication traditional electrical data regenerators are replaced, where possible, by optical amplifiers. Optical amplification has been originally described more than 40 years ago [45]. Photoelectron statistics are also described in [46] but these excellent treatments seem to be not widely known today. In particular, a noise figure definition exists [47] which does not permit exact calculation of the resulting receiver sensitivity. Corrections have been given in [48–50].

3.2.1 *Photon Distributions*

All random variables (RV) in this chapter except for electric fields are nonnegative. The interaction of photons with matter is governed by stimulated emission, absorption, and spontaneous emission. For infinitesimal time increments dt the probability $P(n,t+dt)$ to find n photons in a medium at time $t+dt$ depends on the probabilities to find $n-1$, n, or $n+1$ photons at time t, and the conditional transition probabilities from one of these numbers to n,

$$P(n,t+dt)=P(n|n)P(n,t)+P(n|n-1)P(n-1,t)+P(n|n+1)P(n+1,t). \tag{3.65}$$

For small dt the transition probabilities are

$$\begin{aligned} &P(n|n-1)=((n-1)a+c)dt \qquad P(n|n+1)=(n+1)bdt \\ &P(n|n)=1-P((\neq n)|n)\approx 1-P(n-1|n)-P(n+1|n)=1-(n(a+b)+c)dt \end{aligned} \tag{3.66}$$

with a = stimulated emission rate per photon, b = absorption rate per photon, c = spontaneous emission rate. In the last line other photon numbers did not need not be considered because for $dt\to 0$ the probability of multiple emission/absorption events tends towards zero faster than that of single events. Using $dP(n,t)/dt=(P(n,t+dt)-P(n,t))/dt$, this master equation of photon statistics can be written in differential form as

$$\frac{dP(n,t)}{dt}=-(n(a+b)+c)P(n,t)+((n-1)a+c)P(n-1,t)+(n+1)bP(n+1,t) \tag{3.67}$$

If a particular photon statistic is to persist it must fulfill (3.67) but statistical parameters such as its $\langle n\rangle$ may be time-variable. E.g., if there is only attenuation ($b>0$, $a=c=0$), a Poisson distribution

$$P(n)=e^{-\mu_0}\frac{\mu_0^n}{n!} \tag{3.68}$$

is a solution of (3.67) if its expectation value $\langle n\rangle = \mu_0(t)$ decays exponentially with time, $\mu_0(t) = \mu_0(0)e^{-bt}$. This means a Poisson distribution is conserved under pure attenuation.

The moment generating function (MGF) of a discrete random variable (RV) n is

$$M_n\left(e^{-s}\right) = \left\langle e^{-sn}\right\rangle = \sum_{n=-\infty}^{\infty} P(n)e^{-sn}, \tag{3.69}$$

for a continuous random variable x it is

$$M_x\left(e^{-s}\right) = \left\langle e^{-sx}\right\rangle = \int_{-\infty}^{\infty} p_x(x)e^{-sx}dx. \tag{3.70}$$

The lower summation index / integration boundary is 0 for nonnegative RVs. With $e^{-s} = z^{-1}$, eqn. (3.69) can be inverted by inverse z, while (3.70) is inverted by inverse Laplace transformation.

Adding statistically independent RVs requires convolution of the corresponding PDFs, or multiplication of the corresponding MGFs.

As suggested by the name moment generating function, the MGF allows to obtain all moments via

$$\left\langle x^k\right\rangle = (-1)^k \left.\frac{d^k M\left(e^{-s}\right)}{(ds)^k}\right|_{s=0}. \tag{3.71}$$

Distributions, MGFs, mean values and variances of some discrete and continuous RVs are given in Table 3.2.

Problem: Calculate the MGF of a Poisson distribution, and obtain 1st and 2nd moments $\langle n\rangle$, $\langle n^2\rangle$ from it. Calculate also the standard deviation σ_n.

Solution: $M_n\left(e^{-s}\right) = \sum_{n=0}^{\infty} e^{-sn} e^{-\mu_0}\frac{\mu_0^n}{n!} = e^{-\mu_0}\sum_{n=0}^{\infty}\frac{\left(e^{-s}\mu_0\right)^n}{n!} = e^{-\mu_0}e^{e^{-s}\mu_0} = e^{-\mu_0\left(1-e^{-s}\right)}$

$$\langle n\rangle = -\left.\frac{dM_n\left(e^{-s}\right)}{ds}\right|_{s=0} = \left. e^{-\mu_0\left(1-e^{-s}\right)}\mu_0 e^{-s}\right|_{s=0} = \mu_0$$

$$\left\langle n^2\right\rangle = \left.\frac{d^2 M_n\left(e^{s}\right)}{ds^2}\right|_{s=0} = -\mu_0 \frac{d}{ds}\left.\left(e^{-\mu_0\left(1-e^{-s}\right)}e^{-s}\right)\right|_{s=0}$$

$$= -\mu_0\left.\left(e^{-\mu_0\left(1-e^{-s}\right)}\left(-\mu_0\right)\left(e^{-s}\right)e^{-s} + e^{-\mu_0\left(1-e^{-s}\right)}\left(-e^{-s}\right)\right)\right|_{s=0} = \mu_0(\mu_0+1)$$

$$\sigma^2 = m_2 - m_1^2 = \mu_0(\mu_0+1) - \mu_0^2 = \mu_0 \Rightarrow \sigma = \sqrt{\mu_0}$$

Table 3.2 Some discrete and continuous distributions

Discrete distributions	$P(n)$ ($n \ge 0$)	$M_n(e^{-s})$	$\langle n \rangle$	σ_n^2
Poisson	$e^{-\mu_0} \frac{\mu_0^n}{n!}$	$e^{-\mu_0(1-e^{-s})}$	μ_0	μ_0
Central negative binomial	$\binom{n+N-1}{n} \frac{\mu^n}{(1+\mu)^{n+N}}$	$\frac{1}{(1+\mu(1-e^{-s}))^N}$	$N\mu$	$N\mu(\mu+1)$
Noncentral negative binomial, Laguerre	$\frac{\mu^n e^{-\frac{\mu_0}{1+\mu}}}{(1+\mu)^{n+N}} L_n^{N-1}\left(\frac{-\mu_0}{\mu(1+\mu)}\right)$	$\frac{e^{\frac{-\mu_0(1-e^{-s})}{1+\mu(1-e^{-s})}}}{(1+\mu(1-e^{-s}))^N}$	$\mu_0 + N\mu$	$N\mu(\mu+1) + (2\mu+1)\mu_0$
Continuous distributions	$p_{\tilde{x}}(\tilde{x})$ ($\tilde{x} \ge 0$)	$M_{\tilde{x}}(e^{-s})$	$\langle \tilde{x} \rangle$	$\sigma_{\tilde{x}}^2$
Constant	$\delta(\tilde{x} - \tilde{\mu}_0)$	$e^{-\tilde{\mu}_0 s}$	$\tilde{\mu}_0$	0
Central χ_{2N}^2, Gamma	$\frac{1}{\Gamma(N)} \tilde{\mu}^{-N} \tilde{x}^{N-1} e^{-\tilde{x}/\tilde{\mu}}$	$\frac{1}{(1+\tilde{\mu}s)^N}$	$N\tilde{\mu}$	$N\tilde{\mu}^2$
Noncentral χ_{2N}^2	$\frac{\tilde{x}^{(N-1)/2} e^{-(\tilde{\mu}_0+\tilde{x})/\tilde{\mu}}}{\tilde{\mu}_0^{(N-1)/2} \tilde{\mu}} \cdot I_{N-1}\left(2\sqrt{\tilde{x}\tilde{\mu}_0}/\tilde{\mu}\right)$	$\frac{e^{\frac{-\tilde{\mu}_0 s}{1+\tilde{\mu}s}}}{(1+\tilde{\mu}s)^N}$	$\tilde{\mu}_0 + N\tilde{\mu}$	$N\tilde{\mu}^2 + 2\tilde{\mu}\tilde{\mu}_0$

When a signal passes an optical amplifier the probability $P(n,t)$ of the discrete number of photons n is time-variable. We determine the temporal derivative of the corresponding time-variable MGF $M_n(e^{-s},t)$ [51],

$$\frac{\partial}{\partial t} M_n(e^{-s},t) = \sum_{n=-\infty}^{\infty} e^{-sn} \frac{dP(n,t)}{dt} . \tag{3.72}$$

Here the summation may start at $n=0$ because $P(n,t)=0$ for $n<0$. Using (3.67) we obtain the temporal MGF derivative

$$\begin{aligned} \frac{\partial}{\partial t} M_n(e^{-s},t) &= \sum_{n=0}^{\infty} -e^{-sn}(n(a+b)+c)P(n,t) \\ &+ \sum_{n=1}^{\infty} e^{-sn}((n-1)a+c)P(n-1,t) + \sum_{n=-1}^{\infty} e^{-sn}(n+1)bP(n+1,t) \end{aligned} . \tag{3.73}$$

The summation starts where the first argument of P is nonnegative. 2nd and 3rd terms can be rewritten as $\sum_{n=0}^{\infty} e^{-s(n+1)}(na+c)P(n,t)$, $\sum_{n=0}^{\infty} e^{-s(n-1)} nbP(n,t)$, respectively. This allows to collect $\sum_{n=0}^{\infty} e^{-sn} P(n,t)$ and $\sum_{n=0}^{\infty} n e^{-sn} P(n,t)$, with the result

$$\frac{\partial}{\partial t} M_n\left(e^{-s},t\right) = c\left(e^{-s}-1\right) M_n\left(e^{-s},t\right) - \left(a-be^{s}\right)\left(e^{-s}-1\right) \frac{\partial}{\partial s} M_n\left(e^{-s},t\right). \quad (3.74)$$

The signal reaches the amplifier input at $t=0$, with a photon distribution $P(n,0)$ and a MGF $M_n\left(e^{-s},0\right)$. The solution of (3.74) is

$$M_n\left(e^{-s},t\right) = \left(\frac{a-b}{a-b+a(G-1)\left(1-e^{-s}\right)}\right)^{c/a} M_n\left(\frac{a-b+(bG-a)\left(1-e^{-s}\right)}{a-b+a(G-1)\left(1-e^{-s}\right)},0\right) \quad (3.75)$$

with a time-variable (power) gain

$$G = G(t) = e^{(a-b)t} . \quad (3.76)$$

$M_n\left(e^{-s},t\right) = U^{c/a} M_n\left(e^{-S},0\right)$This means we know the MGF inside or at the output of an optical amplifier as a function of the MGF at its input, if the term e^{-s} is replaced by a more complicated one.

For a transmitted zero there are no photons at the input,

$$P(n,0) = \begin{cases} 1 & n=0 \\ 0 & n \neq 0 \end{cases}, \qquad M_n\left(e^{-s},0\right) = 1 . \quad (3.77)$$

At time t the signal has passed the amplifier and $G = e^{(a-b)t}$ holds. According to (3.75) the output photon MGF is

$$M_n\left(e^{-s},t\right) = \left(1 + \frac{a}{a-b}(G-1)\left(1-e^{-s}\right)\right)^{-c/a} . \quad (3.78)$$

The dependence on t can now be dropped. If there is just one optical mode, stimulated emission is just n times larger than spontaneous emission, hence $c=a$. For N noisy modes

$$N = c/a \quad (3.79)$$

holds. The spontaneous emission factor is

$$n_{sp} = \frac{a}{a-b} . \quad (3.80)$$

It holds $n_{sp} \leq 0$ for $0 \leq a < b$, $G < 1$ and $n_{sp} > 1$ for $a > b$, $G > 1$, with a pole at $a = b$ and the limit $n_{sp} = 1$ for $a \to \infty$. The expectation value of noisy photons in one mode is

$$\mu = n_{sp}(G-1). \tag{3.81}$$

With above substitutions the MGF transformation equation (3.75) becomes

$$M_n\left(e^{-s},t\right) = \left(1+\mu\left(1-e^{-s}\right)\right)^{-N} M_n\left(1-\frac{G\left(1-e^{-s}\right)}{1+\mu\left(1-e^{-s}\right)},0\right), \tag{3.82}$$

and the MGF (3.78) simplifies to

$$M_n\left(e^{-s}\right) = \left(1+\mu\left(1-e^{-s}\right)\right)^{-N}. \tag{3.83}$$

Problem: Calculate the 1st and 2nd moments $\langle n \rangle$, $\langle n^2 \rangle$, and the standard deviation σ_n.

Solution: $\langle n \rangle = -\left.\frac{dM_n\left(e^{-s}\right)}{ds}\right|_{s=0} = N\left(1+\mu\left(1-e^{-s}\right)\right)^{-N-1}\mu e^{-s}\Big|_{s=0} = N\mu$

$$\langle n^2 \rangle = \left.\frac{d^2 M_n\left(e^{s}\right)}{ds^2}\right|_{s=0} = -N\mu\frac{d}{ds}\left(\left(1+\mu\left(1-e^{-s}\right)\right)^{-N-1}e^{-s}\right)\Bigg|_{s=0}$$

$$= -N\mu\left((-N-1)\left(1+\mu\left(1-e^{-s}\right)\right)^{-N-2}\mu\left(e^{-s}\right)e^{-s} + \left(1+\mu\left(1-e^{-s}\right)\right)^{-N-1}\left(-e^{-s}\right)\right)\Bigg|_{s=0}$$

$$= N\mu((N+1)\mu+1)$$

$$\sigma = \sqrt{m_2 - m_1^2} = \sqrt{N\mu((N+1)\mu+1) - N^2\mu^2} = \sqrt{N\mu(\mu+1)}$$

According to Table 3.2 the standard deviation is $\sigma_n = \sqrt{(\mu+1)N\mu} = \sqrt{(\mu+1)\langle n \rangle}$ and is larger a larger fraction of the mean than for the Poisson distribution where it is just $\sigma_n = \sqrt{\mu_0} = \sqrt{\langle n \rangle}$. This is due to the quantum noise of the optical amplifier. It surpasses the quantum noise of a Poisson distribution which is just shot noise. Inverse z transformation of $M_n\left(z^{-1}\right)$ results in a (central) negative binomial distribution

$$P(n) = \binom{n+N-1}{n}\frac{\mu^n}{(1+\mu)^{n+N}}. \tag{3.84}$$

For one mode, $N=1$, this is a Bose-Einstein distribution.

A sufficiently attenuated signal is always Poisson distributed, as will be seen later. For a transmitted one we therefore assume a Poisson distribution at the amplifier input with mean $\tilde{\mu}_0$. With $M_n\left(e^{-s},0\right) = e^{\tilde{\mu}_0\left(e^{-s}-1\right)}$ and (3.82) we find

$$M_n\left(e^{-s},t\right) = \left(1+\mu\left(1-e^{-s}\right)\right)^{-N} e^{\frac{-\mu_0\left(1-e^{-s}\right)}{1+\mu\left(1-e^{-s}\right)}} \tag{3.85}$$

at the amplifier output with $\mu_0 = \tilde{\mu}_0 G$. The corresponding photon distribution is a noncentral negative binomial or Laguerre distribution

$$P(n) = \frac{\mu^n}{(1+\mu)^{n+N}} e^{-\frac{\mu_0}{1+\mu}} L_n^{N-1}\left(\frac{-\mu_0}{\mu(1+\mu)}\right), \tag{3.86}$$

with Laguerre polynomials defined as

$$L_n^\alpha(x) = \frac{1}{n!} e^x x^{-\alpha} \frac{d^n}{dx^n}\left(e^{-x} x^{n+\alpha}\right) = \sum_{m=0}^{n} (-1)^m \binom{n+\alpha}{n-m} \frac{x^m}{m!} . \tag{3.87}$$

There is an important adding property: Two statistically independent RVs n_1, n_2 with the same noise μ and (central or noncentral) negative binomial distributions are added to form a new RV n. It obeys a new negative binomial distribution with $\mu_0 = \mu_{0,1} + \mu_{0,2}$, $N = N_1 + N_2$. This can be verified by multiplying the MGFs or convolving the PDFs.

3.2.2 Noise Figure

It is useful to divide the output noise photon number μ per mode by the amplification G. Thereby one obtains a ficticious input-referred noise photon number $\tilde{\mu} = \mu / G$ per mode which would be amplified by a noise-free amplifier to become the observed output noise. It can also be called an excess noise figure F_z,

$$F_z \equiv \tilde{\mu} = \frac{\mu}{G} = \left(1 - G^{-1}\right) n_{sp} . \tag{3.88}$$

The noise figure itself is

$$F = 1 + F_z = 1 + \left(1 - G^{-1}\right) n_{sp} . \tag{3.89}$$

This noise figure can be called F_{ASE} [49, 52] where ASE stands for amplified spontaneous emission. Due to $n_{sp} \geq 1$ one finds $F \geq 1$ for amplifiers. An ideal amplifier with $n_{sp} = 1$ and amplification $G \to \infty$ possesses the $F = 2$, which can be understood from the following: Without amplifier just the amplitude but not the phase of an optical signal can be measured. With ideal amplifier and a subsequent power splitter both these quantities can be measured simultaneously.

Let us cascade two optical amplifiers. For a Poisson distribution with mean $\tilde{\mu}_0$ at its input the MGF after the first amplifier (index 1) is

$$M_{n,1}\left(e^{-s}, t_1\right) = \left(1 + \mu_1\left(1 - e^{-s}\right)\right)^{-N} e^{\frac{-\mu_{0,1}\left(1 - e^{-s}\right)}{1 + \mu_1\left(1 - e^{-s}\right)}} \tag{3.90}$$

with $\mu_{0,1} = \tilde{\mu}_0 G_1$. Under the assumption that there is only one optical filter, behind all amplifiers and directly in front of the receiver, the mode number N is identical for both amplifiers. The second amplifier carries the index 2. The MGF at its output is

$$M_{n,2}\left(e^{-s}, t_2 + t_1\right) = \left(1 + \mu_2\left(1 - e^{-s}\right)\right)^{-N} M_{n,1}\left(\frac{1 + \left(\mu_2 - G_2\right)\left(1 - e^{-s}\right)}{1 + \mu_2\left(1 - e^{-s}\right)}, t_1\right) \tag{3.91}$$

with $N = \frac{c_i}{a_i}$, $n_{sp,i} = \frac{a_i}{a_i - b_i}$, $G_i = e^{(a_i - b_i)t_i}$, $\mu_i = n_{sp,i}(G_i - 1)$, $i = 1,2$.

This can be rewritten as (3.85) with $t = t_2 + t_1$, $\mu_0 = \mu_{0,1} G_2 = \tilde{\mu}_0 G_1 G_2$ and an expectation value $\mu = \mu_2 + \mu_1 G_2$ for the output noise in one mode. In the cascade

$$F_z = \frac{\mu_1 G_2 + \mu_2}{G_1 G_2} = F_{z1} + \frac{F_{z2}}{G_1}. \tag{3.92}$$

holds. Recursion allows to obtain Friis‘ well known cascading formula

$$F = F_1 + \frac{F_2 - 1}{G_1} + \frac{F_3 - 1}{G_1 G_2} + \ldots \tag{3.93}$$

for a larger cascade (see [48] for a slightly different but equivalent formulation). This is important because optical trunk lines consist alternatively of amplifiers and attenuating fibers. A pure attenuator with $a = 0$, $b > 0$, $n_{sp} = 0$ has a noise figure $F = 1$ and a "gain" $G = e^{-bt} < 1$. E.g., the cascade of an attenuator (index 1) and a subsequent amplifier (index 2) has the noise figure $F = 1 + \frac{F_2 - 1}{G_1}$. If the amplifier (index 1) is in front of the attenuator (index 2), $\mu_2 = 0$ holds. For increasing attenuation, $G_2 \to 0$, we find

$$M_{n,2}\left(e^{-s}, t_2 + t_1\right) = \left(1 + \mu_1 G_2\left(1 - e^{-s}\right)\right)^{-N} e^{\frac{-\mu_{0,1} G_2\left(1 - e^{-s}\right)}{1 + \mu_1 G_2\left(1 - e^{-s}\right)}} \approx e^{-\mu_{0,1} G_2\left(1 - e^{-s}\right)}. \tag{3.94}$$

This means a Laguerre becomes a Poisson distribution if attenuation is large. Since the Poisson distribution is also conserved under attenuation it has a special importance.

An amplifier in which a, b, c vary as a function of the propagation coordinate may be subdivided into infinitesimal sections and treated by (3.93) in which summation is replaced by an integral,

$$F = 1 + \int_0^t \left(\frac{d\left(F_i(\tau) - 1\right)}{d\tau}\right) G^{-1}(\tau) d\tau. \tag{3.95}$$

For (3.95) to be true, the differential $d(F_i(\tau)-1)$ must be the additional noise figure of an infinitesimally short amplifier with group delay $d\tau$. Due to $d\tau \to 0$, $G(\tau)=e^{\int_0^\tau (a(\vartheta)-b(\vartheta))d\vartheta}$ is the total gain including or preceding the infinitesimal amplifier, which itself has the gain $G(\tau)/G(\tau-d\tau)=e^{(a(\tau)-b(\tau))d\tau}$. The derivative $d(F_i(\tau)-1)/d\tau$ assumes the simple value

$$\frac{d(F_i(\tau)-1)}{d\tau}=n_{sp}(\tau)\frac{1-G(\tau-d\tau)/G(\tau)}{d\tau}=\frac{a(\tau)}{a(\tau)-b(\tau)}\frac{1-e^{-(a(\tau)-b(\tau))d\tau}}{d\tau}=a(\tau). \tag{3.96}$$

The integral (3.95) can now be written as

$$F=1+\int_0^t a(\tau)G^{-1}(\tau)d\tau=1+\int_0^t a(\tau)e^{-\int_0^\tau (a(\vartheta)-b(\vartheta))d\vartheta}d\tau. \tag{3.97}$$

For constant coefficients it delivers the already-known value $F=1+\left(1-G^{-1}\right)n_{sp}$.

The longitudinal coordinate L inside the amplifier is $L=\int_0^t v_g d\tau$, where v_g is the group velocity. Due to $dL=v_g dt$ the integration over time can be replaced by an integration over L, if the quantities v_g, a, b are known as a function of L. This is automatically fulfilled for constant v_g.

As an example, consider a fiber which is rendered lossless ($a=b$) by Raman gain with a pump wavelength roughly 50 to 100 nm shorter than the signal wavelength. For a constant attenuation of 0.2 dB/km, the fiber „gain" without pumping ($a=0$) would be $G=e^{-bv_g^{-1}L}=10^{-\frac{1}{10\,\text{dB}}0.2\frac{\text{dB}}{\text{km}}L}$. An $L=50\,\text{km}$ long, Raman-pumped fiber with $a=b$ ideally has the noise figure $F=1+bv_g^{-1}L=3.3$ (if we ignore the difficulty of achieving $a=b$ over such large lengths). For comparison, an ideal amplifier ($n_{sp}=1$) with $G=10$ has $F=1.9$. If it is placed at the input of a 50 km long fiber without Raman gain ($a=0$) the cascade has a noise figure of 1.9. If it is placed at the fiber end the cascade noise figure is 10. In all three cases the total gain equals 1.

Noise figure measurement is easy: The optical amplifier is operated without input signal. Gain saturation effects can be taken into account if the input signal is just switched off during short measurement intervals. Usually G is nearly polarization-insensitive, hence $p=2$ polarization modes of a monomode fiber have to be taken into account. The output signal is passed through an optical filter with bandwidth B_o. Let the measured noise power be P. The quotient $P/(pB_o)$

equals the energy per mode in the frequency domain. The photon number per mode $\mu = \frac{P}{pB_o hf}$ is obtained after dividing it by the photon energy hf. With $F_z = \mu/G$ the noise figure is therefore

$$F = 1 + \frac{P}{pB_o hfG}. \tag{3.98}$$

If a polarizer is placed before the power meter only one polarization mode is evaluated and $p = 1$ holds.

Caution: The hitherto prevailing noise figure definition [47] is based on photon number fluctuations (pnf) in the limit of large photon numbers, $F_{\text{pnf}} = G^{-1} + 2\left(1 - G^{-1}\right)n_{sp} = G^{-1}(1 + 2\mu) = \frac{1}{G}\left(1 + \frac{2P}{pB_o hf}\right)$. With (3.98) we find the conversion formulas $F = 1 + \left(F_{\text{pnf}} - G^{-1}\right)/2$, $F_{\text{pnf}} = 2(F - 1) + G^{-1}$. In an ideal amplifier F_{pnf} assumes the value 2, in an attenuator the value G^{-1}. It is easy to show that F_{pnf} also fulfills the cascading formula (3.93). However, F_{pnf} does not describe the noise of optical amplifiers in a straightforward manner: The calculation of the input-referred number of noise photons per mode $\mu/G = F - 1 = \left(F_{\text{pnf}} - G^{-1}\right)/2$ requires knowledge not only of F_{pnf} but also of the gain *G*! By the way, F_{pnf} also differs from the classical microwave noise figure.

3.2.3 Intensity Distribution

Photons can only be observed by detecting them. When an optical signal is detected the particle aspect causes a granularity, the shot noise. It is comparatively very small if amplification *G* and hence photon numbers *n* are high enough; in particular the standard deviation of a Poisson distribution is the square root of the mean. If granularity is neglected then the photon number *n* which is a discrete RV must be replaced by a continuous RV *x*. For very high *G* the noise generated at the amplifier input dominates the noise at the amplifier output and the shot noise.

Assume the photon distribution can be written as the so-called Poisson transform [46]

$$P(n) = \int_0^\infty p_x(x) e^{-x} \frac{x^n}{n!} dx \tag{3.99}$$

of a PDF $p_x(x)$ that belongs to a nonnegative RV *x*. If we vary the optical power, e.g. by changing the gain *G* of an optical amplifier, $P(n)$ and $p_x(x)$ will depend on G, which means it is not possible to find a limit $\lim_{G\to\infty} p_x(x)$. Yet G mainly scales the x range so that the PDF $p_{\tilde{x}}(\tilde{x})$ of a normalized variable $\tilde{x}=x/G$ depends only weakly on G, and $\lim_{G\to\infty} p_{\tilde{x}}(\tilde{x})$ exists. From equal probabilities $p_x(x)dx = p_{\tilde{x}}(\tilde{x})d\tilde{x}$ it follows $p_x(x)=G^{-1}p_{\tilde{x}}(x/G)$. This allows to write the photon distribution as

$$P(n)=\int_0^\infty p_{\tilde{x}}(\tilde{x})e^{-\tilde{x}G}\frac{(\tilde{x}G)^n}{n!}d\tilde{x}\,. \tag{3.100}$$

The normalization is undone thereby. The quantity x may be called an intensity at the output of the optical amplifier, whereas the normalized variable $\tilde{x}$ is an input-referred intensity. The MGF of $P(n)$ is $M_n\left(e^{-s}\right)=\left\langle e^{-sn}\right\rangle$. After inserting (3.100) the expression $\lim_{G\to\infty} M_n\left(e^{-s/G}\right)=\lim_{G\to\infty}\sum_{n=0}^{\infty}P(n)e^{-(s/G)n}$ becomes

$$\lim_{G\to\infty} M_n\left(e^{-s/G}\right)=M_{\tilde{x}}\left(e^{-s}\right), \tag{3.101}$$

i.e., the MGF of the input-referred intensity $\tilde{x}$. The inverse of (3.100) consists in calculating $M_n\left(e^{-s}\right)$ from $P(n)$, applying (3.101), and backtransforming $M_{\tilde{x}}\left(e^{-s}\right)$ to obtain $p_{\tilde{x}}(\tilde{x})$ (Table 3.3). Eqn. (3.100) is the continuous representation of $P(n)=\sum_i P(n|(\tilde{x}_iG))P(\tilde{x}G=\tilde{x}_iG)$, where the conditional probability $P(n|(\tilde{x}_iG))$ is a Poisson distribution with mean $\tilde{x}_iG$.

For a Laguerre MGF (3.85) we set $\tilde{\mu}=\lim_{G\to\infty}\mu G^{-1}=n_{sp}$, $\tilde{\mu}_0=\lim_{G\to\infty}\mu_0G^{-1}$, and find

$$M_{\tilde{x}}\left(e^{-s}\right)=(1+\tilde{\mu}s)^{-N}e^{\frac{-\tilde{\mu}_0 s}{1+\tilde{\mu}s}}\,. \tag{3.102}$$

The corresponding PDF

$$p_{\tilde{x}}(\tilde{x})=\frac{1}{\tilde{\mu}}\left(\frac{\tilde{x}}{\tilde{\mu}_0}\right)^{\frac{N-1}{2}}e^{-\frac{\tilde{\mu}_0+\tilde{x}}{\tilde{\mu}}}I_{N-1}\left(\frac{2\sqrt{\tilde{x}\tilde{\mu}_0}}{\tilde{\mu}}\right) \tag{3.103}$$

is a noncentral χ^2_{2N} (chi square) PDF with $2N$ degrees-of-freedom ($\tilde{x}\geq 0$). For $\tilde{\mu}_0=0$, or if one starts from the central negative binomial distribution MGF (3.83), the MGF

$$M_{\tilde{x}}\left(e^{-s}\right)=(1+\tilde{\mu}s)^{-N} \tag{3.104}$$

of a central χ^2_{2N} or Gamma PDF

$$p_{\tilde{x}}(\tilde{x}) = \frac{1}{\Gamma(N)} \tilde{\mu}^{-N} x^{N-1} e^{-\tilde{x}/\tilde{\mu}} \tag{3.105}$$

is obtained. It is also found if one inserts $\lim_{|z| \to 0} I_\nu(z) = \frac{1}{\Gamma(\nu+1)} \left(\frac{z}{2}\right)^\nu$ into (3.103). Similar to the addition property of negative binomial distributions, the sum $\tilde{x} = \tilde{x}_1 + \tilde{x}_2$ of two statistically independent RVs $\tilde{x}_1$, $\tilde{x}_2$ with χ^2 PDFs and the same noise $\tilde{\mu} = n_{sp}$ is a new χ^2 PDF with $\tilde{\mu}_0 = \tilde{\mu}_{0,1} + \tilde{\mu}_{0,2}$, $N = N_1 + N_2$, as can be verified by multiplying the corresponding MGFs.

Table 3.3 Eliminating and adding shot noise

$P(n)$:	$\Rightarrow M_n(e^{-s}) \Rightarrow$	$p_{\tilde{x}}(\tilde{x})$:
Poisson distribution, Central negative binomial distribution, Noncentral negative binomial distribution	$M_{\tilde{x}}(e^{-s}) = \lim_{G \to \infty} M_n(e^{-s/G}) \Rightarrow$ Eliminate shot noise by amplification, normalize with respect to G. $\Leftarrow P(n) = \int_0^\infty p_{\tilde{x}}(\tilde{x}) e^{-\tilde{x}G} \frac{(\tilde{x}G)^n}{n!} d\tilde{x} \Leftarrow$ Add shot noise, undo normalization ($x = \tilde{x}G$).	Constant (Dirac function), Central χ^2 distribution, Noncentral χ^2 distribution

For comparison we assume a Poisson distribution with expectation value $\mu_0 = G\tilde{\mu}_0$. Using (3.101) we find

$$M_{\tilde{x}}(e^{-s}) = e^{-\tilde{\mu}_0 s} \tag{3.106}$$

with a corresponding PDF

$$p_{\tilde{x}}(\tilde{x}) = \delta(\tilde{x} - \tilde{\mu}_0). \tag{3.107}$$

It belongs to a constant $\tilde{\mu}_0$. These findings justify assumption (3.99) and can be interpreted: When the light intensity is subject to a detection process, shot noise is added which is Poisson distributed for each value the intensity may assume with a certain probability density. The light intensity may carry just noise (central χ^2), signal and noise (noncentral χ^2), or just signal (constant distribution). The detection, modeled by the Poisson transform, results in central or noncentral negative binomial, or Poisson distributions, respectively. All these can be found in Table 3.2.

Central and noncentral negative binomial distributions with $2N = 4$ are plotted logarithmically for an ideal amplifier ($n_{sp} = 1$) in Fig. 3.17, but for easier comparison normalized probabilities $\log(GP(nG))$ are displayed instead of $\log(P(n))$. A value $\tilde{\mu}_0 = 81.4$ was chosen. The higher the gain $G = 1, 2, 4, \infty$, the higher the noise. For infinite G, χ_4^2 distributions result with a $\text{BER} = 10^{-9}$. The mean photon number for equiprobable 0 and 1 is $\tilde{\mu}_0/2 = 40.7$.

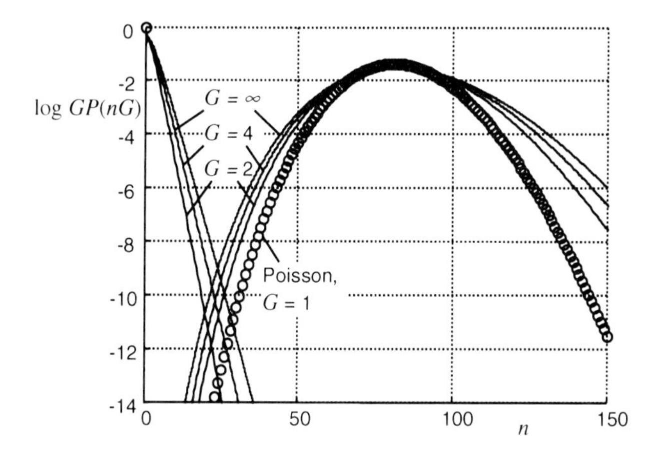

Fig. 3.17 Transition of normalized central and noncentral distributions from Poisson via negative binomial to χ_4^2

In the noise-free case $G = 1$, Poisson distributions (o symbols) result. We calculate the sensitivity of such a hypothetical receiver without optical preamplifier and without thermal noise. Suitably normalized, the decision variable equals the number n of received photons. The decision threshold is chosen as, say, 1/2. So a transmitted 0 can never be falsified to become a 1, and $n \geq 1$ received photons are interpreted as a transmitted 1. The

$$\text{BER} = P_1 P(n = 0) = P_1 e^{-\mu_0} = (1/2) e^{-\mu_0} . \tag{3.108}$$

is the probability $P_1 = 1/2$ of a transmitted 1, multiplied by the conditional probability $P(n = 0)$ of 0 received photons in spite of a transmitted 1 with a photon number expectation value μ_0. For $\text{BER} = 10^{-9}$ an expectation value of

$\mu_0 \approx 20$ photons is needed for the ones. On average half as many suffice. The 10 photons/bit for $\mathrm{BER} = 10^{-9}$ are the quantum limit of a direct-detection optical receiver for binary on-off amplitude shift keying (ASK). It is about 4 times better than with ideal optical preamplifier.

Returning to the χ_4^2 distributions, we compare the obtained data with a simple Gaussian approximation. Table 3.2 yields means and variances of χ^2 distributions which result in

$$Q = \frac{\left(N\tilde{\mu} + \tilde{\mu}_0\right) - N\tilde{\mu}}{\sqrt{N\tilde{\mu}^2 + 2\tilde{\mu}\tilde{\mu}_0} + \sqrt{N\tilde{\mu}^2}}, \quad \mathrm{BER} = \frac{1}{2}\operatorname{erfc}\left(Q/\sqrt{2}\right). \tag{3.109}$$

With $N = 2$, $\tilde{\mu}_0 = 81.4\tilde{\mu}$ a BER = $5.6 \cdot 10^{-9}$ is obtained which is in reasonable agreement with the exact calculation.

The intensity is here the expectation value with respect to the detection process of the photon number. This expectation value normally is a RV itself with respect to ASE, which originates mainly from the amplifier input.

A noncentral χ_{2N}^2 PDF (3.103), (3.105) describes a RV $\chi_{2N}^2 \equiv \tilde{x} = \sum_{i=1}^{2N} u_i^2$ where u_i are independent Gaussian RVs with identical variances $\sigma_{u_i}^2 = \tilde{\mu}/2$. The square sum of the expectation values $\langle u_i \rangle = \sqrt{\tilde{\mu}_{0,i}}$ is $\tilde{\mu}_0 = \sum_{i=1}^{2N} \langle u_i \rangle^2 = \sum_{i=1}^{2N} \tilde{\mu}_{0,i}$.

Proof: It is sufficient to consider a RV $\tilde{x} = z^2$ where z is a Gaussian RV with mean $\sqrt{\tilde{\mu}_0}$, variance $\sigma_z^2 = \tilde{\mu}/2$ and PDF $p_z(z) = \frac{1}{\sqrt{\pi\tilde{\mu}}} e^{-\left(z - \sqrt{\tilde{\mu}_0}\right)^2/\tilde{\mu}}$. Transformation at a square-law device results in

$$\begin{aligned} p_{\tilde{x}}(\tilde{x}) d\tilde{x} &= \sum p_z(z) dz = \sum p_z(z(\tilde{x})) \left| \frac{dz}{d\tilde{x}} \right| d\tilde{x} \\ &= \frac{1}{\sqrt{\pi\tilde{\mu}}} \frac{1}{2\sqrt{\tilde{x}}} \left(e^{-\left(\sqrt{\tilde{x}} - \sqrt{\tilde{\mu}_0}\right)^2/\tilde{\mu}} + e^{-\left(-\sqrt{\tilde{x}} - \sqrt{\tilde{\mu}_0}\right)^2/\tilde{\mu}} \right) d\tilde{x} \end{aligned}.$$

The term in front of $d\tilde{x}$ is the desired $p_{\tilde{x}}(\tilde{x})$. Using $I_n(w) = j^{-n} J_n(jw)$ (Gradstein/Ryshik 8.406 3.), $J_{-1/2}(u) = \sqrt{\frac{2}{\pi u}} \cos u$ (Gradstein/Ryshik 8.464 2.) and $\cos jr = \cosh r = \frac{1}{2}\left(e^r + e^{-r}\right)$ we show that the already-known noncentral χ^2 PDF with one degree-of-freedom ($2N = 1$) is identical with $p_{\tilde{x}}(\tilde{x})$,

$$\frac{1}{\tilde{\mu}}\left(\frac{\tilde{x}}{\tilde{\mu}_0}\right)^{-1/4} e^{-\frac{\tilde{\mu}_0+\tilde{x}}{\tilde{\mu}}} I_{-1/2}\left(\frac{2\sqrt{\tilde{x}\tilde{\mu}_0}}{\tilde{\mu}}\right) = \frac{1}{\tilde{\mu}}\left(\frac{\tilde{x}}{\tilde{\mu}_0}\right)^{-1/4} e^{-\frac{\tilde{\mu}_0+\tilde{x}}{\tilde{\mu}}} j^{1/2} J_{-1/2}\left(j\frac{2\sqrt{\tilde{x}\tilde{\mu}_0}}{\tilde{\mu}}\right)$$

$$= \frac{1}{\tilde{\mu}}\left(\frac{\tilde{x}}{\tilde{\mu}_0}\right)^{-1/4} e^{-\frac{\tilde{\mu}_0+\tilde{x}}{\tilde{\mu}}} j^{1/2} \sqrt{\frac{2}{\pi j \frac{2\sqrt{\tilde{x}\tilde{\mu}_0}}{\tilde{\mu}}}} \cos\left(j\frac{2\sqrt{\tilde{x}\tilde{\mu}_0}}{\tilde{\mu}}\right)$$

$$= \frac{1}{2\tilde{\mu}}\left(\frac{\tilde{x}}{\tilde{\mu}_0}\right)^{-1/4} e^{-\frac{\tilde{\mu}_0+\tilde{x}}{\tilde{\mu}}} \sqrt{\frac{1}{\pi \frac{\sqrt{\tilde{x}\tilde{\mu}_0}}{\tilde{\mu}}}} \left(e^{\frac{2\sqrt{\tilde{x}\tilde{\mu}_0}}{\tilde{\mu}}} + e^{-\frac{2\sqrt{\tilde{x}\tilde{\mu}_0}}{\tilde{\mu}}}\right) = p_{\tilde{x}}(\tilde{x}) .$$

Together with the previously discussed addition property of χ^2 RVs this completes the proof.

Since normalization is arbitrary we understand the light intensity to be the squared magnitude $\left|\underline{\tilde{\mathbf{E}}}\right|^2$ of an electrical field $\underline{\tilde{\mathbf{E}}} = \underline{\mathbf{E}}/\sqrt{G}$ which is normalized with respect to $\sqrt{G}$. $\underline{\tilde{\mathbf{E}}}$ is a carrier with constant amplitude, accompanied by Gaussian zero-mean noise u_i with $\sigma_{\tilde{u}_i}^2 = \tilde{\mu}/2$ in two quadratures in each polarization,

$$\underline{\tilde{\mathbf{E}}}(t) = \left(\left(\sqrt{\tilde{\mu}_0/M} + \tilde{u}_1 + j\tilde{u}_2\right)\underline{\mathbf{e}}_1 + \left(\tilde{u}_3 + j\tilde{u}_4\right)\underline{\mathbf{e}}_2\right)e^{j\omega t} . \tag{3.110}$$

Here $\mathbf{e}_1$ is the unit vector of the signal polarization, and $\underline{\mathbf{e}}_2$ is the orthogonal polarization with $\underline{\mathbf{e}}_1^+\underline{\mathbf{e}}_2 = 0$. For the time being, let $M = 1$. One χ^2-distributed intensity sample

$$\left|\underline{\tilde{\mathbf{E}}}(t)\right|^2 = \left(\sqrt{\tilde{\mu}_0/M} + \tilde{u}_1\right)^2 + \tilde{u}_2^2 + \tilde{u}_3^2 + \tilde{u}_4^2 \tag{3.111}$$

has 4 degrees-of-freedom. If $\underline{\mathbf{e}}_2$ is blocked by a suitably aligned polarizer that is placed behind the optical filter, $\tilde{u}_3 = \tilde{u}_4 = 0$ results, and just 2 degrees-of-freedom remain.

3.2.4 Receivers for Amplitude Shift Keying

We consider now a binary optical ASK receiver (Fig. 3.18). Behind the optical receiver there is an optical filter with an impulse response which is a cosine oscillation having the frequency of the received signal and a rectangular envelope of duration $\tau_1 = 1/B_o$. B_o is the optical bandwidth of that filter. For constant signal, statistically independent normalized fields $\underline{\tilde{\mathbf{E}}}(t + i\tau_1)$ are therefore obtained every time interval τ_1. After photodetection, statistically independent intensity samples $\left|\underline{\tilde{\mathbf{E}}}(t + i\tau_1)\right|^2$ result.

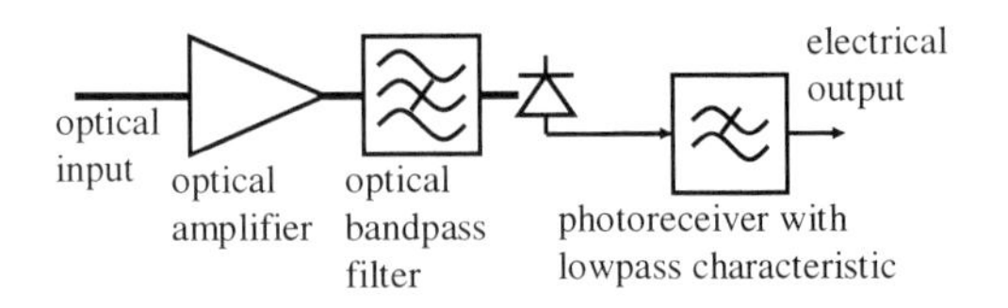

Fig. 3.18 Optical receiver for intensity modulation (amplitude shift keying). A polarizer could be, but normally is not, placed behind the amplifier in order to block noise orthogonal to the signal.

In our model the electrical part of the optical receiver should possess so little thermal noise that it is possible to give it an impulse response equal to a comb of $M=\tau_2/\tau_1$ Dirac pulses with equal amplitudes, spaced by τ_1. In that case the baseband filter is not a lowpass filter but has infinite bandwidth! The Dirac comb filter can in good approximation be exchanged against a filter with a continuous impulse response of length τ_2 or better $\sqrt{\tau_2^2-\tau_1^2}$, unless M is very small. In order to avoid intersymbol interference, τ_2 is of course chosen smaller than one bit duration T, but not much smaller since signal energy would otherwise be lost. Suitably normalized, the signal at the decision circuit input is therefore

$$\tilde{x}=\sum_{i=1}^{M}\left|\underline{\tilde{\mathbf{E}}}(t+i\tau_1)\right|^2 . \tag{3.112}$$

with PDF (3.103). In the hypothetical, noise-free case $\langle\tilde{x}\rangle=\tilde{\mu}_0$ holds. The decision variable has $2N=2pM$ degrees-of-freedom, where the number of polarizations is usually $p=2$. For $\tilde{\mu}_0=0$ the PDF is given by (3.105).

Thermal noise in the electrical part of the optical receiver must also be taken into account. Fig. 3.19 shows $\log P(n)$ for $G=16, 64, 256$, $\tilde{\mu}_0=100$, $2N=8$. However, Gaussian thermal noise with $\sigma=523$ has been added which results in BER $= 7\cdot 10^{-2}$, $1\cdot 10^{-5}$, $2.5\cdot 10^{-10}$, respectively. It is taken into account by convolving the probability distributions or multiplying the corresponding MGFs. A high gain is needed to make the relative thermal noise contribution insignificant. The chosen thermal noise corresponds to a receiver with a bandwidth of 7 GHz and a one-sided thermal noise of $14\,\text{pA}/\sqrt{\text{Hz}}$, suitable to receive 10 Gbit/s, and is calculated as follows: The squared noise current is $\left(14\,\text{pA}/\sqrt{\text{Hz}}\right)^2\cdot 7\,\text{GHz}=1.37\cdot 10^{-12}\,\text{A}^2$. Its standard deviation is $\sqrt{1.37\cdot 10^{-12}\,\text{A}^2}=1.17\cdot 10^{-6}\,\text{A}$. The impulse response is about $1/(2\cdot 7\,\text{GHz})=71\,\text{ps}$ long. The standard deviation of the charge is $1.17\cdot 10^{-6}\,\text{A}\cdot 71\,\text{ps}=8.34\cdot 10^{-17}\,\text{As}$. This is $\sigma=523$ times the electron charge, and each electron is equivalent to a detected photon.

For $G\to\infty$, i.e., χ_{2N}^2 PDFs, the BER vs. $10\log(\tilde{\mu}_0/(2\tilde{\mu}))$, i.e. the mean photon number divided by $\tilde{\mu}=n_{sp}$ and expressed in dB, is shown in Fig. 3.20. It is indeed the signal-to-noise ratio $\text{SNR}=\tilde{\mu}_0/(2\tilde{\mu})$ of the ASK receiver which must be considered: The substitutions $\tilde{\mu}_0 \to a\tilde{\mu}_0$, $\tilde{\mu} \to a\tilde{\mu}$, $\tilde{x} \to a\tilde{x}$ leave a χ^2_{2N} PDF unchanged, except for a magnitude scaling factor $1/a$ which is needed to keep the probability $p_{\tilde{x}}(\tilde{x})d\tilde{x}$ unaffected by the substitutions.

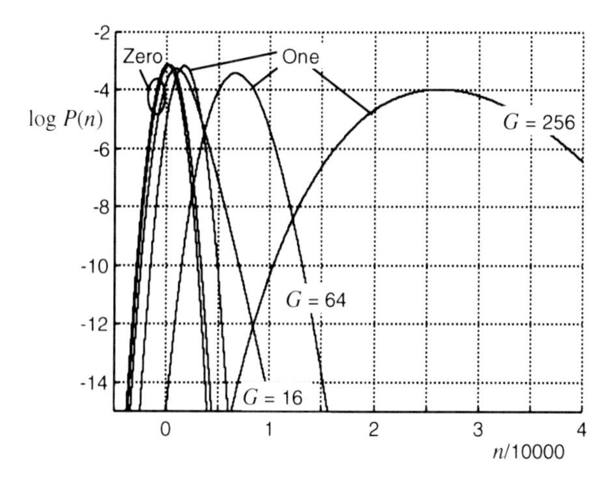

Fig. 3.19 Photon distributions with thermal noise added

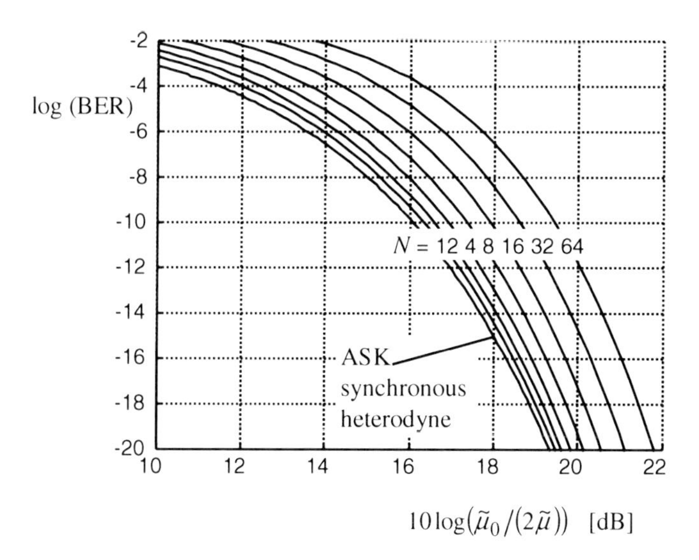

Fig. 3.20 BER vs. SNR with number of modes as a parameter. BER of ASK synchronous heterodyne receiver added for comparison.

Fairly moderate penalties occur for rising N. The BER curves become steeper as N increases. With more detail this behavior is investigated in Fig. 3.21. It shows $10\log(\tilde{\mu}_0/(2\tilde{\mu}))$ vs. N for several constant BER.

A simplistic, not stringent explanation follows: The decision variable for a transmitted one is

$$\begin{aligned}&\sum_{i=1}^{M}\left(\left(\sqrt{\tilde{\mu}_0/M}+\tilde{u}_{i1}\right)^2+\tilde{u}_{i2}^{\,2}+\tilde{u}_{i3}^{\,2}+\tilde{u}_{i2}^{\,2}\right)\\&=\tilde{\mu}_0+\sum_{i=1}^{M}2\sqrt{\tilde{\mu}_0/M}\,\tilde{u}_{i1}+\sum_{i=1}^{M}\left(\tilde{u}_{i1}^{\,2}+\tilde{u}_{i2}^{\,2}+\tilde{u}_{i3}^{\,2}+\tilde{u}_{i2}^{\,2}\right)\end{aligned}. \quad (3.113)$$

The first term on the right side is the desired signal. The second term is signal-ASE beat noise, the third is ASE-ASE beat noise. For increasing optical bandwidth, i.e. rising M, the standard deviation of signal-ASE beat noise stays constant while that of ASE-ASE beat noise rises with $\sqrt{M}$. For large M the ASE-ASE beat noise dominates. For a transmitted zero the first and second terrm disappear but the signal-ASE beat noise of the ones dominates Also, the noise will be Gaussian due to the central limit theorem. In order to keep the BER constant the signal must therefore rise with $\sqrt{M}$. This can be seen in the right part of Fig. 3.21.

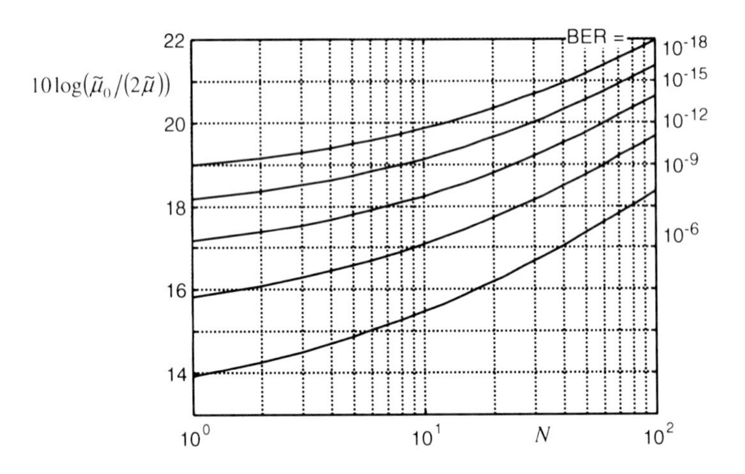

Fig. 3.21 SNR in dB vs. number of modes with BER as a parameter

If there is a finite power ratio between zeros and ones of the optical signal, a transmitted zero will also have a noncentral binomial or χ_{2N}^2 distribution. For the latter case resulting penalties are displayed in Fig. 3.22.

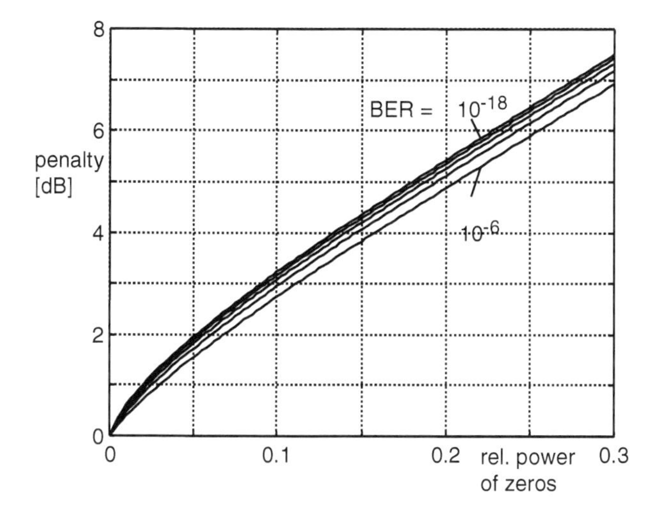

Fig. 3.22 SNR penalty (referred to mean signal power) as a function of power ratio between ones and zeros

Problem: An optical receiver consists of a polarization-independent optical amplifier with $G \to \infty$ and $F = 2$. It is followed by an optical filter having a bandwidth $B_o = 30\,\text{GHz}$, a photodiode and finally an electrical regenerator. The impulse response of photodiode and electrical amplifier is continuous with an approximate length $\tau_2' = 60\,\text{ps}$. Binary 10 Gbit/s NRZ signals are transmitted at $\lambda = 1550\,\text{nm}$. Which mean optical power is approximately needed at the optical amplifier input in order to achieve a BER = 10^{-15} ?

Solution: Since the impulse response does not consist of a Dirac comb we set the length of the "impulse response" that is obtained by the combination of optical filter, photodiode and electrical filter to be $\tau_2 = \sqrt{\tau_1^2 + \tau_2'^2} = \sqrt{B_o^{-2} + \tau_2'^2}$ = 69 ps. Note that in reality it is not an impulse response because the photodiode acts as a squarer! The mode number, with two noisy polarizations, is $N \approx 2\tau_2/\tau_1 \approx 4.1$. Fig. 3.21 yields a BER = 10^{-15} for $10\log(\tilde{\mu}_0/(2\tilde{\mu}))$ = 18.7 and hence $\tilde{\mu}_0/(2\tilde{\mu})$ = 75. For $F = 2$, $\tilde{\mu} = 1$ holds. Since ones and zeros are equally probable, $\tilde{\mu}_0/2$ is the mean photon number averaged over these symbols. The $\tilde{\mu}_0/2 = 75$ photons must arrive in $\tau_2 = 69\,\text{ps}$. The mean power is $P = \frac{hc}{\lambda}\frac{75}{69\,\text{ps}}$ = $1.4 \cdot 10^{-7}$ W or −38.5 dBm.

In reality one reaches roughly $F = 2.5$ or $\tilde{\mu} = 1.5$. This causes a penalty of $10\log\tilde{\mu} = 1.8$ dB. Also, a penalty of 1 ... 3 dB is to be expected due to ISI, nonideal decision circuit (e.g., with hysteresis), and nonideal extinction (see Fig. 3.22). This value depends very much on the hardware. All things considered, a sensitivity of about −35 dBm can be obtained.

3.2.5 Receivers for Differential Phase Shift Keying

Binary differential phase shift keying (DPSK) increases the receiver sensitivity. As will be seen, a recursive modulo-2 addition is needed at the transmitter to create the transmitted symbol $c_i = c_{i-1} \oplus b_i$. According to modulo-2 addition rules the binary data symbol b_i itself is $b_i = c_i \oplus c_{i-1}$. Let now c_i and data symbols b_i be bipolar, ±1. Equations $c_i = c_{i-1} \oplus b_i$ and $b_i = c_i \oplus c_{i-1}$ must hence be replaced by $c_i = b_i c_{i-1}$ and $b_i = c_i c_{i-1}$, respectively. This operation is performed in a Mach-Zehnder interferometer at the receiver side (Fig. 3.23), which is the reason for encoding.

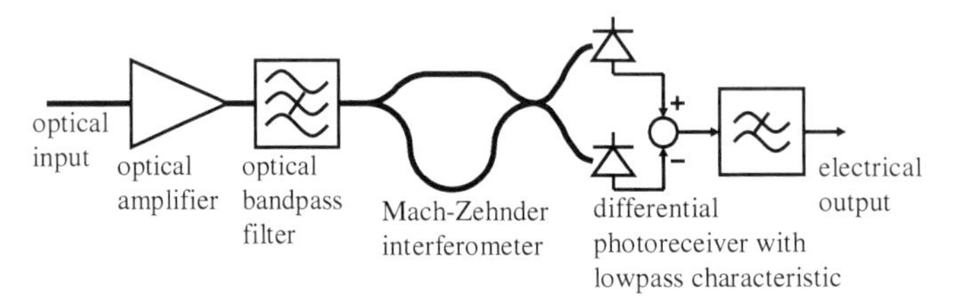

Fig. 3.23 Optical receiver for DPSK signals with Mach-Zehnder interferometer

We have learnt that an optical amplifier adds Gaussian noise in phase and in quadrature to the electric field. Upon detection, shot noise is added to the calculated intensity via the Poisson transform. Here, propagation through the Mach-Zehnder interferometer (MZI) must be considered between the addition of optical amplifier noise and the detection. The wave aspect is to be used, and statistical (in)dependence between the intensities at the two interferometer outputs must of course be investigated. The normalized field at the optical filter output is given by

$$\underline{\tilde{\mathbf{E}}}(t) = \left(\left(c_i\sqrt{\tilde{\mu}_0/M} + \tilde{u}_1 + j\tilde{u}_2\right)\underline{\mathbf{e}}_1 + \left(\tilde{u}_3 + j\tilde{u}_4\right)\underline{\mathbf{e}}_2\right)e^{j\omega t}. \tag{3.114}$$

Both output arms of the splitter in the MZI carry electric fields $\underline{\tilde{\mathbf{E}}}(t)/\sqrt{2}$. In the lower arm the wave is delayed by one symbol duration T. The delay is fine-tuned so that it is also an integer multiple of optical half periods minus 1/4 of a period, which results in $\pm j\underline{\tilde{\mathbf{E}}}(t-T)/\sqrt{2}$. The lower sign is valid for an odd integer, the upper for an even integer, which for simplicity will be assumed from now on. The transfer matrix of the output coupler is $\frac{1}{\sqrt{2}}\begin{bmatrix} 1 & -j \\ -j & 1 \end{bmatrix}$. The output fields at the two photodiodes are therefore $\underline{\tilde{\mathbf{E}}}_1(t) = \frac{1}{2}\left(\underline{\tilde{\mathbf{E}}}(t) + (-j)j\underline{\tilde{\mathbf{E}}}(t-T)\right)$ and $\underline{\tilde{\mathbf{E}}}_2(t) = \frac{1}{2}\left(-j\underline{\tilde{\mathbf{E}}}(t) + j\underline{\tilde{\mathbf{E}}}(t-T)\right)$. For the difference of intensities at the two photodiodes we can write

$$\left|\underline{\tilde{\mathbf{E}}}_1(t)\right|^2 - \left|\underline{\tilde{\mathbf{E}}}_2(t)\right|^2 = \mathrm{Re}\left(\underline{\tilde{\mathbf{E}}}^+(t)\underline{\tilde{\mathbf{E}}}(t-T)\right). \tag{3.115}$$

Polarization matching is achieved by a polarization-independent design of the MZI. In the noiseless case and for real c_i the expression (3.115) equals $c_i c_{i-1}(\tilde{\mu}_0/M)$ as required. Using the definitions

$$\tilde{\mathbf{v}}_k = \mathbf{R}\tilde{\mathbf{u}}_k \qquad \tilde{\mathbf{v}}_k = \begin{bmatrix} \tilde{v}_{1,k} \\ \tilde{v}_{2,k} \end{bmatrix} \qquad \tilde{\mathbf{u}}_k = \begin{bmatrix} \tilde{u}_k(t) \\ \tilde{u}_k(t-T) \end{bmatrix} \qquad \mathbf{R} = \frac{1}{2}\begin{bmatrix} 1 & 1 \\ 1 & -1 \end{bmatrix}$$
$$C_1 = \frac{1}{2}(c_i + c_{i-1}) \qquad C_2 = \frac{1}{2}(c_i - c_{i-1}) \tag{3.116}$$

the fields at the photodiodes can be rewritten ($m = 1,2$) as

$$\underline{\tilde{\mathbf{E}}}_m \propto \left(C_m\sqrt{\tilde{\mu}_0/M} + \tilde{v}_{m,1} + j\tilde{v}_{m,2}\right)\underline{\mathbf{e}}_1 + \left(\tilde{v}_{m,3} + j\tilde{v}_{m,4}\right)\underline{\mathbf{e}}_2 . \tag{3.117}$$

All pairs of RVs $\tilde{u}_k$, $\tilde{u}_l$ ($k,l \in \{1;2;3;4\}$) are pairwise independent, and so are $\tilde{u}_k(t)$, $\tilde{u}_k(t-T)$ because the optical bandpass filter impulse response has a length less than *T*. The same holds for $\tilde{v}_{1,k}$, $\tilde{v}_{2,k}$ because **R** is proportional to an orthogonal matrix and the $\tilde{u}_k(t)$, $\tilde{u}_k(t-T)$ are Gaussian. The intensities at both photodiodes are therefore each χ_4^2-distributed. Note that $\sigma^2_{\tilde{v}_{m,k}} = \sigma^2_{\tilde{u}_k}/2$. This means the term $\tilde{\mu}$ which equals $\sigma^2/2$ of each Gaussian RV must be replaced by $\tilde{\mu}/2$. The difference of these statistically independent intensities is found by multiplying the MGF of one of them by the MGF of the MGF other where s has been replaced by $-s$.

Proof: The difference can be expressed as a sum, $\tilde{x} = \tilde{x}_1 - \tilde{x}_2 = \tilde{x}_1 + \tilde{x}_3$ with $\tilde{x}_3 = -\tilde{x}_2$,

$$p_{\tilde{x}_3}(\tilde{x}_3) = p_{\tilde{x}_2}(-\tilde{x}_3),\ M_{\tilde{x}_3}\left(e^{-s}\right) = \int_{-\infty}^{\infty} p_{\tilde{x}_3}(\tilde{x}_3)e^{-s\tilde{x}_3}d\tilde{x}_3 = \int_{-\infty}^{\infty} p_{\tilde{x}_2}(-\tilde{x}_3)e^{-s\tilde{x}_3}d\tilde{x}_3 = M_{\tilde{x}_2}\left(e^{s}\right).$$

The WDF of statistically independent RVs is obtained by convolution of the individual PDFs,

$$p_{\tilde{x}}(\tilde{x}) = \int_{\tilde{x}=\tilde{x}_1+\tilde{x}_3} p_{\tilde{x}_1,\tilde{x}_3}(\tilde{x}_1, \tilde{x}-\tilde{x}_1)d\tilde{x}_1 = \int_{-\infty}^{\infty} p_{\tilde{x}_1}(\tilde{x}_1)p_{\tilde{x}_3}(\tilde{x}-\tilde{x}_1)d\tilde{x}_1 = p_{\tilde{x}_1}(\tilde{x}) * p_{\tilde{x}_3}(\tilde{x}),\ \text{or}$$

multiplication of the corresponding MGFs,

$$M_{\tilde{x}}\left(e^{-s}\right) = M_{\tilde{x}_1}\left(e^{-s}\right)M_{\tilde{x}_3}\left(e^{-s}\right) = M_{\tilde{x}_1}\left(e^{-s}\right)M_{\tilde{x}_2}\left(e^{s}\right).$$

In the ideal case $1-|C_2| = |C_1| \in \{0;1\}$, and one of the χ_4^2-distributed RVs is central and the other noncentral. A summation happens in the previously defined lowpass filter. As a consequence the MGF of the electrical detection signal is

$$M_{\tilde{x}}\left(e^{-s}\right) = \left(1-(\tilde{\mu}/2)^2 s^2\right)^{-N} e^{\frac{\mp\tilde{\mu}_0 s}{1\pm(\tilde{\mu}/2)s}}, \tag{3.118}$$

where $\tilde{\mu}/2$ has been substituted for $\tilde{\mu}$. The upper signs refer to $b_i = 1$ ($c_i = c_{i-1}$, $|C_1| = 1$), the lower ones to $b_i = -1$ ($c_i = -c_{i-1}$, $|C_1| = 0$). The two possible PDFs are mirror-symmetric. The optimum decision threshold is therefore zero. We now assume the upper signs to hold. The BER can be calculated using

$$\text{BER} = \int_{-\infty}^{0} p_{\tilde{x}}(\tilde{x}) d\tilde{x} \,. \tag{3.119}$$

This can be solved, with the result

$$\text{BER} = e^{-\tilde{\mu}_0/\tilde{\mu}} \frac{1}{2^{2N-1}} \sum_{k=0}^{N-1} (\tilde{\mu}_0/\tilde{\mu})^k \frac{1}{k!} \sum_{n=0}^{N-1-k} \binom{2N-1}{n}. \tag{3.120}$$

Proof (after [53]): Using the inverse Fourier transform, the BER is

$$\begin{aligned}\text{BER} &= \int_{-\infty}^{0} \frac{1}{2\pi} \int_{-\infty}^{\infty} M_{\tilde{x}}\left(e^{-j\omega}\right) e^{j\omega\tilde{x}} d\omega d\tilde{x} = \frac{1}{2\pi} \int_{-\infty-j\varepsilon}^{\infty-j\varepsilon} \frac{1}{j\omega} M_{\tilde{x}}\left(e^{-j\omega}\right) d\omega \\ &= \frac{1}{2\pi} \int_{-\infty-j\varepsilon}^{\infty-j\varepsilon} \frac{1}{j\omega} \left(1 + (\tilde{\mu}/2)^2 \omega^2\right)^{-N} e^{\frac{-j\omega\tilde{\mu}_0}{1+j\omega(\tilde{\mu}/2)}} d\omega\end{aligned}$$

A small positive ε was inserted, which means $e^{j\omega\tilde{x}} \to 0$ for $\tilde{x} \to -\infty$. This allowed the immediate integration over $\tilde{x}$. In the remaining integral we may close the integration contour by a half circle in the mathematically positive sense with a radius approaching infinity because the integrand vanishes for $|\omega| \to \infty$ with sufficient power. We substitute

$$\omega = -\frac{j2}{\tilde{\mu}} \frac{1-p}{1+p}, \quad \frac{d\omega}{dp} = \frac{j}{\tilde{\mu}} \frac{4}{(1+p)^2}$$

and obtain after some manipulations

$$\text{BER} = -e^{-\tilde{\mu}_0/\tilde{\mu}} \frac{1}{2^{2N-1}} \frac{1}{2\pi j} \oint \frac{(1+p)^{2N-1}}{(1-p)p^N} e^{(\tilde{\mu}_0/\tilde{\mu})p} dp$$

The new closed integration contour encloses the poles at $p = 0$ but not the one at $p = 1$, and it runs in the mathematically negative sense.

In the following we let the integration contour run in the mathematically positive sense and therefore drop the leading minus sign. The binomial $(1+p)^{2N-1}$ is also evaluated by a sum.

$$\text{BER} = e^{-\tilde{\mu}_0/\tilde{\mu}} \frac{1}{2^{2N-1}} \sum_{n=0}^{2N-1} \binom{2N-1}{n} \frac{1}{2\pi j} \oint \frac{p^{n-N}}{1-p} e^{(\tilde{\mu}_0/\tilde{\mu})p} dp$$

The remaining integral is evaluated using residues. The upper summation index can be lowered to $N-1$ because a $(N-n)$-fold pole at $p = 0$ exists only for $n = 0,1,...,N-1$. The integral is replaced by $2\pi j R_n$,

$$\mathrm{BER}=e^{-\tilde{\mu}_0/\tilde{\mu}}\frac{1}{2^{2N-1}}\sum_{n=0}^{N-1}\binom{2N-1}{n}R_n\,,$$

where

$$R_n=\frac{1}{(N-n-1)!}\frac{d^{N-n-1}}{dp^{N-n-1}}\left(\frac{1}{1-p}e^{(\tilde{\mu}_0/\tilde{\mu})p}\right)\Bigg|_{p=0}$$

is the corresponding residue. Using the product rule of differentiation, the derivative is evaluated like a binomial,

$$\frac{d^{N-n-1}}{dp^{N-n-1}}\left(\frac{e^{-(\tilde{\mu}_0/\tilde{\mu})p}}{1-p}\right)=\sum_{k=0}^{N-n-1}\binom{N-n-1}{k}\frac{d^k}{dp^k}\left(e^{(\tilde{\mu}_0/\tilde{\mu})p}\right)\frac{d^{(N-n-1)-k}}{dp^{(N-n-1)-k}}\left(\frac{1}{1-p}\right)$$

$$=\sum_{k=0}^{N-n-1}\frac{(N-n-1)!}{k!}e^{(\tilde{\mu}_0/\tilde{\mu})p}(\tilde{\mu}_0/\tilde{\mu})^k\frac{1}{(1-p)^{N-n-k}}\,,$$

$$R_n=\sum_{k=0}^{N-n-1}\frac{1}{k!}(\tilde{\mu}_0/\tilde{\mu})^k\,.$$

While

$$\mathrm{BER}=e^{-\tilde{\mu}_0/\tilde{\mu}}\frac{1}{2^{2N-1}}\sum_{n=0}^{N-1}\binom{2N-1}{n}\sum_{k=0}^{N-n-1}\frac{1}{k!}(\tilde{\mu}_0/\tilde{\mu})^k$$

is a valid expression it may be rewritten as

$$\mathrm{BER}=e^{-\tilde{\mu}_0/\tilde{\mu}}\frac{1}{2^{2N-1}}\sum_{\substack{n,k=0,1,\ldots,N-1\\ n+k\le N-1}}\sum\binom{2N-1}{n}\frac{1}{k!}(\tilde{\mu}_0/\tilde{\mu})^k$$

which after an interchange of summation results in (3.120).

The best possible RX sensitivity results from $N=M=1$,

$$\mathrm{BER}=\frac{1}{2}e^{-\tilde{\mu}_0/\tilde{\mu}}\,. \tag{3.121}$$

However, this requires orthogonally polarized noise to be blocked by a polarizer. A BER = 10^{-9} requires $\tilde{\mu}_0=20\tilde{\mu}$. If there are two noisy polarizations, $N=2M$, $M=1\Rightarrow$

$$\mathrm{BER}=\left(\frac{1}{2}+\frac{1}{8}(\tilde{\mu}_0/\tilde{\mu})\right)e^{-\tilde{\mu}_0/\tilde{\mu}}\,. \tag{3.122}$$

The SNR of the DPSK receiver is SNR = $\tilde{\mu}_0/\tilde{\mu}$. A SNR of 21.9 is required for BER = 10^{-9}. Due to $|c_i|=1$ full optical power is transmitted all the time. For an ideal optical amplifier $\tilde{\mu}=1$ and the SNR equals the number of photons per bit. The SNR for an ideal DPSK receiver is almost 3 dB better than the corresponding SNR = $\tilde{\mu}_0/(2\tilde{\mu})$ = 40.7 for an ideal ASK receiver.

Again we compare to a Gaussian approximation. The difference of the noncentrally and centrally χ_4^2-distributed RVs has a mean equal to the difference of the individual means and a variance equal to the sum of the individual variances. Note also that compared to Table 3.2 $\tilde{\mu}$ must be replaced by $\tilde{\mu}/2$. This yields

$$Q = \frac{2((N\,\tilde{\mu}/2 + \tilde{\mu}_0) - N\,\tilde{\mu}/2)}{2\sqrt{2N(\tilde{\mu}/2)^2 + 2(\tilde{\mu}/2)\tilde{\mu}_0}}, \ \text{BER} = \frac{1}{2}\operatorname{erfc}\left(Q/\sqrt{2}\right). \tag{3.123}$$

$N = 2$, $\tilde{\mu}_0 = 21.9\tilde{\mu}$ results in a BER = $2.4 \cdot 10^{-6}$ which differs a lot from the exact solution.

Fig. 3.24 is similar to Fig. 3.20. For rising mode numbers somewhat larger SNR penalties occur for DPSK compared to ASK. This can be attributed to the larger number of RVs involved: The difference of two photocurrents is taken rather than just one photocurrent.

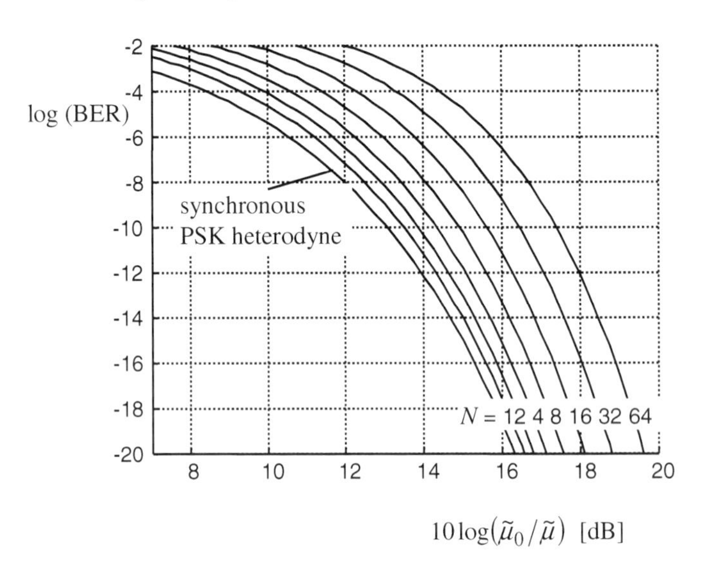

Fig. 3.24 BER of DPSK receiver vs. SNR with number of modes as a parameter. BER of synchronous PSK heterodyne receiver added for comparison.

In order to take shot noise into account the Poisson transform has to be applied to the intensities arriving at each of the photodetectors. Since Poisson transform, subtraction and linear filtering in the baseband part of the receiver are all linear operations it is possible to exchange them and to formulate the result: The decision variable n is the difference of a noncentral and a central negative binomial RV n_1, n_2, each of which carries $2N$ degrees-of-freedom and has a μ half as large as in the ASK case. The probability distribution is $P_n(n) = P_{n_1}(n) * P_{n_2}(-n)$, obtained by discrete convolution. The corresponding MGF $M_n\left(e^{-s}\right) = M_{n_1}\left(e^{-s}\right)M_{n_2}\left(e^{s}\right)$ with $\mu/2$ substituted for μ is

$$M_n\left(e^{-s}\right)=\frac{e^{\frac{-\mu_0\left(1-e^{-s}\right)}{1+(\mu/2)\left(1-e^{-s}\right)}}}{\left(1+(\mu/2)\left(1-e^{-s}\right)\right)^N\left(1+(\mu/2)\left(1-e^{s}\right)\right)^N}. \tag{3.124}$$

If there is no amplifier, $\mu=0$, this becomes a Poisson distribution. The sensitivity limit of a noise-free DPSK receiver, with just one mode, is therefore 20 photons/bit for BER = 10^{-9}. In terms of mean powers this is just half as sensitive as an ideal ASK receiver. Fig. 3.25 shows normalized distributions similar to Fig. 3.17 but referring to DPSK, $|C_1|=1$. To obtain those for $|C_1|=0$ they must be mirrored at the origin $n=0$. For $G=1$ there is just a „noncentral" Poisson distribution because a „central" Poisson distribution has zero photons with unit probability. For higher *G*, until reaching the limit of convolved χ_4^2 distributions there is hardly a change, other than in Fig. 3.17. This is because the noise from both photodiodes partly subtracts in a very beneficial manner.

It remains to be explained why the noise power generated by the optical amplifier is split into equal parts at each photodiode, while the signal is routed without loss either to one or to the other photodiode: The Mach-Zehnder interferometer routes a monochromatic input signal to either arm depending on the phase difference between the arms. Because of the fixed time delay time *T* the phase delay is proportional to ωT. The powers at the two photodiodes are therefore proportional to $\cos^2(\omega T/2+\varphi_{00})$ and $\sin^2(\omega T/2+\varphi_{00})$, where φ_{00} is a constant. The average of these terms over frequency is 1/2.

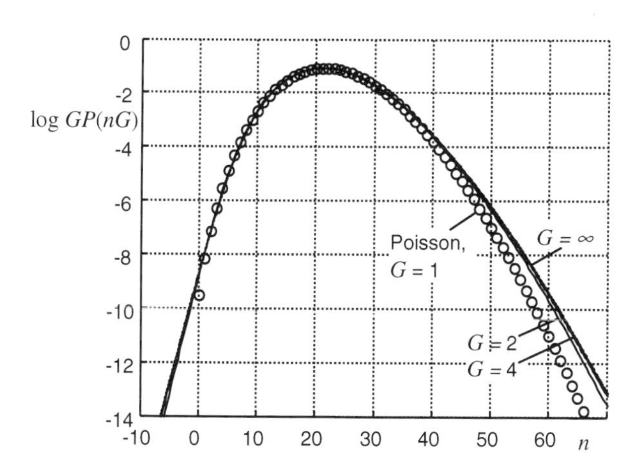

Fig. 3.25 From Poisson via negative binomial to χ_4^2: Transition of the convolution of a noncentral distribution with its central counterpart. Optimum DPSK threshold is at n = 0. Distribution of negated symbol is mirror-symmetric with respect to n = 0.

A possible source of practical problems is a phase error $\Delta\varphi$ between the two interferometer arms. In this case it holds $\underline{\tilde{\mathbf{E}}}_1(t) = \frac{1}{2}\left(e^{j\Delta\varphi/2}\underline{\tilde{\mathbf{E}}}(t) + e^{-j\Delta\varphi/2}\underline{\tilde{\mathbf{E}}}(t-T)\right)$, $\underline{\tilde{\mathbf{E}}}_2(t) = \frac{1}{2}\left(-je^{j\Delta\varphi/2}\underline{\tilde{\mathbf{E}}}(t) + je^{-j\Delta\varphi/2}\underline{\tilde{\mathbf{E}}}(t-T)\right)$. In the noise-free case the intensities at the interferometer outputs are proportional to $\cos^2(\Delta\varphi/2)$, $\sin^2(\Delta\varphi/2)$. Substitutions of the type $\tilde{u}'_{1,3} + j\tilde{u}'_{2,4} = e^{\pm j\Delta\varphi/2}\left(\tilde{u}_{1,3} + j\tilde{u}_{2,4}\right)$ leave the independence and variance of pairs of Gaussian RVs unchanged and allow to replace (3.114) by

$$\underline{\tilde{\mathbf{E}}}(t) = \left(\left(c_i\sqrt{\tilde{\mu}_0/M} + e^{\pm j\Delta\varphi/2}\left(\tilde{u}'_1 + j\tilde{u}'_2\right)\right)\underline{\mathbf{e}}_1 + e^{\pm j\Delta\varphi/2}\left(\tilde{u}'_3 + j\tilde{u}'_4\right)\underline{\mathbf{e}}_2\right)e^{j\omega t} \quad (3.125)$$

Then, in the calculation of $\underline{\tilde{\mathbf{E}}}_{1,2}(t)$, parts of (3.116) must be replaced by

$$\begin{aligned} &\tilde{\mathbf{v}}_k = \mathbf{R}\tilde{\mathbf{u}}'_k \\ &C_1 = \frac{1}{2}\left(e^{j\Delta\varphi/2}c_i + e^{-j\Delta\varphi/2}c_{i-1}\right) \qquad C_2 = \frac{1}{2}\left(e^{j\Delta\varphi/2}c_i - e^{-j\Delta\varphi/2}c_{i-1}\right) \end{aligned} \quad (3.126)$$

For $c_i = c_{i-1}$ this leads to $C_1 \propto \cos\Delta\varphi/2$, $C_2 \propto \sin\Delta\varphi/2$ and finally to the MGFs

$$\begin{aligned} M_{\tilde{x}}\left(e^{-s}\right) &= \left(1-(\tilde{\mu}/2)^2 s^2\right)^{-N} e^{\left(\frac{-\tilde{\mu}_0\cos^2(\Delta\varphi/2)s}{1+(\tilde{\mu}/2)s} + \frac{\tilde{\mu}_0\sin^2(\Delta\varphi/2)s}{1-(\tilde{\mu}/2)s}\right)} \\ M_n\left(e^{-s}\right) &= \frac{e^{\left(\frac{-\mu_0\cos^2(\Delta\varphi/2)\left(1-e^{-s}\right)}{1+(\mu/2)\left(1-e^{-s}\right)} + \frac{-\mu_0\sin^2(\Delta\varphi/2)\left(1-e^{s}\right)}{1+(\mu/2)\left(1-e^{s}\right)}\right)}}{\left(1+(\mu/2)\left(1-e^{-s}\right)\right)^N\left(1+(\mu/2)\left(1-e^{s}\right)\right)^N} \end{aligned} , \quad (3.127)$$

of the continuous RV $\left|\underline{\tilde{\mathbf{E}}}_1\right|^2 - \left|\underline{\tilde{\mathbf{E}}}_2\right|^2$ (intensity difference) and its discrete, not normalized counterpart, respectively. For $c_i = -c_{i-1}$ the distributions are mirrored at the origin. A BER calculation can be found in [53] but is lengthy. Instead, the probability distributions also can be obtained numerically. Results are plotted in Fig. 3.26. For 2 modes, the penalty at a BER = 10^{-9} is kept within 1 dB for phase errors $|\Delta\varphi| \le 0.32$. The synchronous case refers to a carrier phase error $\Delta\varphi$. Synchronous heterodyne receivers will be discussed in chapter 3.3.

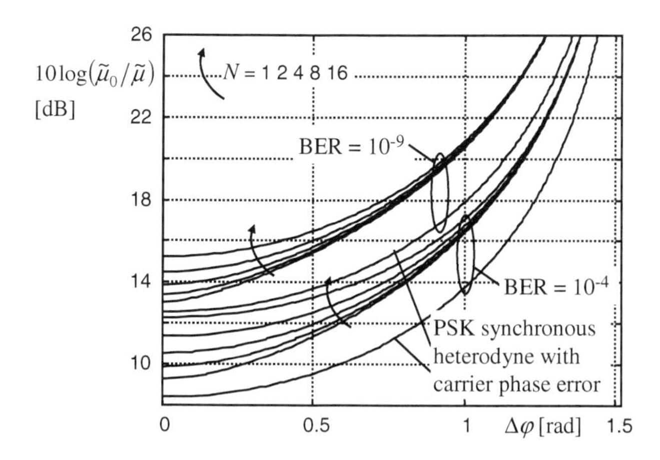

Fig. 3.26 Required SNR vs. interferometer phase error for DPSK receivers, with number of modes N as a parameter. Synchronous case refers to PSK receiver with carrier phase error.

There are various possibilities to generate optical (D)PSK signals. A straightforward way employs a Mach-Zehnder modulator (Fig. 3.27), here schematically shown for X-cut, Y-propagation $LiNbO_3$ with traveling-wave electrodes. The electrical signal causes a differential phase modulation $\Delta\varphi$ between its two arms (see p. 84). Around the operation point $\Delta\varphi = -\pi$ the output electrical field is proportional to $\sin((\pi/2)(V/V_\pi))$, where V_π is the voltage needed for a differential phase shift equal to π between the two arms. A considerable advantage is that the sin function acts as a limiter. Driving voltage undershoot and overshoot have only a reduced effect on the optical field.

The function remains unchanged if the two phase modulation sections are replaced by electroabsorption modulators driven by push-pull signals, as long as the chirp factor of the EAMs is very large so that they act essentially as phase modulators. On the other hand, chirp-free EAMs which work as pure amplitude modulators can also be used in the two arms. The intrinsic path length difference between the two arms must then be 1/2 wavelength. If the transmission factor of one of them is $1/2 + f(V)$, that of the other with complementary driving voltage is $1/2 + f(-V)$. The output electrical field is therefore proportional to $1/2((1/2 + f(V)) - (1/2 + f(-V))) = 1/2(f(V) - f(-V))$. This modulation characteristic is point-symmetric with respect to the origin, which allows to generate symmetric (D)PSK modulation even if the EAM characteristic f is non strictly linear. However, although f saturates for large transmission factors the limiting effect of $1/2(f(V) - f(-V))$ is quite a lot worse than that of $\sin \pi V/V_\pi$. Real EAMs with chirp factors which are neither very large nor zero still allow phase modulation but in order to contain the chirp it is necessary to limit the voltage swing to small values, which prevents limiting.

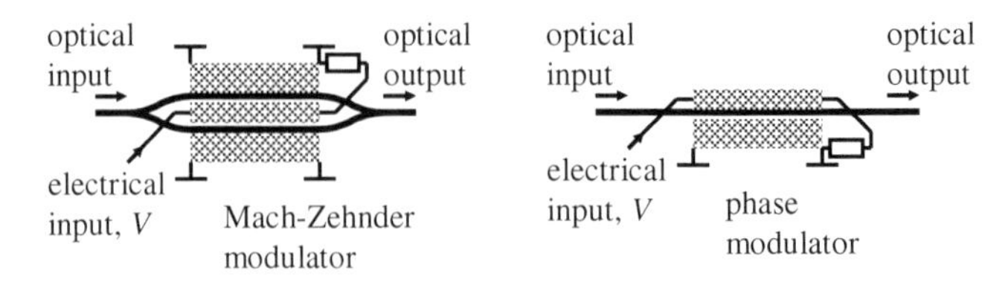

Fig. 3.27 Optical modulators for phase shift keying (PSK) or differential phase shift keying, (DPSK)

The most straightforward way to generate (D)PSK consists in a simple phase modulator, shown for X-cut, Y-propagation $LiNbO_3$. Here the optical phase is proportional to the modulation voltage. The output electrical field is proportional to $e^{j\pi V/V_\pi}$.

When driven by rectangular NRZ driving voltages the optical fields generated by these modulators are identical. However, different spectra result if the driving voltage has finite rise and fall times. For example, linear slopes of length τ can be expressed by

$$k(t)=\begin{cases}1 & \text{for } |t|\le (T-\tau)/2 \\ \dfrac{1}{2}+\dfrac{T/2-|t|}{\tau} & \text{for } |t\pm T/2|\le \tau/2\,. \\ 0 & \text{else}\end{cases} \qquad (3.128)$$

In the Mach-Zehnder modulator the data signal itself is $\underline{a}(t)=\sum_{i=-\infty}^{\infty}\underline{c}_i k(t-iT)$ with a bipolar transmission symbols $c_i=\pm 1$. Due to $\left\langle \underline{c}_i \underline{c}_{i+m}^* \right\rangle = \begin{cases}1 & \text{for } m=0 \\ 0 & \text{for } m\neq 0\end{cases}$, $\underline{v}_m=\begin{cases}1 & \text{for } m=0 \\ 0 & \text{for } m\neq 0\end{cases}$, $N=1$, $\underline{W}_0=0$ the power spectral density is

$$\underline{L}(f)=\frac{1}{T}\left|\underline{S}(f)\right|^2. \qquad (3.129)$$

Here the carrier of the optical signal is suppressed, other than for ASK signals. On the other hand, in the phase modulator

$$\underline{a}(t)=e^{j(\pi/2)\sum_{i=-\infty}^{\infty}\underline{c}_i k(t-iT)} \qquad (3.130)$$

holds, and this gives rise to an additional modulation spectrum including carrier, which is in quadrature to that part of the modulation that is exploited in the receiver. In Fig. 3.28 the chromatic dispersion tolerance is investigated for both schemes. In the presence of chromatic dispersion the phase modulator is found to perform a lot worse than the Mach-Zehnder modulator.

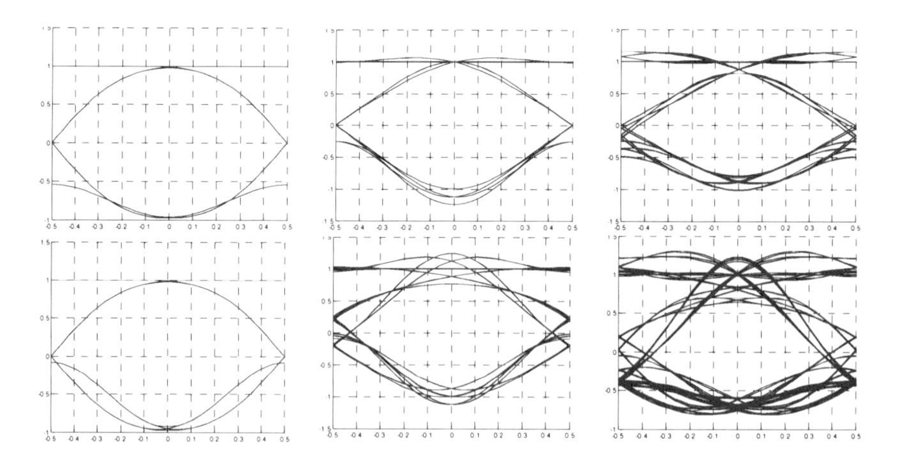

Fig. 3.28 Chromatic dispersion tolerance of DPSK using either a Mach-Zehnder modulator (top) or a phase modulator (bottom). For a data rate of 40 Gbit/s, chromatic dispersion is 0, 34, and 68 ps/nm from left to right.

The combination of DPSK and RZ modulation is useful because it is robust against cross phase modulation. Fig. 3.29 shows experimental waveforms.

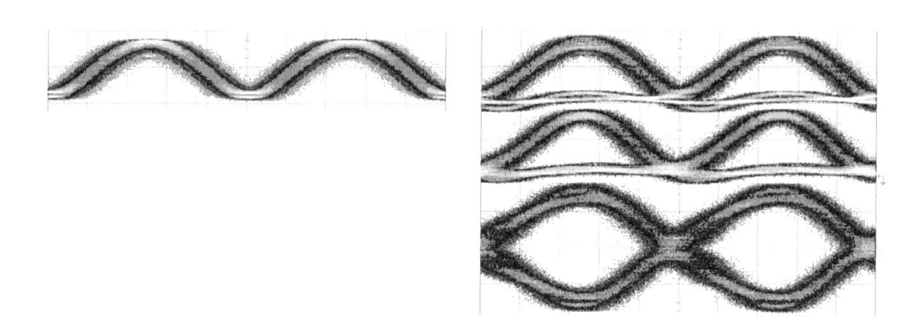

Fig. 3.29 Left: "Eye diagram" of a DPSK-modulated 40Gbit/s RZ signal, detected in a simple photodiode. The phase modulation is of course not visible, only the RZ pulse shape. Right: As before but the signal is fed to a Mach-Zehnder interferometer and is detected in two photodiodes connected to the interferometer outputs (Fig. 3.23). The top two traces are the photocurrents. The bottom trace is their difference, which is sensed by the receiver.

If suitable optical sources were available the receiver of Fig. 3.23 could also be used for frequency shift keying (FSK), in particular with a shortened interferometer delay. DPSK and FSK transmission is also feasible with coherent receivers, with results equivalent to those of the respective direct detection receiver. These and laser linewidth requirements will be discussed later.

Differential quadrature phase shift keying (DQPSK) transmits data in 4 rather than 2 equidistant phasors. Two bit are transmitted per symbol. The demodulation occurs in delay-line interferometers with $\pm\pi/4$ phase differences. The decision variables therefore obey the MGFs (3.127). When optical amplifier noise dominates, the sensitivity is given by Fig. 3.26 with $\Delta\varphi = \pi/4$. For 2 modes the penalty is 4.6 dB with respect to DPSK, at a BER = 10^{-9}. However, since two bit are transmitted per symbol the applicable penalty, referred to one bit stream at twice the symbol rate, is just 1.6 dB. It becomes even smaller at for more modes. This makes DQPSK attractive for a capacity doubling of optical communication systems. Fig. 3.30 shows the calculated BER vs. SNR, also referred to one bit stream at twice the symbol rate.

Similar to Fig. 3.26, Fig. 3.31 shows the SNR increase required to cope with interferometer phase errors in a DQPSK receiver. For 2 modes, the penalty at a BER = 10^{-9} is kept within 1 dB for phase errors $|\Delta\varphi| \leq 0.11$, close to 1/3 of the value for DPSK.

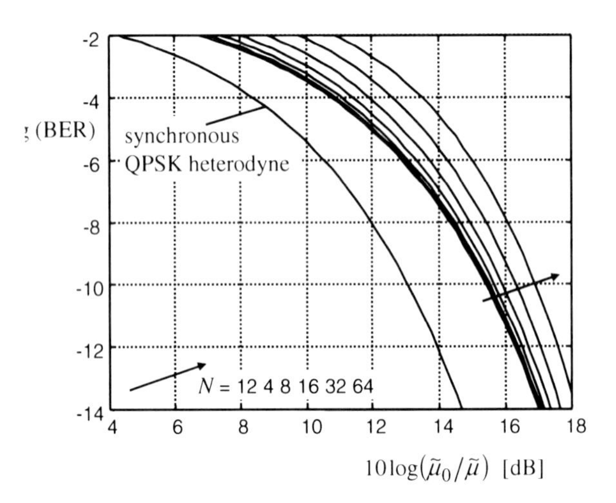

Fig. 3.30 BER of DQPSK receiver vs. SNR with number of modes as a parameter. BER of synchronous QPSK heterodyne receiver added for comparison.

In Fig. 3.32, 40 Gbaud eye diagrams are shown for RZ-DQPSK combined with polarization division multiplex, i.e. a total data rate of 160 Gbit/s per WDM channel.

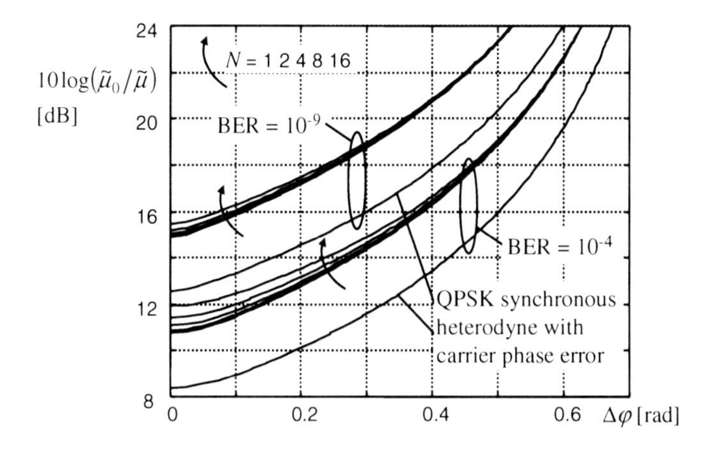

Fig. 3.31 Required SNR vs. interferometer phase error for DQPSK receivers, with number of modes N as a parameter. Synchronous case refers to QPSK receiver with carrier phase error.

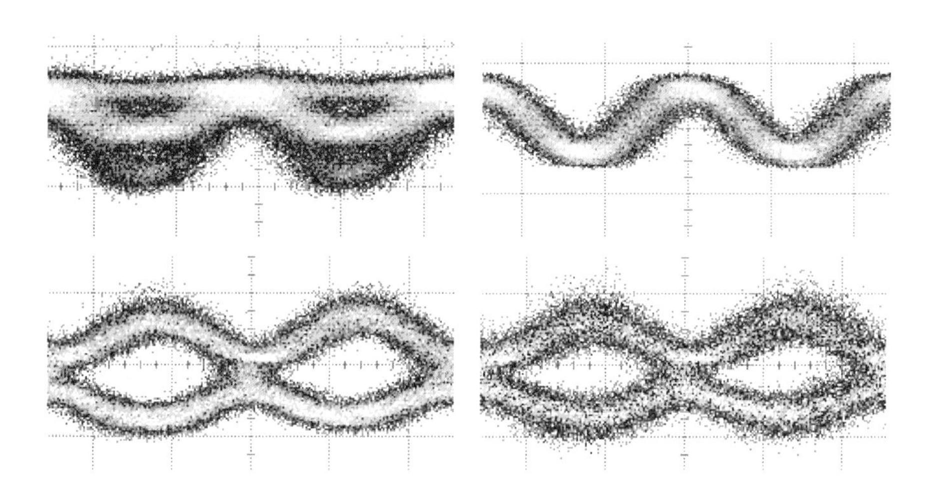

Fig. 3.32 40 Gbaud eye diagrams for RZ-DQPSK combined with polarization division multiplex. Top: ASK "eye diagram" of a DQPSK-modulated 40 Gbit/s NRZ (left) and RZ (right) signal, detected in a photodiode. Bottom: Interferometrically demodulated 40 Gbaud eye diagrams of RZ-DQPSK signals, either at the transmitter (left) or after 324 km of fiber in 4 spans (right). One quadrature is shown in one polarization. Two quadratures and two polarizations were transmitted in 40 WDM channels. Total capacity is therefore 6.4 Tbit/s. The achieved BER was $<10^{-4}$. In the presence of forward error correction this sufficient for quasi error-free transmission, and the corresponding net capacity is 5.94 Tbit/s [54] © 2005 IEEE.

3.3 Coherent Optical Transmission

3.3.1 Receivers with Synchronous Demodulation

Direct detection has so far been the method of choice in commercial optical data transmission systems due to their simplicity and low-cost possibility. Transmission systems using phase shift keying are attractive to increase transmission lengths while keeping amplifier spacing fixed. Of particular interest are coherent receivers. Coherent optical detection could become important in the future because of higher sensitivity and frequency selectivity. These are obtained at the expense of a higher complexity. In addition they allow to perform any required equalization against chromatic or polarization mode dispersions in the electrical domain

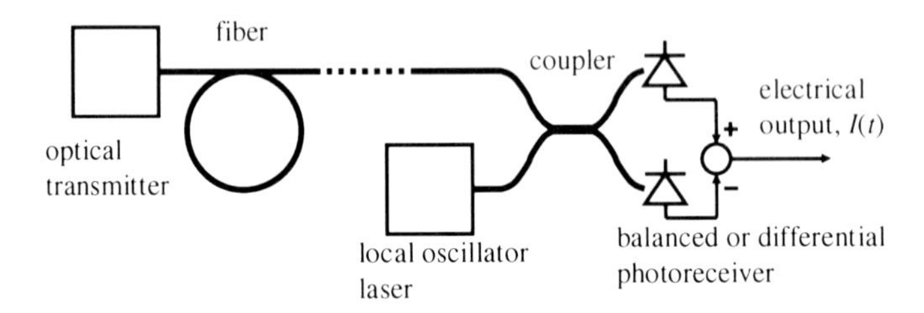

Fig. 3.33 Setup for coherent optical transmission

The difference of a coherent to a direct-detection receiver is as follows: The unmodulated signal of a local oscillator (LO) laser is added to the received signal before photodetection (Fig. 3.33), and the detected interference between these two is processed in the receiver. For an interference to occur the field modes, polarizations and optical frequencies of signal and LO must be identical or almost identical. Field mode matching is achieved automatically if a singlemode directional coupler is used to combine the two optical waves. For polarization matching a polarization control system is needed or a diversity scheme. Signal and LO lasers must be longitudinally single-moded. The LO must be tunable in its frequency in order to match that of the signal. With Jones vectors

$$\underline{\mathbf{E}}_S(t) = \underline{\mathbf{E}}_{S,0} e^{j\omega_S t} \qquad \underline{\mathbf{E}}_{LO}(t) = j\underline{\mathbf{E}}_{LO,0} e^{j\omega_{LO} t} \tag{3.131}$$

for signal (S) and LO, and a coupler transfer matrix $\frac{1}{\sqrt{2}}\begin{bmatrix} 1 & -j \\ -j & 1 \end{bmatrix}$, the output fields at the two photodiodes are $\underline{\mathbf{E}}_1(t) = \frac{1}{\sqrt{2}}\left(\underline{\mathbf{E}}_S - j\underline{\mathbf{E}}_{LO}\right)$ and $\underline{\mathbf{E}}_2(t) = \frac{1}{\sqrt{2}}\left(-j\underline{\mathbf{E}}_S + \underline{\mathbf{E}}_{LO}\right)$, and the corresponding optical powers are

$$P_{1,2} = \frac{1}{2}\left|\underline{\mathbf{E}}_{1,2}\right|^2 = \frac{1}{4}\left(\left|\underline{\mathbf{E}}_S\right|^2 \pm 2\,\mathrm{Re}\left(j\underline{\mathbf{E}}_{LO}^{+}\underline{\mathbf{E}}_S\right) + \left|\underline{\mathbf{E}}_{LO}\right|^2\right). \tag{3.132}$$

When noise is neglected the photocurrent difference is

$$\begin{aligned}
I(t) &= R(P_1 - P_2) = R\,\mathrm{Re}\left(j\underline{\mathbf{E}}_{LO}^{+}\underline{\mathbf{E}}_{S}\right) = R\,\mathrm{Re}\left(\underline{\mathbf{E}}_{LO,0}^{+}\underline{\mathbf{E}}_{S,0}e^{j\omega_{IF}t}\right) \\
&= 2R\sqrt{P_S P_{LO}}\,\frac{\left|\underline{\mathbf{E}}_{LO,0}^{+}\underline{\mathbf{E}}_{S,0}\right|}{\left|\underline{\mathbf{E}}_{S,0}\right|\left|\underline{\mathbf{E}}_{LO,0}\right|}\cos(\omega_{IF}t + \varphi_{IF}) \\
\omega_{IF} &= \omega_S - \omega_{LO} \qquad \varphi_{IF} = \arg\left(\underline{\mathbf{E}}_{LO,0}^{+}\underline{\mathbf{E}}_{S,0}\right) \\
P_S &= \frac{1}{2}\left|\underline{\mathbf{E}}_S\right|^2 \qquad P_{LO} = \frac{1}{2}\left|\underline{\mathbf{E}}_{LO}\right|^2
\end{aligned} \quad . \tag{3.133}$$

This AC signal has an intermediate frequency (IF) equal to the frequency difference between signal and LO. R is the photodiode responsivity. Term $\left|\underline{\mathbf{E}}_{LO,0}^{+}\underline{\mathbf{E}}_{S,0}\right| / \left(\left|\underline{\mathbf{E}}_{S,0}\right|\left|\underline{\mathbf{E}}_{LO,0}\right|\right)$ is the cosine of half the angle between the signal and LO polarizations on the Poincaré sphere. For identical polarizations it is equal to 1, which we will assume from now on.

The local oscillator is strong ($\left|\underline{\mathbf{E}}_{LO}\right| >> \left|\underline{\mathbf{E}}_S\right|$) and unmodulated ($\underline{\mathbf{E}}_{LO} = \text{const.}$). This allows to detect one quadrature of the signal field, just as in synchronous radio frequency communication and in contrast to direct detection optical receivers. The signal field is $\underline{\mathbf{E}}_{S,0}(t) = \underline{\mathbf{E}}_{S,00} \cdot \underline{a}(t)$, with a constant $\underline{\mathbf{E}}_{S,00}$ and $\underline{a}(t)$ given by (3.1). A straightforward and simple scheme consists in amplitude shift keying. In that case $c_i = \sqrt{2}b_i$ ($b_i \in \{0;1\}$) holds. The factor $\sqrt{2}$ was chosen in order to make P_S the mean signal power for equiprobable zeros and ones. The most advantageous modulation format in terms of achievable receiver sensitivity is phase shift keying (PSK). For binary PSK, $c_i = 2b_i - 1 = \pm 1$ is valid. With NRZ pulses, $\underline{a}(t) = c_i$ holds for $|t - iT| < T/2$.

Note that the direct detection term $\left|\underline{\mathbf{E}}_S\right|^2$ is not present in the difference current I. Signal and LO powers result in shot noise in each of the photodiodes, with one-sided power spectral densities equal to $2eRP_{1,2}$. While the powers themselves are correlated because of $P_1 + P_2 = P_S + P_{LO}$ (lossless coupler!) the noise variables themselves are not because $P_{1,2}$ define only the spectral densities. The variances of uncorrelated RVs can be added. Hence the photodiode current difference carries shot noise with a total one-sided power spectral density $2eR(P_S + P_{LO})$. For $\left|\underline{\mathbf{E}}_{LO}\right| >> \left|\underline{\mathbf{E}}_S\right|$ the shot noise contribution from P_S can be neglected. For sufficiently large P_{LO} the (Poissonian) shot noise can be regarded as Gaussian. In the photoreceiver there is also Gaussian thermal noise with a one-sided power spectral density $d\overline{i_{th}^2}/df$. At the output of a matched filter for NRZ signals with impulse response $h(t) = \frac{1}{T}\begin{cases} 1 & |t| \le T/2 \\ 0 & \text{else} \end{cases}$ the noise variance is equal to

$$\begin{aligned} \sigma_{n,out}^2 &= L_n \int_{-\infty}^{\infty} |H(f)|^2 df = L_n \int_{-\infty}^{\infty} |h(t)|^2 dt = \frac{L_n}{T} \\ L_n &= \frac{1}{2}\left(2eR(P_S + P_{LO}) + d\overline{i_{th}^2}/df\right) \end{aligned} \tag{3.134}$$

where L_n is the two-sided power spectral density of frequency-independent (white) noise, and the integral has the value $1/T$ only in this example. For simplicity, we have assumed $d\overline{i_{th}^2}/df$ to be constant.

In a homodyne receiver $\omega_{IF} = 0$ holds. Maximum positive or negative photocurrent difference is achieved for $\varphi_{IF} \in \{0, \pi\}$. This requires an optical phase locked look (PLL) that acts on the local oscillator. If we leave aside the enormous practical difficulties of the OPLL, just a clock and data recovery (CDR) has to be added to the electrical output of Fig. 3.33. It can be composed according to Figs. 3.11 and 3.13. In the homodyne receiver the filtered signal at the decision instant $t_0 = 0$ is

$$s_a = 2R\sqrt{P_S P_{LO}}\, c_i \,. \tag{3.135}$$

This results (approximately for ASK) in

$$\text{BER} = \frac{1}{2}\operatorname{erfc}\left(Q/\sqrt{2}\right) \qquad Q = \frac{\mu_{0,1} - \mu_{0,0}}{\sigma_{n1} + \sigma_{n0}}$$

with

$$Q = \frac{2R\sqrt{2P_S P_{LO}}}{\sqrt{\frac{1}{2T}\left(2eRP_{LO} + d\overline{i_{th}^2}/df\right)} + \sqrt{\frac{1}{2T}\left(2eR(2P_S + P_{LO}) + d\overline{i_{th}^2}/df\right)}} \quad \text{for ASK,}$$

$$Q = \frac{4R\sqrt{P_S P_{LO}}}{2\sqrt{\frac{1}{2T}\left(2eR(P_S + P_{LO}) + d\overline{i_{th}^2}/df\right)}} \quad \text{for PSK.} \tag{3.136}$$

For $P_{LO} \to \infty$ and using $R = \frac{\eta e}{hf}$, where η is the photodiode quantum efficiency, the so-called shot noise limit is achieved,

$$Q=\sqrt{2\eta TP_S/(hf)} \text{ for ASK,} \qquad Q=\sqrt{4\eta TP_S/(hf)} \text{ for PSK.} \tag{3.137}$$

A BER = 10^{-9}, $Q=6$, requires a mean of just $\eta TP_S/(hf)$ = 9 photoelectrons/bit for PSK, and 18 for ASK homodyne receivers.

The LO is matched in frequency, phase and polarization to the signal. In that case the photocurrent difference is given by the scalar and real equation $I(t)=RE_{LO,0}E_{S,0}$ with $P_S=E_{S,0}^2/2$, $P_{LO}=E_{LO,0}^2/2$. We now consider the addition of an optical preamplifier in front of the receiver. The signal power P_S is coupled to the amplifier input. With a noise figure F and a gain G it produces in the optical frequency domain a one-sided optical noise power spectral density $dP/df=hfn_{sp}(G-1)=hf(F-1)G$ in each polarization at its output. As a minor effect, this increases the shot noise in the receiver by an additional optical power $B_o\, dP/df$. The optical bandwidth B_o is that of an optical filter between amplifier and receiver. For simplicity let $B_o >> 1/T$. This means the wanted signal will pass essentially undistorted.

As a major effect the optical amplifier noise will be received like the signal. The lowpass filter in the homodyne receiver passes only frequency components which correspond to optical frequencies in the vicinity of the LO frequency. This means the optically one-sided noise power spectral density dP/df is a two-sided power spectral density in the baseband of the photocurrent difference. 1/2 of this noise, i.e. a two sided power spectral density $\frac{1}{2}dP/df$, is in phase with the optical signal. The homodyne receiver suppresses the other half that is in quadrature to the signal, see (3.133). In line with (3.110), the received electrical field including the major effect of in-phase optical amplifier noise can now be expressed by

$$\left(\sqrt{G}E_S c_i+\tilde{u}(t)\right)e^{j\omega_s t}. \tag{3.138}$$

where the two-sided power spectral density of the noise $\tilde{u}(t)$ is

$$L_{\tilde{u}}=\frac{2}{2}dP/df=hfn_{sp}(G-1)=hf(F-1)G. \tag{3.139}$$

The factor 2 in the numerator applies because of the definition $|\underline{E}|^2=2P$. Due to $I(t)=R\sqrt{2P_{LO}}E_{S,0}$ the two-sided power spectral density of the photocurrent difference is augmented by $2R^2P_{LO}L_{\tilde{u}}$, or the one-sided PSD by $4R^2P_{LO}L_{\tilde{u}}$. In summary, decision variable, noise variance and two-sided noise PSD are now given by

$$s_{out}(t_0)=2R\sqrt{GP_SP_{LO}}\,c_i, \quad \sigma_{n,out}^2=\frac{L_n}{T}, \tag{3.140}$$

$$L_n = \frac{1}{2}\left(4R^2 P_{LO} hf(F-1)G + 2eR((P_S + B_o hf(F-1))G + P_{LO}) + d\overline{i_{th}^2}/df \right). \tag{3.141}$$

For PSK, Q is given by

$$Q = \sqrt{\frac{2TP_S}{hf(F-1) + \frac{2eR(P_S + B_O hf(F-1))}{4R^2 P_{LO}} + \frac{2eR}{4GR^2} + \frac{d\overline{i_{th}^2}/df}{4GR^2 P_{LO}}}}. \tag{3.142}$$

In the limit $P_{LO} \to \infty$, $G \to \infty$,

$$Q = \sqrt{\frac{2TP_S}{hf(F-1)}}. \tag{3.143}$$

Here the quantum limit for an ideal optical amplifier ($F = 2$) is $TP_S/(hf) = 18$ photons/bit for a BER = 10^{-9}. At the optical amplifier input the sensitivity is worse than at the input of a receiver without optical amplifier! For ASK the calculation is similar, yielding a mean of 36 photons/bit.

Next consider a heterodyne receiver (Fig. 3.34) where there is $\omega_{IF} \neq 0$. It is useful to choose π/ω_{IF} to be an integer multiple of the symbol duration T.

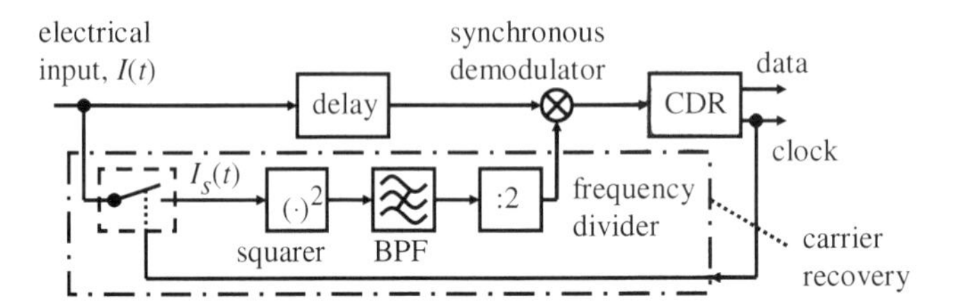

Fig. 3.34 IF and baseband parts of synchronous PSK heterodyne receiver, to be connected to the electrical output of Fig. 3.33

We write the photocurrent difference in a form that shows the PSK or ASK modulation,

$$\begin{aligned} I(t) &= 2R\sqrt{P_S P_{LO}} \frac{\left| \underline{\mathbf{E}}_{LO,0}^+ \underline{\mathbf{E}}_{S,00} \right|}{\left| \underline{\mathbf{E}}_{S,00} \right| \left| \underline{\mathbf{E}}_{LO,0} \right|} c_i \cos(\omega_{IF} t + \tilde{\varphi}_{IF}) \\ \tilde{\varphi}_{IF} &= \arg\left(\underline{\mathbf{E}}_{LO,0}^+ \underline{\mathbf{E}}_{S,00} \right) \end{aligned}. \tag{3.144}$$

Again we assume full polarization matching. This IF signal is connected to one input of a multiplier that serves as a synchronous demodulator. To the other input a carrier signal $2\cos(\omega_{IF}t+\hat{\varphi}_{IF})$ is connected. Ideally, $\hat{\varphi}_{IF}=\tilde{\varphi}_{IF}$ holds. The multiplication product is lowpass-filtered in the clock and carrier recovery (CDR) before being sampled by the decision circuit. Ideally the matched filter impulse response is $h(t)=\frac{1}{T}\begin{cases}1 & |t|\leq T/2\\ 0 & \text{else}\end{cases}$. In that case the signal at the decision instant $t_0=0$ is

$$s_a=\left(h(t)*\left(I(t)2\cos(\omega_{IF}t+\hat{\varphi})\right)\right)\Big|_{t=t_0}=2R\sqrt{P_S P_{LO}}\,c_i\,. \tag{3.145}$$

$I(t)$ is accompanied by zero-mean, white Gaussian noise $n(t)$ that is statistically independent of $I(t)$. At the decision instant, after demodulation and lowpass-filtering, the noise is

$$n_a=\left(h(t)*\left(n(t)2\cos(\omega_{IF}t+\hat{\varphi})\right)\right)\Big|_{t=t_0}=\left(\left(h(t)2\cos(\omega_{IF}t-\hat{\varphi}-\omega_{IF}t_0)\right)*n(t)\right)\Big|_{t=t_0}\,. \tag{3.146}$$

Since the noise is white, its variance is

$$\sigma_{n_a}^2=L_n\int_{-\infty}^{\infty}\left|h(t)2\cos(\omega_{IF}t-\hat{\varphi}-\omega_{IF}t_0)\right|^2dt=\frac{2}{T}L_n\,. \tag{3.147}$$

with L_n given in (3.134). Equal signal amplitude and doubled noise variance result in Q being $1/\sqrt{2}$ as large as in (3.136), (3.137). In the shot noise limit, a mean of $\eta TP_S/(hf)$ = 18 photoelectrons/bit for PSK, and 36 for ASK heterodyne receivers, respectively, is needed for BER = 10^{-9}.

Alternatively, with a different normalization, we can state the difference of a heterodyne versus a homodyne receiver as follows: The electrical signal power is only 1/2 as large, the needed electrical bandwidth is twice as large but only half of the noise power, i.e. the portion in phase with the signal, influences the detection. So there is a 3dB sensitivity disadvantage of a heterodyne versus a homodyne receiver when both these operate without optical amplifier.

We return to (3.147). In the presence of optical amplifiers we must insert the noise PSD from (3.141). Compared to $\sigma_{n,a}^2=L_n/T$ in the homodyne receiver the noise variance in the heterodyne receiver $\sigma_{n,a}^2=2L_n/T$ is doubled (with respect to the signal power after demodulation). A possible quantum limit (BER = 10^{-9}) is therefore 36 photons/bit for PSK and 72 for ASK. However, note that in the optical domain not only ω_S but also the image frequency $2\omega_{LO}-\omega_S$ will contribute to an IF $|\omega_S-\omega_{LO}|$. At least if $|\omega_S-\omega_{LO}|>>2\pi/T$ holds there will be negligible signal energy at $2\omega_{LO}-\omega_S$. Therefore we can cut out frequencies around $2\omega_{LO}-\omega_S$ by an optical filter. This halves the variance of heterodyned noise. If B_o reflects the changed optical bandwidth we can write for the heterodyne receiver with image rejection filter

$$L_n = \frac{1}{2}\left(2R^2 P_{LO} hf(F-1)G + 2eR((P_S + B_o hf(F-1))G + P_{LO}) + d\overline{i_{th}^2}/df \right). \tag{3.148}$$

In WDM environments the image band can easily be suppressed by an arrayed waveguide demultiplexer. The image band suppression reduces the quantum limit (related to the amplifier input) to the shot noise limit of the heterodyne receiver alone. See also Table 3.4.

Problem: One of the photodiodes in Fig. 3.33 is removed, and the lossless coupler transmits a fraction $(1-a)$ of the incident signal power to the remaining photodiode. Calculate the shot noise limit of a PSK heterodyne receiver, and the quantum limit if there is an additional optical preamplfier.

Solution: The coupler transfer matrix is $\begin{bmatrix} \sqrt{1-a} & \mp j\sqrt{a} \\ \mp j\sqrt{a} & \sqrt{1-a} \end{bmatrix}$. Field and power at the photodiode are

$\underline{\mathbf{E}}_1(t) = \sqrt{1-a}\,\underline{\mathbf{E}}_S \mp j\sqrt{a}\,\underline{\mathbf{E}}_{LO}$ and $P_1 = \frac{1}{2}\left((1-a)|\underline{\mathbf{E}}_S|^2 \pm 2\sqrt{(1-a)a}\,\mathrm{Re}\left(j\underline{\mathbf{E}}_{LO}^{+}\underline{\mathbf{E}}_S\right) + a|\underline{\mathbf{E}}_{LO}|^2 \right)$,

respectively. The signal amplitude is multiplied by $\sqrt{(1-a)a}$ with respect to the balanced photoreceiver, and the part of the shot noise variance which is due to P_{LO} is multiplied by a. In the shot noise limit Q is therefore multiplied by $\sqrt{(1-a)a}/\sqrt{a}$. For a given BER, $(1-a)^{-1/2}$ as much signal power is needed at the receiver input. In the limit $a \to 0$ the sensitivity can be the same as for the balanced photoreceiver of Fig. 3.33, but a much larger P_{LO} is required.

If there is an optical amplifier the quantum noise limit (which refers to the amplifier input) is not influenced by the single-ended photoreceiver. This is because the amplifier noise, which determines the sensitivity, is attenuated in the coupler just as much as the wanted signal.

So far we have considered an optical signal with just one quadrature, i.e. $\mathrm{Im}\, c_i = 0$. If we choose instead a signal with complex $\underline{c}_i$ reception will not be altered as long as $\mathrm{Re}\, \underline{c}_i$ stays unchanged. This means it is possible to transmit the information capacity by using also the other quadrature. The most interesting case is quadrature phase shift keying (QPSK). Each symbol

$$\underline{c}_i = (2b_{i,1} - 1) + j(2b_{i,2} - 1). \tag{3.149}$$

carries information from two data bits $b_{i,1}, b_{i,2}$ (Fig. 3.35 left). It is possible to create the symbols in one phase modulator (Fig. 3.27 right) driven by a quaternary signal with Gray encoding. However, unwanted ISI generation in the modulator makes this method little advisable. It is more advantageous to implement eqn. (3.149) directly. This is possible with two Mach-Zehnder modulators placed between two Y forks (Fig. 3.35 right). One of the arms must have an extra $\pi/2$ phase shift with respect to the other. Half of the power is therefore lost because it excites an odd output waveguide mode.

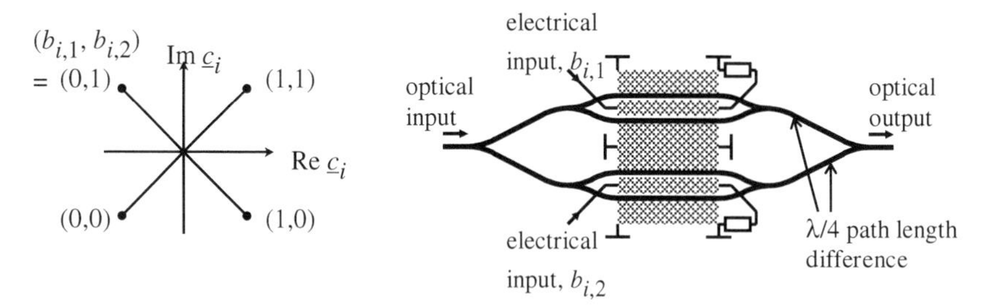

Fig. 3.35 Phasor of QPSK signal (left) and modulator for its generation (right)

In the case of the heterodyne receiver the electrical signal is split to the inputs of two demodulators (Fig. 3.36). Different carriers $2\cos(\omega_{IF}t+\hat{\varphi}_{IF})$ and $2\sin(\omega_{IF}t+\hat{\varphi}_{IF})$ are connected to the respective carrier input of the demodulators. The CDR contains two decision circuits and a PLL with a common clock VCO. If the LO is strong, thermal noise does not play any role in the receiver. This means that electrical power splitting does not change the sensitivity. In terms of photoelectrons/bit the QPSK heterodyne receiver is as sensitive as the PSK one. If the bit rate is doubled by adding the second quadrature then of course twice as many photoelectrons/symbol are needed because each symbol contains now 2 rather than 1 bit. The same holds for the signal photons necessary at an optical preamplifier input.

Electrical splitting is not possible in the homodyne receiver because due to $\omega_{IF}=0$ information from one quadrature is lost. Essentially, a QPSK homodyne receiver is therefore composed of two PSK homodyne receivers (Fig. 3.37) [55].

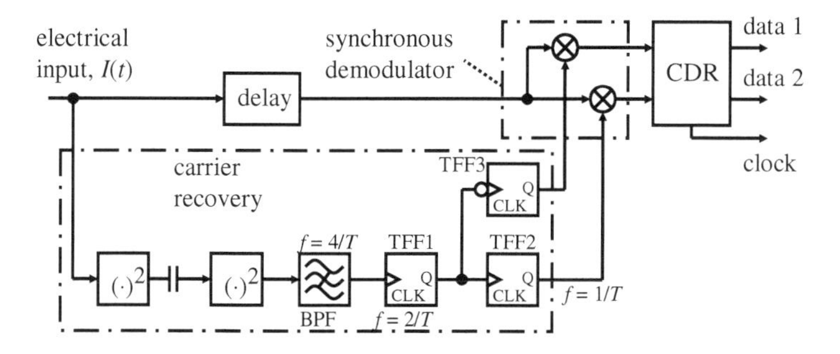

Fig. 3.36 IF and baseband parts of synchronous QPSK heterodyne receiver, to be connected to the electrical output of Fig. 3.33

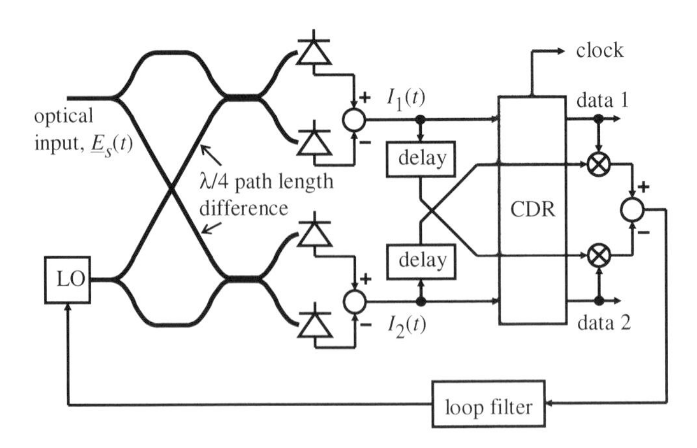

Fig. 3.37 QPSK homodyne receiver. © 2005 IEEE.

An optical power splitter is needed which divides the available signal power equally to either receiver. Under the assumption of otherwise equal path lengths in the two receivers the local oscillator signals must be out of phase by $\pi/2$ with respect to each other. The arrangement of two Y-forks, two couplers, $\pi/2$ phase shift in the interconnections, photodiodes and subtractors may be called an optoelectronic 90° hybrid. The optical waveguides and couplers can be implemented for example in $LiNbO_3$ [56]. For long-term usage the $\pi/2$ phase difference needs to be stabilized [57]. In-phase and quadrature interference can also be detected in fiber couplers, see pp. 228, 230.

Problem: Signal and local oscillator are connected with 45° linear and circular polarization, respectively, to the inputs of the fiber coupler. Calculate the differences $P_5 - P_7$ and $P_6 - P_8$ between the optical powers at the outputs of the two x-y polarization beam splitters (PBS).

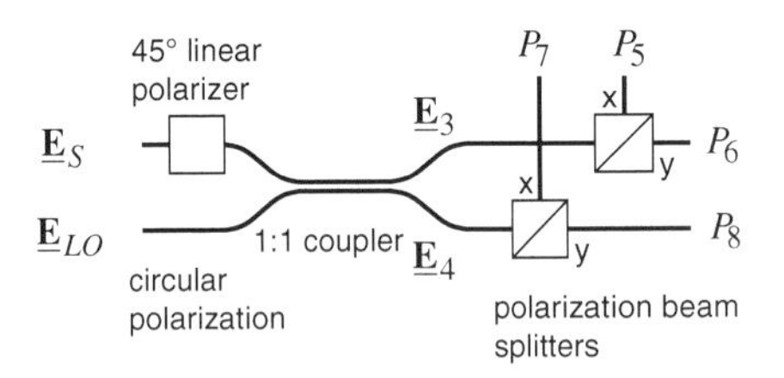

Solution: We start with the Jones vectors $\underline{\mathbf{E}}_S = \underline{E}_S \frac{1}{\sqrt{2}}\begin{bmatrix}1\\1\end{bmatrix}$, $\underline{\mathbf{E}}_{LO} = \underline{E}_{LO} \frac{1}{\sqrt{2}}\begin{bmatrix}j\\1\end{bmatrix}$. The 1:1 coupler has a polarization-independent transfer matrix $\frac{1}{\sqrt{2}}\begin{bmatrix}1 & -j\\-j & 1\end{bmatrix}$ and yields output Jones vectors $\underline{\mathbf{E}}_3 = \frac{1}{2}\begin{bmatrix}\underline{E}_S + \underline{E}_{LO}\\ \underline{E}_S - j\underline{E}_{LO}\end{bmatrix}$, $\underline{\mathbf{E}}_4 = \frac{1}{2}\begin{bmatrix}-j\underline{E}_S + j\underline{E}_{LO}\\ -j\underline{E}_S + \underline{E}_{LO}\end{bmatrix}$. The power be $P = \frac{1}{2}|\underline{E}|^2$. The PBS pass only the respective polarization components, which yields

$P_{5,7} = \frac{1}{8}\left(\left|\underline{E}_S\right|^2 \pm 2\mathrm{Re}\left(\underline{E}_S \underline{E}_{LO}^*\right) + \left|\underline{E}_{LO}\right|^2\right)$, $P_{6,8} = \frac{1}{8}\left(\left|\underline{E}_S\right|^2 \pm 2\,\mathrm{Im}\left(\underline{E}_S \underline{E}_{LO}^*\right) + \left|\underline{E}_{LO}\right|^2\right)$. The power differences $P_5 - P_7 = \frac{1}{2}\mathrm{Re}\left(\underline{E}_S \underline{E}_{LO}^*\right)$, $P_6 - P_8 = \frac{1}{2}\mathrm{Im}\left(\underline{E}_S \underline{E}_{LO}^*\right)$ are in quadrature to each other [58]. The large sensitivity against variations of the received signal polarization can be reduced by inserting a 45° linear polarizer in the signal path before the fiber coupler [59].

Due to the splitter the needed number of photoelectrons/bit doubles in the QPSK homodyne receiver with respect to the PSK homodyne receiver. When there is an optical preamplifier the quantum noise limit of a PSK homodyne receiver, which is only as good as that of a PSK heterodyne receiver, is not further decreased because under the assumption of large optical gain the power splitting does not matter. QPSK homodyne is therefore ideally as sensitive as QPSK heterodyne.

Quaternary synchronous ASK transmission is possible but is as uninteresting as binary synchronous ASK. If the considerable effort for synchronous demodulation is undertaken then the highest possible receiver sensitivity should be achieved, and this requires (Q)PSK, not ASK.

More (but not more sensitive) receiver configurations will be discussed in connection with carrier recovery.

3.3.2 Carrier Recovery

Synchronous demodulation is only possible if the carrier is available in the receiver. For a homodyne receiver the local oscillator generates the carrier.

In a heterodyne receiver the LO just transponds the signal into a lower frequency band (microwaves). An electrical carrier signal must be generated. For carrier recovery of a PSK heterodyne receiver consider again Fig. 3.34. Let us first assume that the switch is replaced by a through connection. The electrical heterodyne signal $I(t) \propto c_i \cos\left(\omega_{IF} t + \tilde{\varphi}_{IF}\right)$ is squared. This results in a signal proportional to $I^2(t) \propto 1 + \cos\left(2\omega_{IF} t + 2\tilde{\varphi}_{IF}\right)$ which, however, contains a lot of noise. A subsequent bandpass filter (BPF) centered at $2\omega_{IF}$ removes so much of the noise that it becomes possible to feed the frequency-doubled signal to a frequency divider. If no counting errors occur the frequency divider output signal will be proportional to $\mathrm{sgn}\left(\cos\left(\omega_{IF} t + \hat{\varphi}_{IF}\right)\right)$ where the replacement of $\tilde{\varphi}_{IF}$ by $\hat{\varphi}_{IF}$ reflects phase distortions of bandpass filter and frequency divider. The rectangular shape of this carrier does not matter; it will just cause a few harmonics after demodulation which are eliminated by subsequent lowpass filtering.

The larger the laser linewidths the larger will be the temporal fluctuations of $\tilde{\varphi}_{IF}$. This means that a delay line before the signal input of the demodulator is essential to cancel the group delay occurring in the carrier recovery. The temporal derivative $d\tilde{\varphi}_{IF}/dt$ is an offset to the IF ω_{IF}. The BPF must be so wide that the resulting broadened spectrum can pass, otherwise $\hat{\varphi}_{IF}$ will momentarily differ from $\tilde{\varphi}_{IF}$. However, if the BPF is wide the SNR of its output signal is poor and it can not be excluded that the frequency divider will every now and then fail to change its state when it should. This means all subsequent data will change polarity, and this is clearly not acceptable. It is therefore helpful to differentially encode the transmitted signal, and to differentially decode the data at the output of the CDR [60]. Thereby counting errors of a static frequency divider will be transformed into adjacent double errors of the differentially decoded signal. This error rate doubling reduces the sensitivity advantage of PSK over DPSK transmission systems.

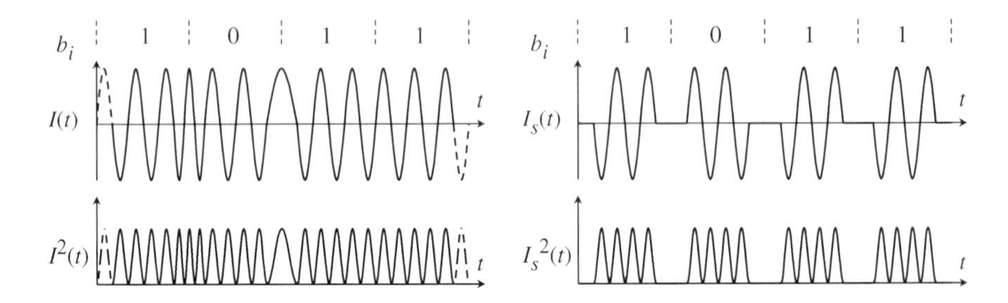

Fig. 3.38 Signals in PSK heterodyne receiver with phase modulator at the transmitter

The generation of optical PSK signals has already been discussed on p. 210. A Mach-Zehnder modulator keeps the optical signal strictly to one quadrature. During the transition periods between subsequent symbols the signal just disappears. The situation is different for phase modulators. Depending on the sign of the transition the IF will momentarily increase or decrease (Fig. 3.38 left).

The same is be true for the frequency-doubled signal $I^2(t)$ after the squarer. In quadrature to the frequency-doubled carrier it contains a modulated signal with zero mean. The total momentary IF offset $d\tilde{\varphi}_{IF}/dt$ contains not only a symmetric part due to phase modulation but also by a more slowly variable part caused by laser phase noise which differs from zero over longer times although its mean is also zero.

After subtraction of a mean group delay at $2\omega_{IF}$ the frequency response of a symmetric BPF can in the vicinity of $2\omega_{IF}$ be approximated by $e^{-(a+jb)(\omega-2\omega_{IF})^2}$. The phase noise part of $d\tilde{\varphi}_{IF}/dt$ has the effect that perturbations caused by the modulation induced part of $d\tilde{\varphi}_{IF}/dt$ during 01 and 10 data symbol transitions can not cancel out. At the BPF output this results in a slow residual phase noise, $\hat{\varphi}_{IF} \neq \tilde{\varphi}_{IF}$. A solution to this problem is to cut out the phase transition periods by the switch at the input of the carrier recovery branch [60] (Fig. 3.34, with switch inserted). The data itself will only be known quite some time later at the output of the CDR. The only possibility to achieve this is therefore cut out all periods where phase transitions may occur. The recovered clock signal is used to that purpose. The signal after the switch is $I_s(t)$ (Fig. 3.38 right). Its square $I_s^2(t)$ contains only one quadrature of the frequency-doubled carrier. The periodic modulation of the envelope does not matter because it is removed by the BPF. Since the clock frequency is highly stable its phase noise is negligible, and the feedback to the switch is hence permitted.

For QPSK heterodyne receivers the feedforward carrier recovery scheme can be cascaded (Fig. 3.36). After the first squarer the DC part is removed by AC coupling. The resulting signal $\cos(2\omega_{IF}t + 2\tilde{\varphi}_{IF})$ is squared again and then contains an AC term $\cos(4\omega_{IF}t + 4\tilde{\varphi}_{IF})$. For QPSK signals any of the signal phases $\pi/4, 3\pi/4, 5\pi/4, 7\pi/4$ are possible. After quadrupling all these are modulo 2π equal to π. The signal $\cos(4\omega_{IF}t + 4\tilde{\varphi}_{IF})$ is now bandpass-filtered in the vicinity of $4\omega_{IF}$. After that its frequency is divided by 4, using cascaded toggle flip flops. There are two T flip flops with opposite triggering polarity in the second divider stage in order to generate the two orthogonal recovered carriers proportional to $\text{sgn}(\cos(\omega_{IF}t + \hat{\varphi}_{IF}))$ and $\text{sgn}(\sin(\omega_{IF}t + \hat{\varphi}_{IF}))$, respectively.

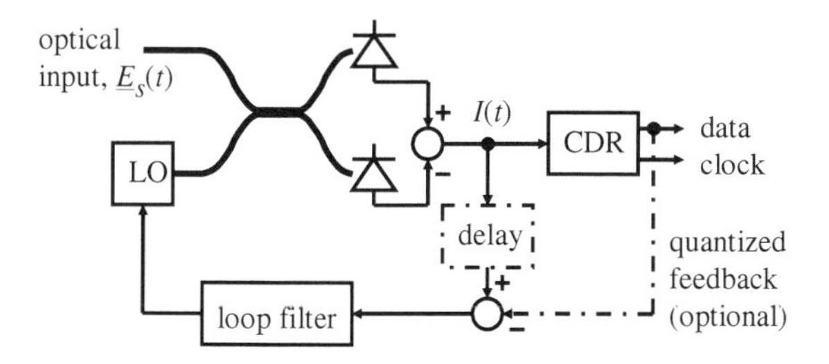

Fig. 3.39 Homodyne receiver with optical phase-locked loop (OPLL) suitable for PSK signals with residual carrier

In PSK homodyne receivers the optical receiver must serve as a phase detector of an optical phase-locked loop (OPLL). Unfortunately the IF signal $\propto \cos\tilde{\varphi}_{IF}$ requires $\tilde{\varphi}_{IF} \in \{0; \pi\}$ for maximum sensitivity. For operation as a phase detector, however, $\tilde{\varphi}_{IF} \approx \pm\pi/2$ is required. There are several solutions to this problem. A straightforward way (Fig. 3.39) is to send a carrier which is in quadrature to the wanted signal. Most easily this is accomplished by using a pure phase modulator for encoding instead of a Mach-Zehnder modulator, as has been discussed on p. 210. The phase detector signal is fed to a loop filter – usually a proportional integral controller – which in turn is connected to a LO with a frequency control input. OPLL design does not differ from electrical PLL design in principle. A particular difficulty lies in the fact that the optical oscillators (signal and LO lasers) have much larger linewidths than electrical oscillators. While in principle this can be overcome by a sufficiently large OPLL bandwidth there are severe practical limits imposed by unwanted loop delay. In particular, any fiber will act as a delay line. Standard DFB lasers with MHz linewidths can therefore not be used as signal and LO lasers. Rather, external cavity lasers with kHz linewidths are needed to implement the scheme of Fig. 3.39. These are costly. For example, a QPSK receiver with an OPLL having a realistic loop delay of 100 symbol durations has been predicted to tolerate a laser sum linewidth / symbol rate ratio of $\sim 5 \cdot 10^{-6}$ [61].

Another difficulty is that in addition to the phase information that is needed only for the OPLL the homodyne signal $I(t)$ of course contains the data signal. Owing to the random nature of data, ones (or zeros) may occur in larger numbers temporarily. In the frequency domain this results in a power spectral density that is non-zero in the vicinity of the carrier. This data crosstalk into the OPLL can be overcome by quantized feedback: The signal itself is known after regeneration. Suitably scaled, it is subtracted from the phase detector signal. The data signal at the CDR output contains no random noise, so the SNR can not be degraded. However, a delay is needed to equalize that of the CDR. Although the additional delay decreases the achievable loop frequency further quantized feedback is still beneficial because it improves the SNR in the OPLL.

Phase detection and carrier recovery in a PSK homodyne receiver is also possible if there is no residual carrier, i.e. for modulation in a Mach-Zehnder modulator. For this purpose we can retain the setup of Fig. 3.37, but data recovery is needed only for one data symbol. The homodyne signals in the two branches are

$$I_1(t) \propto \operatorname{Re}\left(\underline{c}_i e^{j(\omega_{IF} t + \tilde{\varphi}_{IF})}\right) \qquad I_2(t) \propto \operatorname{Re}\left(-j\underline{c}_i e^{j(\omega_{IF} t + \tilde{\varphi}_{IF})}\right), \tag{3.150}$$

the recovered data signals are $\operatorname{Re}\underline{c}_i = 2b_{i,1} - 1$ and $\operatorname{Im}\underline{c}_i = 2b_{i,2} - 1$, respectively. Assume $\omega_{IF} = 0$. The upper multiplier forms the correlation product $I_2(t)\operatorname{Re}\underline{c}_i \propto \left(\sin\tilde{\varphi}_{IF}\operatorname{Re}\underline{c}_i + \cos\tilde{\varphi}_{IF}\operatorname{Im}\underline{c}_i\right)\operatorname{Re}\underline{c}_i$. For binary PSK $\operatorname{Im}\underline{c}_i = 0$ holds. With $\operatorname{Re}^2\underline{c}_i = 1$ for NRZ PSK signals the correlation product is $I_2(t)\operatorname{Re}\underline{c}_i \propto \sin\tilde{\varphi}_{IF}$, and this is the needed phase detection signal. Loop filter and LO, acting as a VCO, complete the OPLL. A delay element at one multiplier input is needed to make up for the delay inside the CDR.

The complete Fig. 3.37 lends itself also for carrier detection in a QPSK homodyne receiver. At the subtractor output the term

$$\begin{aligned} &I_2(t)\operatorname{Re}\underline{c}_i - I_1(t)\operatorname{Im}\underline{c}_i \\ &\propto \begin{pmatrix} \sin\tilde{\varphi}_{IF}\operatorname{Re}\underline{c}_i \\ +\cos\tilde{\varphi}_{IF}\operatorname{Im}\underline{c}_i \end{pmatrix}\operatorname{Re}\underline{c}_i - \begin{pmatrix} \cos\tilde{\varphi}_{IF}\operatorname{Re}\underline{c}_i \\ -\sin\tilde{\varphi}_{IF}\operatorname{Im}\underline{c}_i \end{pmatrix}\operatorname{Im}\underline{c}_i = 2\sin\tilde{\varphi}_{IF} \end{aligned} \tag{3.151}$$

is found as a phase detection signal. Due to uncorrelated data, $\left\langle \operatorname{Im}\underline{c}_i \operatorname{Re}\underline{c}_i \right\rangle = 0$, a single correlator would suffice in principle but this would throw up a patterning problem similar as discussed in the context of Fig. 3.39.

An advantage of the heterodyne QPSK receiver over its homodyne counterpart is that the carrier recovery is a feed-forward process whereas OPLLs with usual long loop delays impose extremely stringent laser linewidth requirements. On the other hand, homodyne receivers allow to do all processing in the baseband, which is an enormous advantage at high bit rates. Also, no unwanted image frequency exists for homodyne receivers.

It is possible to retain the best from both worlds, i.e., to combine the respective advantages of the Figs. 3.36 and 3.37. Such receivers may be called intradyne [62, 55]. The IF is not equal to zero but is in general a lot smaller than the symbol clock frequency. It is all the same possible to set $\omega_{IF}=0$ because a non-zero IF can still be expressed by a time-variable difference phase $\tilde{\varphi}_{IF}$ between signal laser and local oscillator laser. The two photocurrents of Fig. 3.37 can be understood to be a complex photocurrent

$$\underline{I}(t)=I_1(t)+j\cdot I_2(t)\propto \underline{c}\cdot e^{j\tilde{\varphi}_{IF}} \tag{3.152}$$

which serves as an IF signal. This QPSK signal can be processed according to Fig. 3.40. The data symbol $\underline{c}$ is defined in (3.149). Due to $\underline{c}^4=-4$ the quadrature phase shift keying is wiped off if $\underline{I}$ is raised to the 4th power. A frequency-doubled I&Q signal $\underline{I}^2$ is obtained by the operations $\operatorname{Re}\underline{I}^2 := \operatorname{Re}^2\underline{I}-\operatorname{Im}^2\underline{I}$, $\operatorname{Im}\underline{I}^2 := 2\operatorname{Re}\underline{I}\cdot\operatorname{Im}\underline{I}$ where := is an assignment in a signal flow diagram. This is repeated to obtain $\operatorname{Re}\underline{I}^4 := \operatorname{Re}^2\underline{I}^2-\operatorname{Im}^2\underline{I}^2$, $\operatorname{Im}\underline{I}^4 := 2\operatorname{Re}\underline{I}^2\cdot\operatorname{Im}\underline{I}^2$. Then the frequency-multiplied carrier $\underline{I}^4\propto -e^{j4\tilde{\varphi}_{IF}}$ is bandpass-filtered. In the intradyne case this means that both signals $\operatorname{Re}\underline{I}^4$, $\operatorname{Im}\underline{I}^4$ are lowpass-filtered.

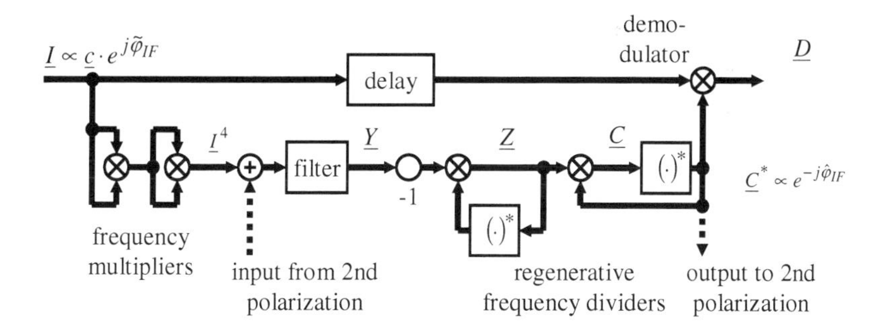

Fig. 3.40 QPSK intradyne receiver with feedforward carrier recovery scheme based on regenerative frequency dividers [55]. © 2005 IEEE.

Filtering removes broadband noise but alters the phase angle. In complex representation the filtered signal is $\underline{Y}\propto -e^{j4\hat{\varphi}_{IF}}$ where $\hat{\varphi}_{IF}$ ideally would be identical to $\tilde{\varphi}_{IF}$. The filtered signal needs to be frequency-divided by 4. This is possible by implementing the operations $\underline{Z}=(-\underline{Y})^{1/2}$, $\underline{C}=\underline{Z}^{1/2}$. $\underline{C}\propto e^{j\hat{\varphi}_{IF}}$ is the recovered complex carrier. These signal processing units can be called regenerative intradyne divide-by-2 frequency dividers. E.g., in the second stage the complex output signal $\underline{C}$ is being assigned the value $\underline{C} := \underline{Z} \cdot \underline{C}^*$. For the two quadratures this means $\mathrm{Re}\,\underline{C} := \mathrm{Re}\,\underline{Z} \cdot \mathrm{Re}\,\underline{C} + \mathrm{Im}\,\underline{Z} \cdot \mathrm{Im}\,\underline{C}$, $\mathrm{Im}\,\underline{C} := \mathrm{Im}\,\underline{Z} \cdot \mathrm{Re}\,\underline{C} - \mathrm{Re}\,\underline{Z} \cdot \mathrm{Im}\,\underline{C}$. In order to stabilize $|\underline{C}|$ the associated linear combiners may be slightly overdriven. Else the equations may be modified by an amplitude control term, for example $\underline{C} := \underline{Z} \cdot \underline{C}^* \cdot \int \left(1 - |\underline{C}|^2\right) dt$. In the time-discrete regime (only there!) the phase of $\underline{C} := \underline{Z} \cdot \underline{C}^*$ is not asymptotically stable. For modeling purposes it is therefore useful to set $\underline{C}_n := (1/2)\left(\underline{A}_n + \underline{A}_{n-1}\right)$, $\underline{A}_n := \underline{Z}_n \cdot \underline{C}_{n-1}^*$, where adjacent samples with indices $n-1$, n are one symbol duration T apart. This also helps to model a realistic lowpass behavior.

Since there is no OPLL implemented, the IF will in general be unequal zero. In order to exploit a maximum of phase noise tolerance an automatic frequency control (AFC) system for the IF is needed. It may act on temperature or a current of the local oscillator laser. A suitable intradyne frequency discriminator is seen as a part of Fig. 3.44. It derives a frequency error signal $\mathrm{Im}\left(\underline{Y}(t) \cdot \underline{Y}^*(t-\tau)\right)$ from the frequency-multiplied, filtered carrier $\underline{Y}$, and τ is a delay time. The frequency error signal may be integrated and then passed on to the LO.

The complex IF signal $\underline{I}$ must be delayed by the group delay that is suffered in the carrier recovery branch in order to minimize phase noise influence. The demodulated data signal is $\underline{D} := \underline{I} \cdot \underline{C}^* \propto \underline{c} \cdot e^{j\Delta\varphi}$ where $\Delta\varphi = \tilde{\varphi}_{IF} - \hat{\varphi}_{IF}$ is the residual phase error. The decision variables of the two quadratures are $\mathrm{Re}\,\underline{D}$, $\mathrm{Im}\,\underline{D}$.

It is tempting to implement this intradyne receiver digitally. In that case analog-to-digital converters digitize the photodetected signals and all subsequent processing occurs in a digital integrated circuit. A corresponding scheme is described in 3.3.5 and [63]. It allows synchronous QPSK transmission with DFB lasers [64, 65].

A similar setup is possible for 3-phase IF signals [66]: These are generated from 2-phase IF signals $\mathrm{Re}\,\underline{I}$, $\mathrm{Im}\,\underline{I}$ by the operations

$$i_1 = \mathrm{Re}\,\underline{I}, \; i_{2,3} = -(1/2)\mathrm{Re}\,\underline{I} \pm \left(\sqrt{3}/2\right)\mathrm{Im}\,\underline{I}, \tag{3.153}$$

or usually in a receiver having a symmetric 3×3 coupler with transfer matrix (2.409).

Problem: Signal and local oscillators are connected to fibers 1 and 2 of the coupler with transfer matrix (2.409). Show that three photodiodes connected to the output fibers 4, 5, 6 (see Fig. on p. 105) yield 3-phase photocurrents.

Solution: The output fields are

$$\begin{bmatrix} \mathbf{E}_4 \\ \mathbf{E}_5 \\ \mathbf{E}_6 \end{bmatrix} = \mathbf{T} \begin{bmatrix} \mathbf{E}_S \\ \mathbf{E}_{LO} \\ 0 \end{bmatrix} = e^{j(-Ml+\pi/18)} \frac{j}{\sqrt{3}} \begin{bmatrix} \mathbf{E}_S + e^{-j2\pi/3}\mathbf{E}_{LO} \\ e^{-j2\pi/3}\mathbf{E}_S + \mathbf{E}_{LO} \\ e^{-j2\pi/3}\mathbf{E}_S + e^{-j2\pi/3}\mathbf{E}_{LO} \end{bmatrix}. \tag{3.154}$$

The photocurrents are $i_{1,2,3} = \frac{R}{2}\left|\mathbf{E}_{4,5,6}\right|^2$,

$$\begin{aligned} i_1 &= (R/6)\left(\left|\mathbf{E}_S\right|^2 + 2\,\mathrm{Re}\left(e^{-j2\pi/3}\mathbf{E}_S^+\mathbf{E}_{LO}\right) + \left|\mathbf{E}_{LO}\right|^2\right) \\ i_2 &= (R/6)\left(\left|\mathbf{E}_S\right|^2 + 2\,\mathrm{Re}\left(e^{j2\pi/3}\mathbf{E}_S^+\mathbf{E}_{LO}\right) + \left|\mathbf{E}_{LO}\right|^2\right) \\ i_3 &= (R/6)\left(\left|\mathbf{E}_S\right|^2 + 2\,\mathrm{Re}\left(\mathbf{E}_S^+\mathbf{E}_{LO}\right) + \left|\mathbf{E}_{LO}\right|^2\right) \end{aligned} \tag{3.155}$$

as requested. Supplied with photodetectors, this coupler may be called an optoelectronic 120° hybrid. Its advantage compared to the 90° hybrid of Fig. 3.37 is that the hybrid phase is intrinsically stable and does not depend on fiber length differences.

To find the best possible inverse operation we consider shot and thermal noises which accompany each of $i_{1,2,3}$. These have equal variances and are statistically independent in $i_{1,2,3}$. More generally, consider m signals A_k ($k = 1...m$), each of them accompanied by a zero-mean noise variable n_k. These can be arranged in the vectors $\mathbf{A}$ and $\mathbf{n}$, respectively. All n_k are statistically independent and have equal variances σ^2. We want to add all signals with suitable weighting by a vector a composed of weights a_i, to form a sum $\mathbf{a}^T(\mathbf{A}+\mathbf{n})$ with maximized signal-to-noise ratio (SNR). The signal power is $\left|\mathbf{a}^T\mathbf{A}\right|^2$, the noise power is $|\mathbf{a}|^2\sigma^2$. The SNR is proportional to $\left|\mathbf{a}^T\mathbf{A}\right|^2 \Big/ \left(|\mathbf{A}|^2|\mathbf{a}|^2\right)$. This term reaches the value 1, which is its maximum, for $\mathbf{a} = c\mathbf{A}$ where c is a constant. The weights of $\mathrm{Re}\,\underline{I}$ in the expressions for $i_{1,2,3}$ are 1, –1/2, –1/2. The weights of $\mathrm{Im}\,\underline{I}$ in $i_{1,2,3}$ are 0, $\sqrt{3}/2$, $-\sqrt{3}/2$. We choose c = 2/3 and find

$$\mathrm{Re}\,\underline{I} = (2i_1 - i_2 - i_3)/3, \quad \mathrm{Im}\,\underline{I} = (i_2 - i_3)/\sqrt{3}. \tag{3.156}$$

The expressions with maximized SNRs contain $\mathrm{Re}\,\underline{I}$, $\mathrm{Im}\,\underline{I}$ alone, and are the formulas needed to invert (3.153). If $i_{1,2,3}$ are given, then the subsequent application of (3.156) and (3.153) results in a new set $i_{1,2,3}$ which can be written by the redefinition

$$[x_1 \quad x_2 \quad x_3]^T := [x_1 \quad x_2 \quad x_3]^T - (1/3)(x_1 + x_2 + x_3). \tag{3.157}$$

with $x = i$. This means that the new $i_{1,2,3}$ are zero-mean variables.

It is now possible to derive an equation for the calculation of a 3-phase product $\underline{W} = \underline{U}\underline{V}$ of two 3-phase signals $\underline{U}, \underline{V}$, using $\mathrm{Re}\underline{U} = (2u_1 - u_2 - u_3)/3$, $\mathrm{Im}\underline{U} = (u_2 - u_3)/\sqrt{3}$, $\mathrm{Re}\underline{V} = (2v_1 - v_2 - v_3)/3$, $\mathrm{Im}\underline{V} = (v_2 - v_3)/\sqrt{3}$, $\underline{W} = \underline{U}\underline{V}$, $w_1 = \mathrm{Re}\underline{W}$, $w_{2,3} = -(1/2)\mathrm{Re}\underline{W} \pm (\sqrt{3}/2)\mathrm{Im}\underline{W}$. The first signal is

$$\begin{aligned} w_1 &= \mathrm{Re}\underline{U}\,\mathrm{Re}\underline{V} - \mathrm{Im}\underline{U}\,\mathrm{Im}\underline{V} \\ &= \frac{2}{9}(2u_1v_1 - u_1v_2 - u_1v_3 + 2u_2v_3 - u_2v_1 - u_2v_2 + 2u_3v_2 - u_3v_1 - u_3v_3). \end{aligned} \tag{3.158}$$

This complicated expression can be simplified if we subtract from each 3-phase signal its mean before calculating $\underline{W} = \underline{U}\underline{V}$. So we implement (3.157) for at least one of the signals $\underline{U}$, $\underline{V}$. This can be done directly, without going through (3.156) and (3.153). After such pre-conditioning one may apply $u_1 + u_2 + u_3 = 0$ or $v_1 + v_2 + v_3 = 0$ in (3.158) and obtains

$$w_1 = (2/3)(u_1v_1 + u_2v_3 + u_3v_2). \tag{3.159}$$

The other expressions

$$w_2 = (2/3)(u_3v_3 + u_1v_2 + u_2v_1), \quad w_3 = (2/3)(u_2v_2 + u_3v_1 + u_1v_3) \tag{3.160}$$

are found in a similar way. A particularly simple possibility to obtain them is to write $\left(e^{\pm j2\pi/3}\underline{W}\right) = \left(e^{\pm j2\pi/3}\underline{U}\right)\underline{V}$. This means that a cyclic index shift of w_k and u_k with unchanged indices of v_k is allowed in (3.159), from which follows (3.160). Application of (3.157) has an effect on shot and thermal noises but not on optical amplifier noise which can be considered as an unwanted part of the complex signal. If optical amplifier noise dominates, $u_1 + u_2 + u_3 = 0$ holds anyway and application of (3.157) is no longer necessary. But even if shot and thermal noises dominate $u_1 + u_2 + u_3 = 0$ holds no longer. If (3.159), (3.160) are calculated without prior application of (3.157) a moderate penalty will be suffered, as will be discussed later in the similar case of a 3-phase intradyne receiver for DPSK signals. A simpler, hence better alternative to the application of (3.157) is to redefine

$$[x_1 \quad x_2 \quad x_3]^T := \left(1/\sqrt{3}\right)[x_1 - x_3 \quad x_2 - x_1 \quad x_3 - x_2]^T. \tag{3.161}$$

The sum of the redefined variables vanishes as required. At the same time the complex variable has been phase shifted by $\pi/6$. In the noiseless case this amounts to the substitution $\underline{X} := e^{j\pi/6}\underline{X}$. Both (3.157) and (3.161) yield 3-phase signals with only 2 instead of 3 independent noise variables.

Problem: Signal and local oscillator are connected to fibers 1 and 2 of the coupler with transfer matrix (2.409). Show that photodiodes connected to two of the output fibers 4, 5, 6 (see Fig. on p. 105) yield photocurrents with 90° phase difference. Which advantages and disadvantages exist compared to the 90° hybrid of Fig. 3.37?

Solution: The output fields are

$$\begin{bmatrix} \mathbf{E}_4 \\ \mathbf{E}_5 \\ \mathbf{E}_6 \end{bmatrix} = \mathbf{T} \begin{bmatrix} \mathbf{E}_S \\ \mathbf{E}_{LO} \\ 0 \end{bmatrix} = e^{j\xi} \frac{1}{\sqrt{5}} \begin{bmatrix} \mathbf{E}_S + (-1-j)\mathbf{E}_{LO} \\ (-1-j)\mathbf{E}_S + \mathbf{E}_{LO} \\ (-1-j)\mathbf{E}_S + (-1-j)\mathbf{E}_{LO} \end{bmatrix}. \quad (3.162)$$

The photocurrents are $i_{1,2,3} = \frac{R}{2}\left|\mathbf{E}_{4,5,6}\right|^2$,

$$\begin{aligned} i_1 &= (R/10)\left(\left|\mathbf{E}_S\right|^2 + 2\mathrm{Re}\left((-1-j)\mathbf{E}_S^+\mathbf{E}_{LO}\right) + 2\left|\mathbf{E}_{LO}\right|^2\right) \\ i_2 &= (R/10)\left(2\left|\mathbf{E}_S\right|^2 + 2\mathrm{Re}\left((-1+j)\mathbf{E}_S^+\mathbf{E}_{LO}\right) + \left|\mathbf{E}_{LO}\right|^2\right) \\ i_3 &= (R/10)\left(2\left|\mathbf{E}_S\right|^2 + 2\mathrm{Re}\left(\mathbf{E}_S^+\mathbf{E}_{LO}\right) + 2\left|\mathbf{E}_{LO}\right|^2\right) \end{aligned} . \quad (3.163)$$

The interference terms in $i_{1,2}$ have equal amplitudes and 90° phase difference, while i_3 is of no use and needs not be detected. The advantage of this optoelectronic 90° hybrid is that only two photoreceivers are required and no arithmetic function. Furthermore, like for the 120° hybrid, no phase needs to be stabilized. The disadvantages are that 2/5 of both light powers is wasted to fiber output 6, and that it is not possible to suppress the direct detection terms. In coherent receivers where optical amplifier noise dominates this may be well permissible.

QPSK combined with polarization multiplex quadruples channel capacity over state-of-the-art intensity-modulated systems but some modifications become necessary. The channel capacity can be doubled by adding the information bits $b_{3,4} = \pm 1$ in the second polarization mode. The data symbol is

$$\underline{\mathbf{c}} = \begin{bmatrix} (2b_1 - 1) + j(2b_2 - 1) \\ (2b_3 - 1) + j(2b_4 - 1) \end{bmatrix}. \quad (3.164)$$

Both vector components can be recovered separately in a polarization diversity (and intradyne) receiver with a common local oscillator. Details will be discussed later. Suitable processing provided, there are two independent intradyne signals with equal carrier phases. They can be written as a complex Jones vector $\begin{bmatrix} \underline{I}_1 & \underline{I}_2 \end{bmatrix}^T \propto \underline{\mathbf{c}} \cdot e^{j\varphi'}$. For carrier recovery each polarization diversity branch needs a frequency multiplier as defined above. The phase-aligned frequency-multiplied carriers $\underline{I}_1^4$, $\underline{I}_2^4 \propto -e^{j4\tilde{\varphi}_{IF}}$ of both branches are added before being passed through the lowpass filters and the regenerative frequency dividers. This addition is quite advantageous because it increases the carrier recovery SNR by 3dB.

For calculation of the laser linewidth tolerance we neglect frequency divider errors for the time being and proceed as follows: The decision variable in one decision branch of a QPSK receiver subject to the phase error $\Delta\varphi$ is for example $\mathrm{Re}\,\underline{D} = (2b_1 - 1)\cos\Delta\varphi - (2b_2 - 1)\sin\Delta\varphi$. The resulting BER is

$$\mathrm{BER} = \int_{-\infty}^{\infty} (1/2)\mathrm{erfc}\left(Q_0 \begin{pmatrix} \cos\Delta\varphi \\ -(2b_1-1)(2b_2-1)\sin\Delta\varphi \end{pmatrix} \Big/ \sqrt{2} \right) \cdot p_{\Delta\varphi}(\Delta\varphi) \cdot d\Delta\varphi \quad (3.165)$$

where $p_{\Delta\varphi}(\Delta\varphi)$ is the PDF of $\Delta\varphi$. In an ideal receiver $Q_0 = \sqrt{2(\text{photons/bit})}$ holds where the number of photons/bit holds either at the intradyne receiver input or at the input of a preceding ideal optical amplifier. Ideally, 18 photons/bit needed for a BER = 10^{-9}. The phase difference $\tilde{\varphi}_{IF}(t) - \tilde{\varphi}_{IF}(t-\tau)$ accumulated during a time slot τ has a Gaussian distribution with the variance $2\pi|\tau|\Delta f$ where Δf is the IF linewidth. The PDF $p_{\Delta\varphi}(\Delta\varphi)$ has been determined by a time-discrete system simulation over 10^5 symbols. The lowpass filter was simulated as a transversal filter of finite length with equal coefficients. The assignments in the frequency divider equations have been implemented with a one symbol delay. $p_{\Delta\varphi}(\Delta\varphi)$ was inserted into (3.165) to evaluate the BER. The true PDF and its Gaussian approximation yield similar results. For Fig. 3.41 we use the Gaussian approximation of $p_{\Delta\varphi}(\Delta\varphi)$ because it predicts the BER with less random fluctuations. Missing points indicate that $|\Delta\varphi| > \pi/4$ occurred, which means that a frequency division error has occurred in the carrier recovery. Nevertheless a non-vanishing number of frequency division errors will not preclude operation because they will result in single bit errors if the data streams $b_{1...4}$ are differentially encoded, and differentially decoded after decision in the receiver. In order to combine acceptable sensitivity and large phase noise tolerance for BER = 10^{-9} in a single-polarization QPSK system the lowpass filter width is chosen as $0.0625/T$ where T is the symbol duration. An intermediate frequency (IF) linewidth / symbol rate ratio $\Delta f \cdot T$ of at least 0.0005 seems to be tolerable (Fig. 3.41 top). A polarization multiplexed QPSK system is also simulated. A lowpass filter width equal to $0.125/T$ yields a tolerable $\Delta f \cdot T$ of at least 0.001 (Fig. 3.41 bottom). Note however, that a complete system evaluation requires frequency division errors to be assessed, also for lower BER. So the importance of Fig. 3.41 is limited.

Straightforward simplifications of the above allow to evaluate binary PSK systems: $2b_2 - 1 = 2b_4 - 1 = 0$, N = 2, only one frequency doubler, no need to invert the frequency-multiplied signal due to $\underline{c}^2 = 1$, only one frequency halver, $|\Delta\varphi| > \pi/2$ indicates division error. The optimum sensitivity is again 18 photons/bit. The tolerable $\Delta f \cdot T$ value is at least 0.005 (0.01) for the single (dual) polarization case, using lowpass filter widths of $0.0625/T$ ($0.125/T$).

The feedforward carrier recovery for QPSK/BPSK receivers is extremely laser linewidth tolerant compared to the alternative of implementing an OPLL. Frequency halvers implemented as regenerative frequency dividers make the scheme applicable for baseband type intradyne receivers. These are attractive at 40 Gbaud, thereby enabling 160 Gbit/s transmission by dual polarization QPSK.

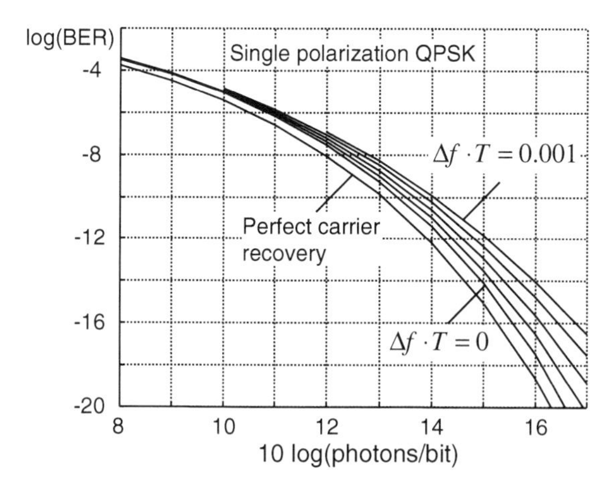

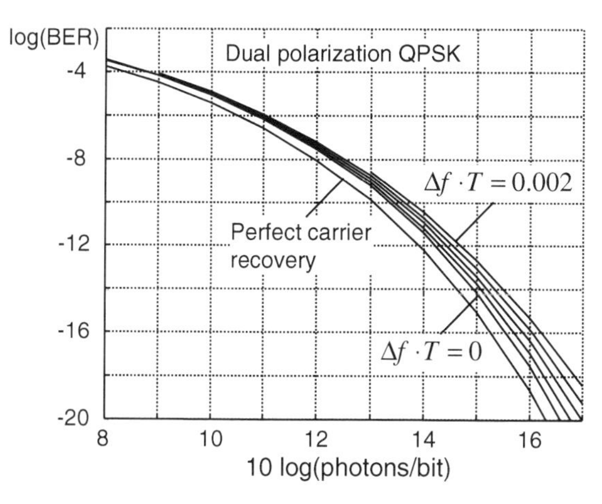

Fig. 3.41 log(BER) vs. 10 log(photons/bit) for various IF linewidth / symbol rate ratios in QPSK single (top) and dual (bottom) polarization receivers. The frequency division errors are neglected (see text). In each case, the leftmost curve holds for an ideal receiver with perfect carrier recovery [55]. © 2005 IEEE.

3.3.3 Receivers with Asynchronous Demodulation

The main advantages of coherent optical systems are

- superior sensitivity (in case of synchronous detection),
- improved selectivity (because electrical filtering works the same way as optical filtering in a direct detection receiver) and
- fast tunability (depending on the properties of the LO).

However, the laser linewidth and architectural requirements of synchronous demodulation are quite stringent. Easier to implement is asynchronous demodulation. As a first example consider a DPSK heterodyne receiver (Fig. 3.42 left).

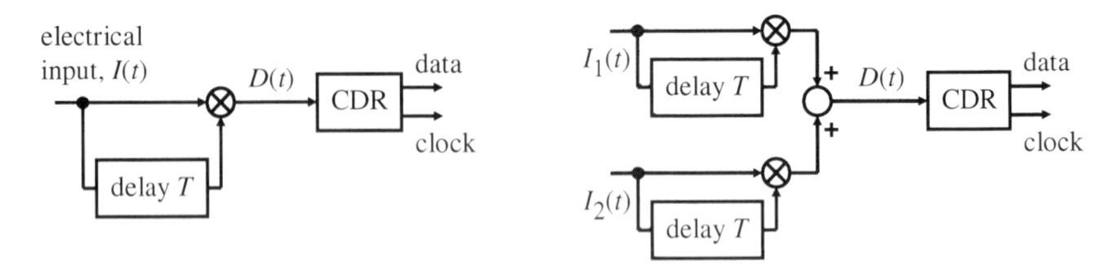

Fig. 3.42 IF and baseband parts of asynchronous DPSK heterodyne receiver, to be connected to the electrical output of Fig. 3.33 (left). Phase diversity counterpart thereof, to be connected to $I_1(t)$, $I_2(t)$ of Fig. 3.37 (right).

The heterodyne signal $I(t)$ and a copy of it, $I(t-T)$, that is delayed by one symbol duration T are multiplied. The IF must be chosen such that $\omega_{IF}T = k\pi$ where k is an integer. The multiplier output signal has to be lowpass-filtered to remove spectral components around $2\omega_{IF}$, unless the CDR does this automatically due to its lowpass behavior by design. Just like the electrical field in (3.114) the heterodyne signal can be written as a complex signal with noise,

$$\underline{I}(t) = \left(c_i\sqrt{\tilde{\mu}_0/M} + \tilde{u}_1 + j\tilde{u}_2\right)e^{j\omega_{IF}t}. \tag{3.166}$$

The real, lowpass-filtered output signal of the delay-line multiplier is then

$$D(t) \propto \operatorname{Re}\left(\underline{I}(t)\underline{I}^*(t-T)\right) = \operatorname{Re}\underline{I}(t)\operatorname{Re}\underline{I}(t-T) + \operatorname{Im}\underline{I}(t)\operatorname{Im}\underline{I}(t-T), \tag{3.167}$$

just like the mean of the real transported electrical power is $(1/2)\operatorname{Re}\left(\underline{U}\,\underline{I}^*\right)$. This is equivalent to (3.115). If there is an optical preamplifier with $G \to \infty$, receiver noise becomes insignificant. $D(t)$ is the difference of noncentral and central χ^2-distributed independent RVs. The BER is given by (3.120). However, the polarization-sensitive nature of the heterodyne receiver blocks the noise in one polarization ($N = M$) – and also at least partly the signal, should the LO polarization be misaligned. According to (3.121) the best achievable receiver sensitivity (quantum limit, $N = 1$) for BER = 10^{-9} is $\tilde{\mu}_0 = 20\tilde{\mu}$ at the optical amplifier input.

Let us also calculate the shot noise limit of the DPSK heterodyne receiver including receiver noise. In order to stick to our model an IF bandpass filter is now needed before the delay-line multiplier, just as in the PSK heterodyne receiver without optical preamplifier. Since the PSK heterodyne receiver had the same IF noise statistics whether operated alone or with an optical preamplifier of infinite gain, and since the shot noise limit in the first was equal to the quantum limit of the latter case, the DPSK heterodyne BER and shot noise limit are also the same as the optically preamplified DPSK heterodyne BER and shot noise limit, i.e. 20 photoelectrons/bit or photons/bit for $N = 1$.

Fig. 3.42 right shows a setup that is equivalent to the left side but works in the baseband in an intradyne receiver. These are sometimes called phase diversity receivers. The in phase and quadrature currents $I_1(t)$, $I_2(t)$ can be regarded as an analytical heterodyne signal $\underline{I}(t) = I_1(t) + jI_2(t)$ with a low IF that can (but need not) be zero. Lowpass filtering requirements are considerably relaxed because the doubled IF $2\omega_{IF}$ is suppressed due to $c_i \cos\tilde{\varphi}(t) c_{i-1} \cos\tilde{\varphi}(t-T)$ $+ c_i \sin\tilde{\varphi}(t) c_{i-1} \sin\tilde{\varphi}(t-T) = c_i c_{i-1} \cos(\tilde{\varphi}(t) - \tilde{\varphi}(t-T))$. If there is no or just insignificant phase noise, $\cos(\tilde{\varphi}(t) - \tilde{\varphi}(t-T)) \approx 1$ holds as desired.

The scheme can also be modified for 3-phase signals $i_{1,2,3}$. Ref. [67] discusses this and other coherent receiver types with asynchronous demodulation. One needs to determine

$$D(t) \propto (2/3)(i_1(t) i_1(t-T) + i_2(t) i_2(t-T) + i_3(t) i_3(t-T)). \tag{3.168}$$

In the noiseless case $i_k(t) = c_i \cos(\tilde{\varphi}(t) + (k-1)2\pi/3)$, $i_k(t-T)$ $= c_{i-1} \cos(\tilde{\varphi}(t-T) + (k-1)2\pi/3)$ ($k = 1,2,3$) this results in $D(t) \propto c_i c_{i-1}$. Each product $i_k(t) i_k(t-T)$ can be written as $\frac{1}{4}\left((i_k(t) + i_k(t-T))^2 - (i_k(t) - i_k(t-T))^2\right)$. If independent zero-mean noise variables accompany the signal then $D(t)$ is the difference of a non-central and a central χ^2 variable, just like in (3.115) but with 3/2 times as many degrees-of-freedom. Each mode has here 3 degrees-of-freedom. If the associated moderate penalty is not tolerable it is sufficient to apply (3.161) (or (3.157)) before calculating $D(t)$. These transformations leave the ratio of total signal power divided by the variance of one of the noise independent noise variables unchanged; only the number of degrees-of-freedom reduces to 2/3.

Seen from another point-of-view, if one implements (3.167) using (3.156) and 3-phase signals then one ends up with a linear combination of 9 products to be formed. The expression simplifies into (3.168) if the sum of the 3-phase signals vanishes.

Consider now a true phase modulation (3.130) in conjunction with the trapezoidal pulse of (3.128). The angular optical frequency deviation is the derivative thereof,

$$\begin{aligned} \Delta\omega(t) &= \sum_{i=-\infty}^{\infty} c_i \frac{\pi}{2} \frac{dk(t-iT)}{dt} \\ &= \begin{cases} \pm\pi/\tau & \text{for } |t - (i+1/2)T| \le \tau/2,\ c_i = \pm 1,\ c_{i-1} = -c_i \\ 0 & \text{else} \end{cases}. \end{aligned} \tag{3.169}$$

Each frequency deviation pulse causes a phase change by $\pm\pi$. Clearly, the existence of both positive and negative frequency deviations will broaden the optical spectrum more than necessary. For low bit rates it is not necessary to employ an optical phase modulator. Rather, a semiconductor laser can be directly modulated [68]. The frequency response of the frequency modulation caused by the pump current modulation of a semiconductor laser is highly non-uniform, however. There is a fast electronic contribution to it, and a slow (but much stronger) thermally induced contribution. This means that the step response will start with a frequency step that not only diminuishes but eventually acquires opposite polarity and much more strength. Therefore it is advisable to use a modulation format with a DC-free modulation current. This is possible with the bipolar frequency modulation scheme (3.169).

Some multielectrode lasers have a very large electronic frequency modulation effect that is even a lot stronger than the thermal component. In that case the step response is a frequency step that diminuishes somewhat as the thermal effect sets on, but keeps the same polarity. In such cases it would be possible to choose only frequency deviation pulses of like polarity,

$$\Delta\omega(t)=\begin{cases}\pi/\tau & \text{for } \left|t-(i+1/2)T\right|\le\tau/2,\ c_i=-c_{i-1} \\ 0 & \text{else}\end{cases} \quad (3.170)$$

without significant eye opening impairment. This narrows the optical spectrum considerably. Note that the upper case corresponds to $b_i=-1$ due to $b_i=c_i c_{i-1}$, and the lower case applies also for $b_i=1$. Differential encoding is no more necessary because the optical phase is the integral of the angular optical frequency. Another practical problem, and reason for spectral broadening, is the generation of short pulses of duration τ for $b_i=-1$. It would be convenient to choose $\tau=T$ which means the frequency deviation becomes an NRZ signal.

Instead of a frequency deviation that is either positive or zero with equal probabilities it is more instructive to describe the scheme by a bipolar frequency deviation

$$\Delta\omega(t)=b_i m\pi/T \quad (3.171)$$

where m is a frequency modulation index and $b_i=\pm1$ is bipolar. This is called continuous-phase frequency shift keying (CPFSK). In our special case, $m=1/2$, it is called minimum shift keying (MSK). Note that the bipolar definition of the frequency deviation changes the mean IF by $m\pi/T$. For an unmodulated signal at the new center IF the phase delay of the delay line is therefore such that the mean multiplier output signal equals zero. A serious disadvantage of MSK is that it takes a complete symbol duration T until the necessary phase difference of π has built up between the signal and its delayed replica. This situation is worsened by any IF filtering or optical bandpass filtering and by any baseband filtering behind the delay-line multiplier. Since filtering is indispensable in practice there is in general substantial intersymbol interference associated with MSK. Fortunately it is possible to circumvent this problem. The required π phase shift is achieved in a time $T/(2m)$, and it is sufficient to choose $m>1/2$. Of course the frequency deviation (3.171) must still lead to a maximally positive or negative demodulator output signal. As a consequence the new delay time τ must be chosen equal to $\tau = T/(2m)$, and the mean IF must be chosen such that for an unmodulated signal at the mean IF there is a phase delay of $\pm\pi/2$ between the signals at the multiplier input (Fig. 3.43).

A CPFSK phase diversity receiver with zero mean IF can not be built with the setup of Fig. 3.42 right since with respect to the changed mean IF the demodulator needs to perform the operation

$$D(t) \propto \operatorname{Im}\left(\underline{I}(t)\underline{I}^*(t-\tau)\right) = \operatorname{Im}\underline{I}(t)\operatorname{Re}\underline{I}(t-\tau) - \operatorname{Re}\underline{I}(t)\operatorname{Im}\underline{I}(t-\tau). \tag{3.172}$$

Fortunately a slight circuit modification allows indeed to implement this operation for $\underline{I}(t) = I_1(t) + jI_2(t)$ (Fig. 3.44) [59].

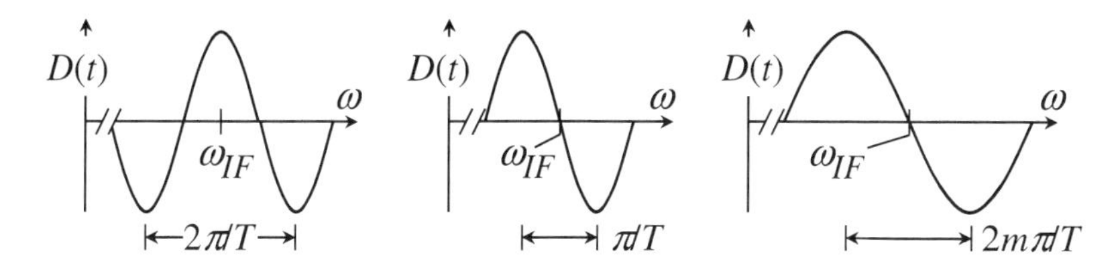

Fig. 3.43 Delay-line demodulator characteristics for DPSK (left), MSK (middle) and general CPFSK (right) assuming unmodulated signals

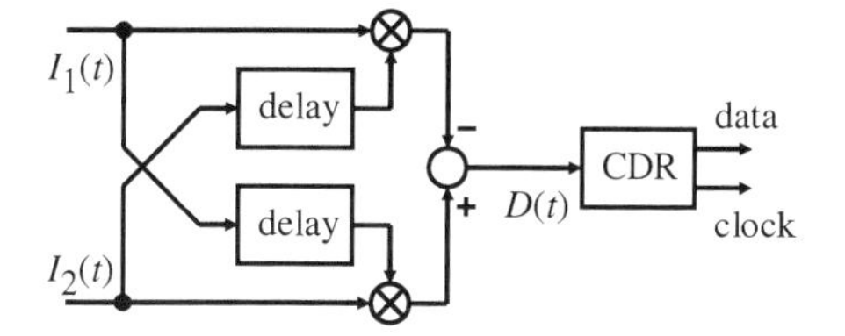

Fig. 3.44 Baseband part of CPFSK phase diversity receiver, to be connected to $I_1(t)$, $I_2(t)$ of Fig. 3.37 (right)

The highest possible sensitivity of CPFSK transmission is achieved for $m \approx 0.7$. It is about 1...2 dB worse than that of DPSK transmission. For large modulation indices m the optical bandwidth requirement becomes excessive and the sensitivity is about 3 dB worse than that of DPSK. In terms of mean power it is then identical to that of ASK (while the required peak power is only half as large). Note that all what has been said about CPFSK is also applicable to direct detection receivers with a Mach-Zehnder interferometer and a differential photoreceiver.

There has been much hope for the success of coherent CPFSK systems with data rates from 100 Mbit/s to 10 Gbit/s, both for trunk lines and multichannel access systems until about 1991. Since then coherent technology has been progressively abandoned. Nowadays the prospects of coherent technology lie in WDM trunk line technology with synchronous demodulation of (Q)PSK signals at data rates of at least 10 Gbit/s, possibly combined with polarization division multiplex. Polarization diversity receivers with asynchronous demodulation can inherently tolerate endless signal polarization changes.

Table 3.4 summarizes ideal receiver sensitivities for various modulation and detection schemes. Practically achievable sensitivities are of course worse, especially for the asynchronous demodulation schemes where a compromise has to be found between low mode number and low intersymbol interference.

Table 3.4 Mean required number of photons or photoelectrons per bit (quantum and shot noise limit, respectively) for BER = 10^{-9} of various modulation and detection schemes. Technically interesting cases are typed in boldface. Preamp. means optical preamplifier.

Demodulation	asynchronous				synchronous		
Modulation format	ASK	CPFSK	DPSK	DQPSK	ASK	PSK	QPSK
Direct detection with noise-free receiver	~10		~20				
Direct detection with preamp., 2 (1) polarizations	**40.7** (38)	≥ 31 (≥ 29)	**21.9** (20)	**~32** (**~31**)			
Heterodyne or phase diversity detection without or with preamp. (with preamp. but no image rejection)	38 (76)	≥ 29 (≥ 58)	20 (40)	~31 (~62)	36 (72)	18 (36)	**18** (36)
Homodyne detection with (without) preamp.					36 (18)	18 (9)	18 (18)

3.3.4 Laser Linewidth Requirements

In order to take phase noise into account the optical signal phasor must be multiplied by $e^{j\varphi_S(t)}$. This will influence the BER. Since usually the linewidth is much smaller than the symbol rate it is reasonable to assume a $\varphi_S(t)$ that is constant over each symbol duration. In an ideal PSK homodyne receiver with fixed LO phase the BER would be equal to $\mathrm{BER} = (1/2)\mathrm{erfc}\left(Q/\sqrt{2}\right)$ with $Q = Q_0 \cos\varphi_S(t)$ with $Q_0 = \sqrt{4\eta T P_S/(hf)}$ for PSK homodyne. In reality the OPLL takes care that the LO phase follows the signal phase as closely as possible. What matters is therefore just a phase difference $\varphi(t)$. The phase difference dependent BER is

$$\mathrm{BER}(\varphi) = (1/2)\mathrm{erfc}\left(Q_0 \cos\varphi/\sqrt{2}\right), \tag{3.173}$$

and the averaged BER which is of interest is

$$\mathrm{BER} = \int_{-\infty}^{\infty} p_\varphi(\varphi)\mathrm{BER}(\varphi)d\varphi\,. \tag{3.174}$$

In the homodyne receiver

$$\begin{aligned}&\varphi(t)=\varphi_S(t)-\varphi_{LO}(t)=\varphi_1(t)-\varphi_2(t)\\&\varphi_1(t)=\varphi_S(t)-\varphi_{LO,0}(t)\qquad \varphi_2(t)=\varphi_{LO}(t)-\varphi_{LO,0}(t)\end{aligned}\tag{3.175}$$

holds. In the free-running, unlocked state no correction $\varphi_2(t)$ is being imposed by an OPLL on the free-running LO phase angle $\varphi_{LO,0}(t)$. Each phase difference $\Delta\varphi_k(t,\tau)=\varphi_k(t)-\varphi_k(t-\tau)$ ($k=\{S,\ LO,0\}$) has a Gaussian distribution with a variance $\sigma^2_{\Delta\varphi_k(t,\tau)}=2\pi|\tau|\Delta f_k$ proportional to $|\tau|$, where Δf_k is the respective laser linewidth. The distribution of $-\Delta\varphi_{LO}(t,\tau)$ is identical to that of $\Delta\varphi_{LO}(t,\tau)$. Since the two $\Delta\varphi_k(t,\tau)$ are also independent RVs the distribution of $\Delta\varphi_1(t,\tau)=\Delta\varphi_S(t,\tau)-\Delta\varphi_{LO,0}(t,\tau)$ is also Gaussian, and its variance is

$$\sigma^2_{\Delta\varphi_1(t,\tau)}=\sigma^2_{\Delta\varphi_S(t,\tau)}+\sigma^2_{\Delta\varphi_{LO,0}(t,\tau)}=2\pi|\tau|\Delta f\qquad \Delta f=\Delta f_S+\Delta f_{LO}\,.\tag{3.176}$$

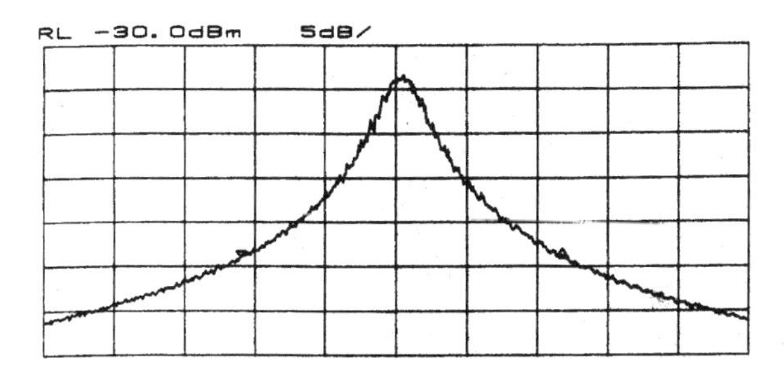

Fig. 3.45 Beat signal of two 1550 nm lasers (~20 mW) with long cavities (600 and 800 μm). Center IF: 1130 MHz. Horizontal span: 2 MHz/div. Vertical span: 5 dB/div. Measured total –20 dB linewidth is 9 MHz, corresponding to a –3 dB linewidth of ~0.9 MHz.

A heterodyne signal without phase locking, which can be seen as a complex quantity $\underline{I}(t)=R\underline{\mathbf{E}}^+_{LO,0}\underline{\mathbf{E}}_{S,0}e^{j\omega_{IF}t}\propto e^{j\varphi_1(t)}$, has therefore a sum linewidth Δf equal to the sum of the individual linewidths. Standard DFB lasers typically exhibit linewidths of 1...20MHz. Over wide ranges, the linewidth scales inversely with the power and inversely with the square of the cavity length but good sidemode suppression and single-mode yield require short cavities. A very narrow beat linewidth is shown in Fig. 3.45.

As long as $|\varphi(t)|<<1$ holds the error signal is proportional to the true phase error $\varphi(t)$, and any noise which accompanies it is independent of it and Gaussian. It is therefore possible to apply linear PLL theory.

For $\tau \geq \tau_2 \geq 0$ the autocorrelation function $l_{\Delta\varphi_1}(\tau_2)$ of $\Delta\varphi_1(t,\tau)$ is

$$\begin{aligned} l_{\Delta\varphi_1}(\tau_2) &= \langle \Delta\varphi_1(t,\tau)\Delta\varphi_1(t-\tau_2,\tau)\rangle \\ &= \langle (\Delta\varphi_1(t,\tau_2)+\Delta\varphi_1(t-\tau_2,\tau-\tau_2))(\Delta\varphi_1(t-\tau_2,\tau-\tau_2)+\Delta\varphi_1(t-\tau,\tau_2))\rangle \end{aligned} . \tag{3.177}$$

The three different RVs on the right side are all zero-mean Gaussian and pairwise independent. Only the product of the two like ones survives the expectation value operation, $l_{\Delta\varphi_1}(\tau_2) = \left\langle \Delta\varphi_1^2(t-\tau_2,\tau-\tau_2)\right\rangle = 2\pi|\tau-\tau_2|\Delta f$. For $\tau_2 \geq \tau \geq 0$ the autocorrelation function vanishes because there is no temporal overlap. Taking symmetry into account, this results in

$$l_{\Delta\varphi_1}(\tau_2) = \begin{cases} 2\pi(|\tau|-|\tau_2|)\Delta f & \text{for } |\tau_2| \leq |\tau| \\ 0 & \text{else} \end{cases} . \tag{3.178}$$

This results in a two-sided power spectral density

$$L_{\Delta\varphi_1}(f) = 2\pi\Delta f \frac{\sin^2 \pi f|\tau|}{\pi^2 f^2} . \tag{3.179}$$

With $d\varphi_1(t)/dt = \lim_{\tau\to 0} \Delta\varphi_1(t,\tau)/\tau$ the power spectral density of $d\varphi_1(t)/dt$ with $d\varphi_1(t)/dt$ ○—● Ω_1 is

$$|\Omega_1|^2 \equiv L_{d\varphi_1/dt}(f) = \lim_{\tau\to 0} \tau^{-2} L_{\Delta\varphi_1}(f) = 2\pi\Delta f . \tag{3.180}$$

The unavoidable noise added in the estimation of the true phase error $\varphi(t)$ is modeled by a Gaussian estimation error $\varphi_e(t)$. For example, in a heterodyne receiver half the noise power is in phase, half is in quadrature with the signal. In the shot noise limit this yields the two-sided power spectral density

$$L_{\varphi_e} = \frac{1}{Q^2/T} = \frac{hf}{2\eta P_S} . \tag{3.181}$$

According to p. 180,

$$\begin{aligned} &\varphi_2(t) = h_o(t) * (\varphi_1(t) - \varphi_2(t) + \varphi_e(t)) \quad \text{○—●} \quad \Phi_2 = H_o(\Phi_1 - \Phi_2 + \Phi_e), \\ &\Phi_1 - \Phi_2 = \frac{1}{1+H_o}\Phi_1 - \frac{H_o}{1+H_o}\Phi_e \\ &= \frac{1}{1+H_o}\frac{\Omega_1}{j2\pi f} - \frac{H_o}{1+H_o}\Phi_e \end{aligned} \tag{3.182}$$

holds where $H_o = \frac{1}{j2\pi f} K_0 H_R(f) K_d$ is the transfer function of the open PLL. This corresponds to

$$L_{\varphi_1-\varphi_2} = \left|\frac{1}{1+H_o}\right|^2 \frac{|\Omega_1|^2}{|2\pi f|^2} + \left|\frac{H_o}{1+H_o}\right|^2 L_{\varphi_e}$$
$$= \left|j2\pi f + K_0 H_R(f) K_d\right|^{-2} \left(2\pi\Delta f + \left|K_0 H_R(f) K_d\right|^2 L_{\varphi_e}(f)\right). \tag{3.183}$$

The variance of $\varphi_1 - \varphi_2$ is $\sigma^2_{\varphi_1-\varphi_2} = \int_{-\infty}^{\infty} L_{\varphi_1-\varphi_2}(f)df$. The phase error is stationary because it has a constant variance

$$\sigma_\varphi^2 = \sigma^2_{\varphi_1-\varphi_2} = \int_{-\infty}^{\infty} \frac{2\pi\Delta f + \left|K_0 H_R(f) K_d\right|^2 L_{\varphi_e}(f)}{\left|j2\pi f + K_0 H_R(f) K_d\right|^2} df . \tag{3.184}$$

All involved RVs are Gaussian because they result from linear operations (convolution) on the Gaussian RV $\varphi_1(t)$. In addition to serious constraints resulting from unwanted time delay in the PLL the controller transfer function H_R may be chosen so as to minimize σ_φ^2. While a small loop bandwidth minimizes the influence of the phase error detection noise φ_e a large loop bandwidth is needed to keep the contribution of the combined laser linewidth Δf sufficiently low. Both PLL dead time and $L_{\varphi_e}(f)$ depend very much on the receiver architecture. While a PLL-based carrier recovery was chosen as an example the BER calculation is not restricted to it as long as one is able to find σ_φ^2.

Now it is possible to evaluate (3.174),

$$\mathrm{BER} = \int_{-\infty}^{\infty} \frac{1}{\sqrt{2\pi}\sigma_\varphi} e^{-\varphi^2/\left(2\sigma_\varphi^2\right)} (1/2)\,\mathrm{erfc}\left(Q_0 \cos\varphi/\sqrt{2}\right) d\varphi . \tag{3.185}$$

An analytical expression is derived in [69]. For any non-zero value of σ_φ^2 there exists a BER floor, i.e., a constant $\mathrm{BER} = \int_{\cos\varphi<0} \frac{1}{\sqrt{2\pi}\sigma_\varphi} e^{-\varphi^2/\left(2\sigma_\varphi^2\right)} d\varphi$ for $Q_0 \to \infty$. In order to keep Fig. 3.46 comparable to Fig. 3.24 a system with optical preamplifier has been chosen as a reference. Data holds therefore both for homodyne and heterodyne receivers. It also holds for the photoelectrons/bit of a heterodyne receiver without optical preamplifier.

The laser linewidth has to chosen low enough to allow a controller design which keeps the BER floor low enough. It is useful to keep it so low that the sensitivity for a given BER is degraded only negligibly, say, by 0.5 dB. For a BER = 10^{-9} the corresponding permissible phase error variance is about 0.03. Note that our calculation was not strictly accurate. In particular, the phase error detection will have a sine-like characteristic which saturates for large φ. This means that large φ which cause particularly large $\mathrm{BER}(\varphi)$ will be corrected not as fast as in the linear case, and this may increase the true phase noise penalty.

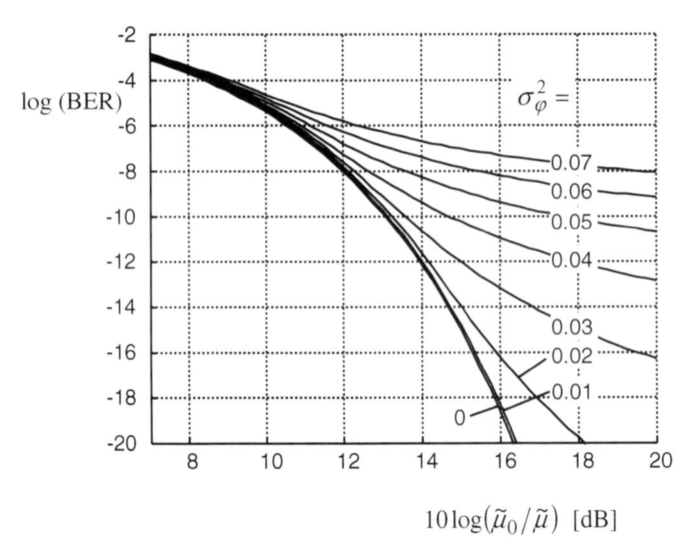

Fig. 3.46 BER performance for synchronous PSK in the presence of phase noise with a variance as indicated. Horizontal axis refers to SNR at optical preamplifier input for PSK heterodyne or homodyne.

As one tries to increase receiver sensitivity and fiber capacity by choosing an appropriate modulation and demodulation scheme the demands on laser linewidth increase dramatically. Data for a PSK homodyne receiver with residual carrier (Fig. 3.39) is not given; it is even worse than that of a Costas loop.

A QPSK system with the same optical power per data channel, i.e. unchanged Q_0, and the phasor (3.149) one synchronously demodulated data channel can be expressed by $\mathrm{Re}\left(\underline{c}_i e^{-j\varphi}\right) = \cos\varphi \,\mathrm{Re}\,\underline{c}_i + \sin\varphi \,\mathrm{Im}\,\underline{c}_i$, hence $Q = Q_0(\cos\varphi \pm \sin\varphi)$ is valid. The distribution of φ is symmetrical with respect to zero, so it is possible to choose one arbitrary sign. The

$$\mathrm{BER} = \int_{-\infty}^{\infty} \frac{1}{\sqrt{2\pi}\sigma_\varphi} e^{-\varphi^2/\left(2\sigma_\varphi^2\right)} (1/2)\,\mathrm{erfc}\left(Q_0(\cos\varphi + \sin\varphi)/\sqrt{2}\right) d\varphi\,. \tag{3.186}$$

is evaluated in Fig. 3.47. Tolerable phase noise variance and laser linewidths are about 1/10 of those permissible for PSK.

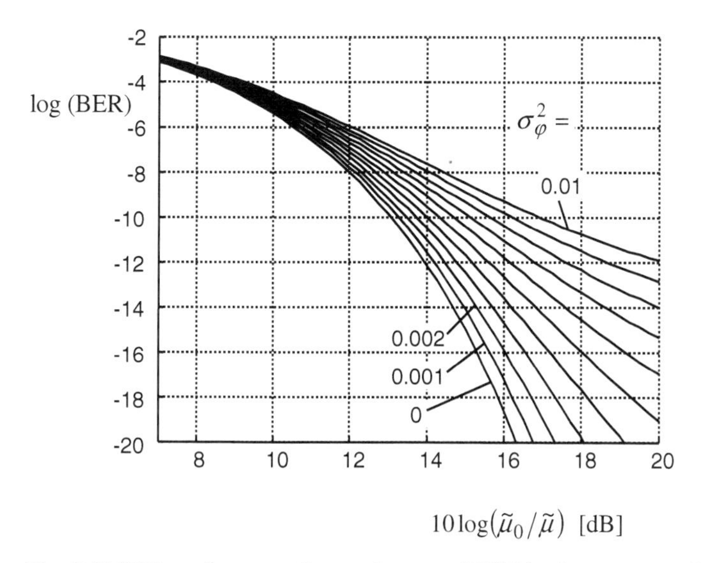

Fig. 3.47 BER performance for synchronous QPSK in the presence of phase noise with a variance as indicated. Horizontal axis refers to SNR at optical preamplifier input for each of the two data channels.

Table 3.5 Approximate values of permissible products of total laser linewidth times symbol duration for small degradation at BER = 10^{-9}. Synchronous demodulation assumes feedforward carrier recovery. Values for synchronous demodulation can approximately be doubled in case of polarization division multiplex.

	asynchronous demodulation			synchronous demodulation	
	ASK	CPFSK $m = 0.7$	DPSK	PSK	QPSK
Direct detection or heterodyne or intradyne / phase diversity with optical preamplifier	~0.1	0.01... 0.02	~0.004		
Heterodyne or intradyne / phase diversity detection				~0.005	~0.0005

While it is outside our present scope to completely evaluate various carrier recovery schemes and the corresponding phase noise tolerance, Table 3.5 lists some permissible products of total laser linewidth times symbol rate.

In order to calculate the phase noise tolerance of DPSK its BER must first be calculated for the case $\underline{c}_i\underline{c}_{i-1}^* = \pm e^{j\Delta\varphi(t,T)}$. This means the delay-line multiplier behaves similar to an ideal carrier recovery with a dead time of one symbol duration. The corresponding phase error variance is as low as $\sigma_{\Delta\varphi}^2 = 2\pi T\Delta f$. An analytical calculation of $\mathrm{BER}(\Delta\varphi)$ is possible for DPSK and CPFSK [53] but the result is not fully straightforward to evaluate.

For comparison we lay out the path for a numerical BER evaluation. Due to $\mathrm{Im}\left(\underline{c}_i \underline{c}_{i-1}^*\right) \neq 0$ (3.118) must now be replaced by

$$M_{\tilde{x}}\left(e^{-s}\right) = \left(1-(\tilde{\mu}/2)^2 s^2\right)^{-N} e^{\frac{-|C_1|^2 \tilde{\mu}_0 s}{1+(\tilde{\mu}/2)s} + \frac{|C_2|^2 \tilde{\mu}_0 s}{1-(\tilde{\mu}/2)s}}$$

$$= \left(1-(\tilde{\mu}/2)^2 s^2\right)^{-N} e^{\tilde{\mu}_0 s \frac{-\mathrm{Re}\left(c_i c_{i-1}^*\right) + (\tilde{\mu}/2)s}{1-(\tilde{\mu}/2)^2 s^2}} \qquad \left(|c_i|^2 = |c_{i-1}|^2 = 1\right). \tag{3.187}$$

The corresponding PDF is obtained by inverse transformation and $\mathrm{BER}(\Delta\varphi)$ by integration of the PDF tail. One can set $\cos\Delta\varphi = \mathrm{Re}\left(c_i c_{i-1}^*\right)$. Insertion of $\mathrm{BER}(\Delta\varphi)$ into $\mathrm{BER} = \int_{-\infty}^{\infty} p_{\Delta\varphi}(\Delta\varphi)\mathrm{BER}(\Delta\varphi)d\Delta\varphi$ with a zero-mean Gaussian $\Delta\varphi$ yields the final result. For $N = 2$ modes we may consider $\sigma_\varphi^2 \approx 0.025$ as the tolerable maximum (Fig. 3.48). The delay time $\tau = T$ equals one bit duration. According to $\sigma_\varphi^2 = 2\pi|\tau|\Delta f$ this corresponds to a linewidth times bit duration product of $\Delta f \cdot T \approx 0.004$ which is consistent with [68].

Numerical system modeling has the great advantage that any other nonideal effect (e.g., multiplier distortions, linear and nonlinear optical signal distortion) can be easily included. Disadvantage is of course the computational effort.

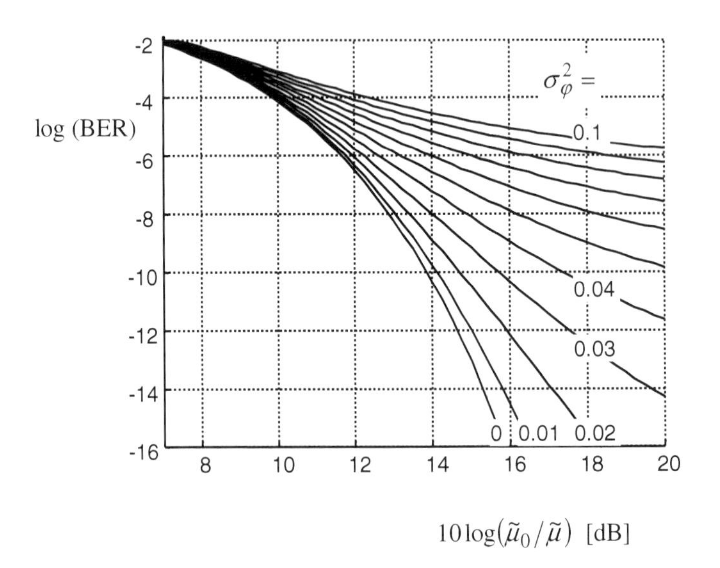

Fig. 3.48 BER performance for asynchronous DPSK with $N = 2$ modes in the presence of phase noise with a variance as indicated. Horizontal axis refers to SNR at optical preamplifier input.

3.3.5 Digital Coherent QPSK Receiver

The foregoing details the principle of coherent optical receivers. Analog circuitry is difficult to implement at high speeds. It is advantageous to build the receiver in such a way that the detected signals are digitized and then processed digitally. As one example we consider a polarization-multiplexed QPSK transmission system with intradyne receiver, feedforward carrier recovery and decision-directed electronic polarization control [63]. Subsequently we will discuss possible improvements and alternatives.

Fig. 3.49 shows a QPSK polarization division multiplex transmission system. The transmitter laser (TX) signal is split, QPSK-modulated in two branches and recombined with two (p = 1, 2) orthogonal polarizations in a polarization beam splitter (PBS). Two bipolar data bits d_{p1}, d_{p2} are to be transmitted in each polarization, four per symbol in total. d_{p1}, d_{p2} are Gray-encoded into a data quadrant number $n_{d,p}$ (Table 3.6). It is differentially encoded to form an encoded quadrant number $n_{c,p}$ with $n_{c,p}(i)=(n_{d,p}(i)+n_{c,p}(i-1))\mathrm{mod}\,4$. This number defines the quadrant of the transmitted complex symbol $\underline{c}_p = \mathrm{Re}\,\underline{c}_p + j\,\mathrm{Im}\,\underline{c}_p = (\pm 1 \pm j)/\sqrt{2}$ (p = 1, 2; Table 3.6), the real and imaginary parts of which drive the respective QPSK modulator. It holds $|\underline{c}_p|^2 = 1$.

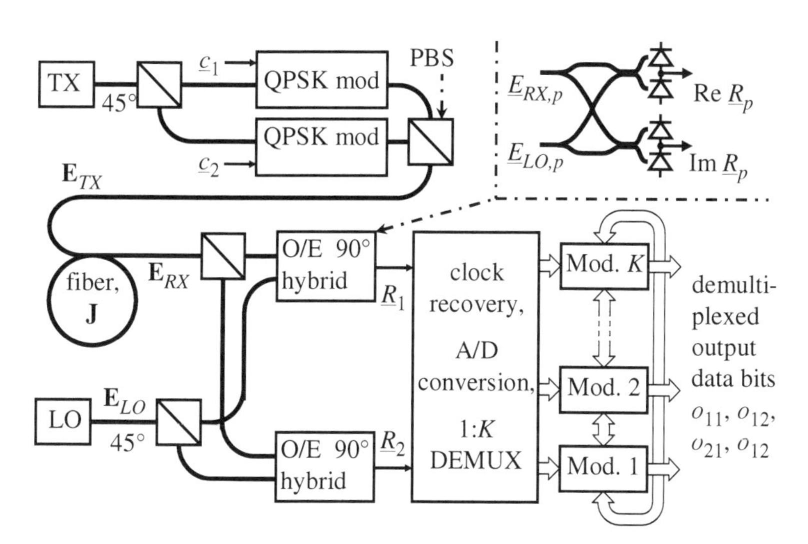

Fig. 3.49 Polarization multiplexed synchronous QPSK transmission system with digital receiver. Inset: Optoelectronic 90° hybrid. © 2005 IEEE.

Table 3.6 Bipolar bits vs. quadrant number. © 2005 IEEE.

d_{p1}, Re $\underline{c}_p$, Re $\underline{r}_p$, o_{p1}	d_{p2}, Im $\underline{c}_p$, Im $\underline{r}_p$, o_{p2}	$n_{d,p}$, $n_{c,p}$, $n_{r,p}$, $n_{o,p}$
$1/\sqrt{2}$	$1/\sqrt{2}$	0
$-1/\sqrt{2}$	$1/\sqrt{2}$	1
$-1/\sqrt{2}$	$-1/\sqrt{2}$	2
$1/\sqrt{2}$	$-1/\sqrt{2}$	3

The transmitted electric field $\mathbf{E}_{TX}$ is proportional to the vector $\mathbf{c} = [\underline{c}_1 \quad \underline{c}_2]^T$. The received electric field $\mathbf{E}_{RX} = \mathbf{J}\mathbf{E}_{TX}$ has been multiplied by the fiber Jones matrix $\mathbf{J}$.

The receiver features a 45°-polarized local oscillator laser (LO). Received and LO signals are each split into orthogonal polarizations, and the fields $\underline{E}_{RX,p}, \underline{E}_{LO,p}$ of each polarization p are detected in optoelectronic 90° hybrids (inset of Fig. 3.49). The in-phase and quadrature (I&Q) signals can be understood to be real and imaginary parts Re $\underline{R}_p$, Im $\underline{R}_p$ of a complex received intermediate frequency (IF) signal $\underline{R}_p$, and these form together a received IF signal vector $\mathbf{R} = [\underline{R}_1 \quad \underline{R}_2]^T = \mathbf{J} \cdot \mathbf{c} \cdot e^{j\varphi''}$. Angle φ'' is the phase difference between signal and LO lasers. The IF is chosen to be near or at zero.

The clock can be recovered in an extra intensity modulation direct detection receiver or from the electrical I&Q signals. After clock recovery the I&Q signals of both polarization branches are sampled at the symbol rate, digitized and demultiplexed 1:K to a low symbol rate where complex digital functions can be implemented in CMOS. The K data streams are processed in K modules (Mod.). The modules also have to communicate among each other, as indicated by the inputs and outputs terminated by circles in Figs. 3.50 and 3.51. The ith vector sample $\mathbf{R}$ enters the kth module (Fig. 3.50), where $k = i$ mod K. It is multiplied by a complex Jones matrix $\mathbf{M}$ to form a polarization-separated IF vector sample $\mathbf{X}(i) = [\underline{X}_1(i) \quad \underline{X}_2(i)]^T = \mathbf{MR}(i)$. Ideally, $\mathbf{MJ}$ is proportional to the unity matrix and it holds

$$\mathbf{X} = \mathbf{c} \cdot e^{j\varphi'} . \tag{3.188}$$

Angle $\varphi' - \varphi''$ is the phase shift introduced by the matrix product $\mathbf{MJ}$.

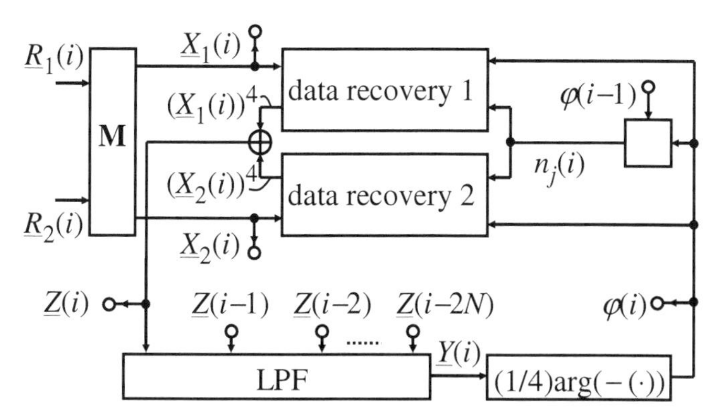

Fig. 3.50 Block diagram of kth ($k = i \bmod K$) digital signal processing module after the demultiplexer. Here and in Fig. 3.51, the inputs and outputs terminated by a circle may or must interface with other modules for carrier recovery. © 2005 IEEE.

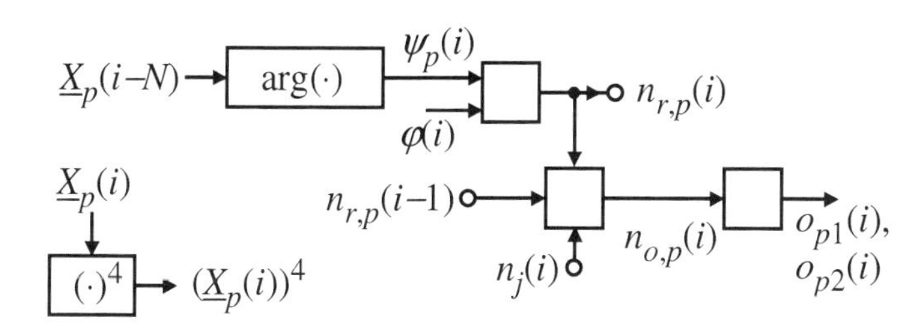

Fig. 3.51 Block diagram of "data recovery p" (polarization p = 1, 2). Signal $\underline{X}_p(i-N)$ is provided by another (not the kth) module. $n_{r,p}(i)$, $n_{r,p}(i-1)$ assure communication between data recoveries of neighbor modules. $o_{p1}(i)$, $o_{p2}(i)$ are output data bits (Fig. 3.49). © 2005 IEEE.

The components $\underline{X}_p$ (p = 1, 2) are processed in data recoveries (Fig. 3.51). Sample $\underline{X}_p(i)$ is raised to the 4th power. Due to $\underline{c}_p^4 = -4$ the modulation is thereby eliminated [70]. The resulting frequency-quadrupled carrier components $\underline{X}_p^4 \propto -e^{j4\varphi'}$ of both polarizations are added,

$$\underline{Z} = \underline{X}_1^4 + \underline{X}_2^4 \propto -e^{j4\varphi'} . \tag{3.189}$$

A further noise suppression results from a filtering of $\underline{Z}$, in a lowpass filter because the IF is at or near zero. A good filter may take the sum

$$\underline{Y}(i) = \sum\nolimits_{m=0}^{2N} \underline{Z}_p(i-m) \tag{3.190}$$

of the $2N+1$ most recent samples of the frequency-quadrupled carrier $\underline{Z}$. Most of them come from adjacent modules. The group delay in this filter (LPF) equals N symbols. $N = 2$ is a fairly good choice [70].

The filter also alters the phase angle. $\underline{Y} \propto -e^{j4\varphi}$ holds, where $4\varphi(i)$ would ideally be equal to $4\varphi'(i-N)$. The next step is to divide phase and frequency of $\underline{Y}$ by a factor of 4 in order to recover the carrier phase. The best choice is probably a 2D-lookup table which calculates carrier phase samples $\varphi(i) = (1/4)\arg(-\underline{Y}(i))$.

In each of the data recoveries ($p = 1, 2$), the phase angle $\psi_p(i) = \arg \underline{X}_p(i-N)$ of the received signal is calculated, with a delay equal to that experienced in the lowpass filter. For demodulation, an integer $n_{r,p}(i)$ which fulfills

$$n_{r,p}(i)\pi/2 \le \psi_p(i) - \varphi_p(i) < (n_{r,p}(i)+1)\pi/2 \tag{3.191}$$

is simply determined. It may be called a received quadrant number because $\underline{n}_{r,p}(i) = n_{c,p}(i-N)$ holds for $|\varphi(i) - \varphi'(i-N)| < \pi/4$.

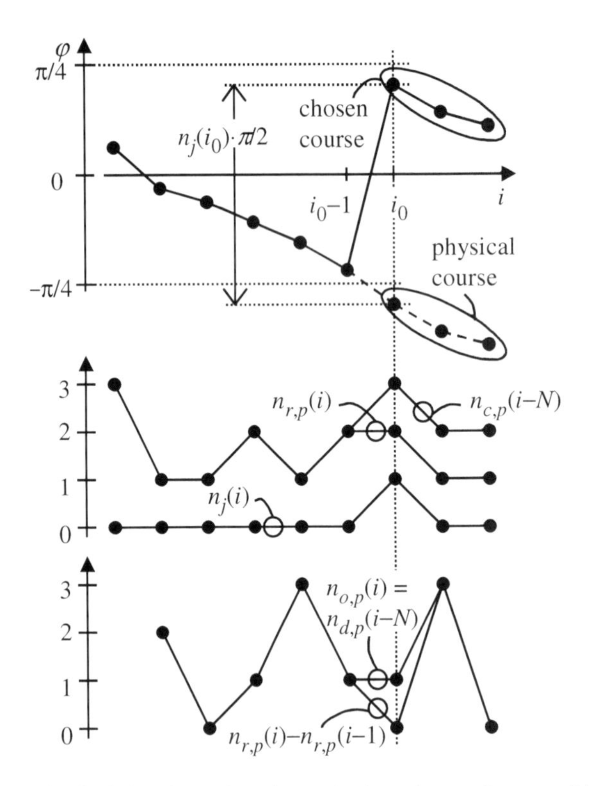

Fig. 3.52 Quadrant phase jump, its detection, and successful correction. © 2005 IEEE.

There is a 4fold ambiguity in the calculation of $\varphi(i)$, and $n_{r,p}(i)$ does therefore not contain all needed information. Choosing $\varphi(i)$ as close as possible to $\varphi(i-1)$ could solve the problem but is practically impossible because the correct quadrant can not be selected within one symbol duration. Instead, let $|\varphi(i)| \le \pi/4$ be always chosen. For correction of arising errors the kth module must detect whether φ has jumped by an integer multiple of $\pi/2$. It determines a quadrant jump number $n_j(i)$ for this purpose. This is an integer which fulfills

$$|\varphi(i) - \varphi(i-1) - n_j(i)\pi/2| < \pi/4 . \tag{3.192}$$

All quadrant numbers and operations are understood to be valid modulo 4 because the angle functions are periodic. Fig. 3.52 shows a quadrant phase jump at $i = i_0$ and its detection. The value $n_j(i_0) \ne 0$ indicates that all $n_{r,p}(i_+)$ with $i_+ \ge i_0$ carry an unwanted offset $-n_j(i_0)$. It is not possible to correct all $n_{r,p}(i_+)$ accordingly because this would have to be done at the symbol rate.

But the differential quadrant encoding at the transmitter side allows to calculate an output quadrant number

$$n_{o,p}(i) = (n_{r,p}(i) - n_{r,p}(i-1) + n_j(i)) \mathrm{mod}\, 4 = n_{d,p}(i-N) . \tag{3.193}$$

It is identical to the delayed data quadrant number, and it yields the output data bits $o_{p1}(i)$, $o_{p2}(i)$ (Table 3.6). These are equal to the delayed data bits $d_{p1}(i-N)$, $d_{p2}(i-N)$. The successful correction of the quadrant jump at $i = i_0$ is illustrated in the bottom of Fig. 3.52.

Coherent interaction occurs only between equally polarized lightwaves. Due to the random and time-variable birefringence of standard fibers some kind of polarization handling is needed [71].

Without loss of generality we assume horizontal transmitter polarization, or horizontal and vertical for polarization division multiplex. If polarization-dependent loss (PDL) is excluded the fiber can be modeled by a differential phase retardation δ_1, a subsequent circular retardation δ_2 and a final differential phase retardation δ_3. This way the overall Jones matrix is proportional to

$$\begin{bmatrix} 1 & 0 \\ 0 & e^{-j\delta_3} \end{bmatrix} \begin{bmatrix} \cos\delta_2/2 & -\sin\delta_2/2 \\ \sin\delta_2/2 & \cos\delta_2/2 \end{bmatrix} \begin{bmatrix} 1 & 0 \\ 0 & e^{-j\delta_1} \end{bmatrix} . \tag{3.194}$$

One polarization handling approach is polarization control of the incoming signal or the LO. In the latter case the rest of the coherent receiver must be sufficiently polarization-independent. Due to the long fiber length temperature-dependent polarization changes may be so large that the received signal polarization marches several or many times around the Poincaré sphere. If the polarization control system is capable of tracking sufficiently slow but arbitrarily large polarization fluctuation without any interruption it is called endless [11].

In Fig. 3.49 we use another approach, polarization diversity. The incoming signal is split into two orthogonal polarization components and each of them is received with permanently aligned LO polarization. The electrical signals are combined prior to the decision circuit.

A purely electronic polarization handling scheme in a polarization diversity receiver with synchronous demodulation for QPSK polarization multiplex transmission is also possible. We consider the case of decision-directed control and discuss its digital implementation [63]. This scheme is easily implemented and is being used in practice [72]. The coefficients of the polarization control matrix **M** can and should be the same in all modules, but they must be updated to track changes of **J**. If a perfect estimate $\langle \mathbf{Q} \rangle$ of the matrix product **MJ** is available, polarization can be controlled electronically and penalty-free by applying the iterative formula

$$\mathbf{M} := \langle \mathbf{Q} \rangle^{-1} \mathbf{M} \qquad \text{(zero-forcing algorithm).} \tag{3.195}$$

For $\langle \mathbf{Q} \rangle \to \mathbf{1}$, where **1** is the unity matrix, the inverse can be approximated by

$$\langle \mathbf{Q} \rangle^{-1} = (\mathbf{1} - (\mathbf{1} - \langle \mathbf{Q} \rangle))^{-1} \approx \mathbf{1} + (\mathbf{1} - \langle \mathbf{Q} \rangle) = \mathbf{1} + \langle \mathbf{1} - \mathbf{Q} \rangle . \tag{3.196}$$

In practice it is therefore allowed to set

$$\mathbf{M} := (\mathbf{1} + \langle \mathbf{1} - \mathbf{Q} \rangle)\mathbf{M} = \mathbf{M} + \langle \mathbf{1} - \mathbf{Q} \rangle \mathbf{M} . \tag{3.197}$$

An unknown matrix **N** can be estimated by correlating an input vector **A** having independent zero-mean elements with the output vector NA,

$$\langle \mathbf{N} \rangle = \left\langle (\mathbf{NA})\mathbf{A}^{+} \right\rangle . \tag{3.198}$$

Matrix **Q** is obtained as follows: The polarization-separated IF vector **X**, suitably delayed, is multiplied by $e^{-j\varphi}$ to get rid of phase noise. The result contains phase jumps and is differentially coded. Therefore it must be correlated not with the recovered data itself but with the received data vector $\mathbf{r} = [\underline{r}_1 \quad \underline{r}_2]^T$ whose elements are defined in Table 3.6 by the received quadrant numbers $n_{r,p}$,

$$\mathbf{Q}(i) = \mathbf{X}(i-N) \cdot e^{-j\varphi(i)} \mathbf{r}(i)^{+} . \tag{3.199}$$

Once the algorithm has converged it holds $\mathbf{X}(i-N) \cdot e^{-j\varphi(i)} = \mathbf{r}(i)$, $\mathbf{Q}(i) = \mathbf{r}(i)\mathbf{r}(i)^{+}$, $\langle \mathbf{Q} \rangle = \mathbf{1}$. This means that the term $\mathbf{1} - \mathbf{Q}$ is very noisy; only the average $\langle \mathbf{1} - \mathbf{Q} \rangle = \mathbf{0}$ vanishes. Therefore it is better to replace $\mathbf{1} - \mathbf{Q}$ by $\mathbf{E} = \mathbf{r}(i)\mathbf{r}(i)^{+} - \mathbf{X}(i-N) \cdot e^{-j\varphi(i)} \mathbf{r}(i)^{+} = \left(\mathbf{r}(i) - \mathbf{X}(i-N) \cdot e^{-j\varphi(i)}\right) \cdot \mathbf{r}(i)^{+}$.

Finally, better than to wait for a perfect estimate $\langle \mathbf{1} - \mathbf{Q} \rangle$ or $\langle \mathbf{E} \rangle$ is it to choose a low control gain g with $0 < g << 1$ and to update **M** immediately as

$$\mathbf{M} := \mathbf{M} + g\mathbf{T} , \tag{3.200}$$

This means $\mathbf{M}(i+1)=\mathbf{M}(i)+g\mathbf{T}(i)$, with

$$\mathbf{T}(i)=\mathbf{E}(i)\mathbf{M}(i)=\left(\mathbf{r}(i)-\mathbf{X}(i-N)\cdot e^{-j\varphi(i)}\right)\cdot\mathbf{r}(i)^{+}\mathbf{M}(i). \tag{3.201}$$

System simulations in the presence of PDL and with random start values always correctly recover the carrier and **M**, and indicate that $g=10^{-3}$ is more than adequate. To allow for realtime operation, the $\mathbf{E}(i)$ in (3.201) may be accumulated before an updating step of **M** (3.200) is undertaken.

The control time constant equals roughly $1/g$ updating intervals. Since the symbol duration is so short it should be sufficient in most applications to base the updating of **M** on matrices **T** obtained in only one of the H signal processing modules. This reduces electronic integrated circuit floorspace. The calculations can be cast into the blocks of Fig. 3.53. These must be added in one of the signal processing modules. Index l is the M-fold subsampled symbol index at which the chosen module is clocked. The associated control time constant is H/g symbol durations, e.g. 1.6 μs for 10 Gbaud operation, $g=10^{-3}$, and $H=16$.

In principle, the updating of **M** also recovers the carrier, but the practical contribution of **M** to carrier recovery is negligible because the updating is slow.

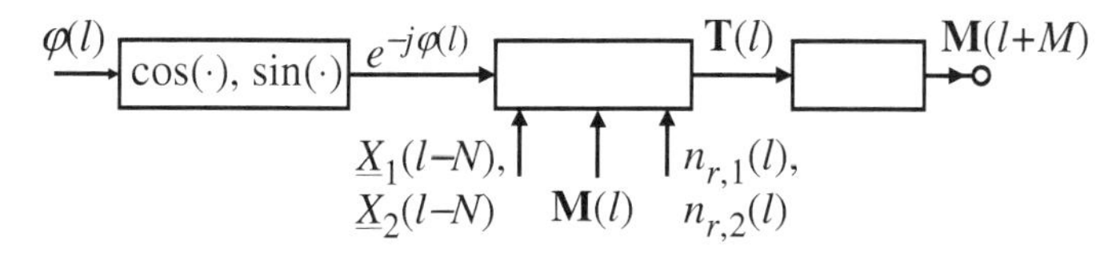

Fig. 3.53 Updating of the polarization control matrix **M**. © 2005 IEEE.

So far it holds $\mathbf{X}=\mathbf{MR}$. The received data vector may also be written as $\mathbf{r}=\left\lfloor e^{-j\varphi}\mathbf{X}\right\rfloor_D$ where $\lfloor\cdot\rfloor_D$ is the rounding to the nearest of the symbols defined in Table 3.6. If the I&Q signals of the same polarization have unequal amplitudes or crosstalk due to 90° hybrid imbalance and similar effects, then the polarization control algorithm must be generalized: We define a real-valued received IF signal vector $\mathbf{R}_r$. The elements of $\mathbf{R}_r$ are the actual signals of the four photoreceivers contained in Fig. 3.49. Ideally it holds $\mathbf{R}_r=\left[\operatorname{Re}\underline{R}_1 \quad \operatorname{Im}\underline{R}_1 \quad \operatorname{Re}\underline{R}_2 \quad \operatorname{Im}\underline{R}_2\right]^T$. A real polarization-separated IF vector $\mathbf{X}_r=\mathbf{M}_r\mathbf{R}_r$ is computed. $\mathbf{M}_r$ is a real 4×4 matrix. The real-valued received data vector $\mathbf{r}_r=\left[\operatorname{Re}\underline{r}_1 \quad \operatorname{Im}\underline{r}_1 \quad \operatorname{Re}\underline{r}_2 \quad \operatorname{Im}\underline{r}_2\right]^T$ corresponding to **r** is $\mathbf{r}_r=\left\lfloor\begin{bmatrix}\mathbf{L}(-\varphi) & \mathbf{0}\\ \mathbf{0} & \mathbf{L}(-\varphi)\end{bmatrix}\mathbf{X}_r\right\rfloor_D$ with carrier rotation matrix $\mathbf{L}(\varphi)=\begin{bmatrix}\cos\varphi & -\sin\varphi\\ \sin\varphi & \cos\varphi\end{bmatrix}$. The real-valued polarization control matrix is updated as

$$\mathbf{M}_r := \mathbf{M}_r + g\mathbf{T}_r, \tag{3.202}$$

$$\mathbf{T}_r(i) = \left(\mathbf{r}_r(i) - \begin{bmatrix} \mathbf{L}(-\varphi(i)) & \mathbf{0} \\ \mathbf{0} & \mathbf{L}(-\varphi(i)) \end{bmatrix} \mathbf{X}_r(i-N) \right) \cdot \mathbf{r}_r(i)^T \mathbf{M}_r(i). \tag{3.203}$$

Closer analysis reveals that the resulting polarization control gain is halved compared to (3.200), (3.201). Intuitively this can be understood by the fact that $\mathbf{M}_r$ contains twice as many degrees-of-freedom as $\mathbf{M}$. We have found now a generalized algorithm that can form suitable linear combinations of data signals so that crosstalk is minimized. And indeed, with 3 real input and 2 real output signals such an algorithm (still with term $\mathbf{1}-\mathbf{Q}$ instead of $\mathbf{E}$) has been used to transform the signals of a single-polarization receiver with symmetric 3×3 coupler into the I&Q domain [66]. Ideally the transformation would be given by (3.156), practically it may need to be slightly different since there are not ideal three-phase input signals.

As an alternative to the decision-directed polarization control scheme there is also the so-called constant modulus algorithm (CMA), which demultiplexes the two polarization channels in a non-data-aided approach. A detailed derivation of the CMA can be found in [73], its adaptation for polarization control in a coherent optical receiver is explained in [74]. The CMA exploits the fact that polarization cross-talk causes the amplitude of the signal to fluctuate. Thus by minimizing these fluctuations and forcing the complex signal of each recovered polarization onto the unity circle a perfect separation of two polarization channels can be achieved. The required error signal is calculated as

$$\mathbf{T} = \begin{bmatrix} 1-\left|\underline{X}_1\right|^2 & 0 \\ 0 & 1-\left|\underline{X}_2\right|^2 \end{bmatrix} \mathbf{X}\mathbf{R}^+ \tag{3.204}$$

and the polarization control matrix is incrementally updated by (3.200).

Intuitively the CMA can be explained as follows: Quantity $g\mathbf{T}\mathbf{R} = g\begin{bmatrix} 1-\left|\underline{X}_1\right|^2 & 0 \\ 0 & 1-\left|\underline{X}_2\right|^2 \end{bmatrix}\mathbf{X}$ is added to the original term $\mathbf{X} = \begin{bmatrix} \underline{X}_1 \\ \underline{X}_2 \end{bmatrix} = \mathbf{M}\mathbf{R}$, eventually enforcing constant moduli $\left|\underline{X}_{1,2}\right| \to 1$.

A disadvantage of the CMA is that in general $\langle \mathbf{M} \rangle \neq \mathbf{J}^{-1}$ holds, because it only minimizes the cross-talk between the two polarization channels but does not recover the phase offset between them. Separate carrier recoveries are therefore needed for the two polarizations.

If one wants to use one carrier recovery for both polarization channels then the differential phase needs to be compensated. This is possible if (3.200), (3.204) are modified [75] as

$$\mathbf{M} := \mathbf{M} + g(\mathbf{T} + \mathbf{U}), \qquad \mathbf{U} = \begin{bmatrix} -j & 0 \\ 0 & j \end{bmatrix} \frac{\Delta\zeta}{2} \mathbf{M}, \tag{3.205}$$

$$\Delta\zeta = \left(\left(\arg \underline{X}_1 - \arg \underline{X}_2 + \frac{\pi}{4} \right) \bmod \frac{\pi}{2} \right) - \frac{\pi}{4} . \tag{3.206}$$

Term $\Delta\zeta$ is the differential phase between $\underline{X}_1$ and $\underline{X}_2$, taken modulo $\pi/2$ in the interval $[-\pi/4, \pi/4[$. The phases $\arg \underline{X}_{1,2}$ are determined anyway in the demodulation process. The original term $\mathbf{X} = \mathbf{MR}$ will be augmented by $g\mathbf{UR} = g\begin{bmatrix} -j\Delta\zeta/2 & 0 \\ 0 & j\Delta\zeta/2 \end{bmatrix}\mathbf{X}$, which is equivalent to a multiplication of $\underline{X}_{1,2}$ by $e^{\mp jg\Delta\zeta/2}$, thereby decreasing $\Delta\zeta$ to its $(1-g)$-fold. Using $|\underline{X}_{1,2}| \to 1$ and $\arctan x \approx x$ for $|x| << 1$, the numerical determination of $\Delta\zeta$ can be trivialized.

The various polarization control algorithms are compared in Fig. 3.54 as a function of polarization control gain g [76]. Performance is worst for the original decision-directed algorithm (ODDA) [63] with $\mathbf{T} = (\mathbf{1} - \mathbf{Q})\mathbf{M}$ employed in (3.200) and $\mathbf{Q}$ given by (3.199). For a specified penalty, CMA and DPC-CMA allow to increase the control gain to a higher value. However, the modified decision-directed algorithm (MDDA) (3.200), (3.201) performs best.

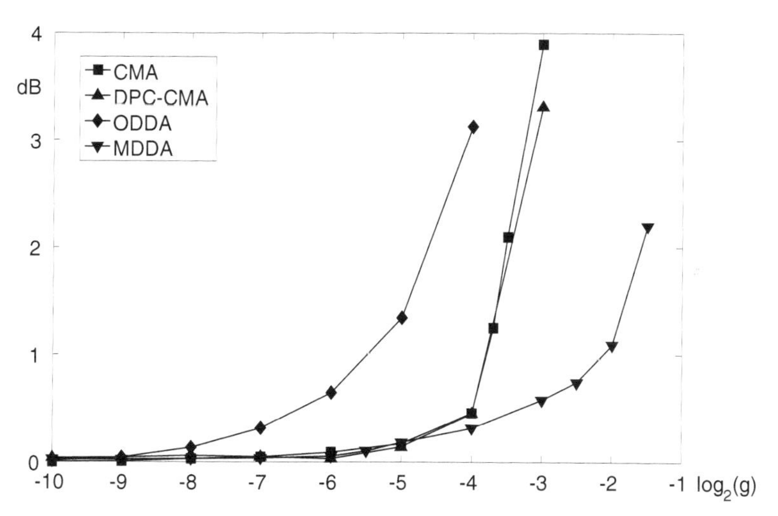

Fig. 3.54 Sensitivity penalty at BER = 10^{-3} of various QPSK polarization control algorithms vs. control gain g, for $\Delta f \cdot T_S = 1 \cdot 10^{-4}$, $\mathbf{MJ} = \mathbf{1}$. Each data point corresponds to 200,000 symbols. © 2010 IEEE.

Important improvements and simplifications are possible for the digital QPSK carrier phase recovery. In a parallelized digital signal processing unit (DSPU) it causes much effort to provide estimated phase values for each received symbol. In order to reduce this effort, block phase estimators generate only a single estimation value for K symbols processed in parallel [77, 78]. But they are more sensitive to phase noise than sliding-window estimators.

An algorithm which needs no multiplications and is fully parallelizable and phase noise tolerant was developed by Hoffmann [79, 80]. We start (similar to $\mathbf{X} = \mathbf{c} \cdot e^{j\varphi'}$ (3.188)) with a single-polarization signal

$$X(k) = c(k) \cdot e^{j\varphi'(k)} + n(k) \tag{3.207}$$

where $c(k)$ is the complex QPSK symbol, $\varphi'(k)$ the carrier phase and $n(k)$ additive complex noise, all for sample k. Generating a modulation-free signal like $X^4(k)$ is the first task. But taking the 4th power of complex signal samples (3.189) is not optimum. Rather, the 4th power $e^{j4\psi(k)}$ of the corresponding phasor is a better choice [81, 77]. The positive angle

$$\psi(k) = \arg(X(k)) \bmod 2\pi = (\gamma(k) + \varphi'(k) + \nu(k)) \bmod 2\pi \tag{3.208}$$

contains also $\nu(k)$ as the noise contribution from $n(k)$ and the QPSK modulation $\gamma(k) = \arg(c(k))$. The latter is removed by the modulo $\pi/2$ operation

$$\vartheta(k) = \psi(k) \bmod (\pi/2) \tag{3.209}$$

where it holds $e^{j4\psi(k)} = e^{j4\vartheta(k)}$. In digital signal processing a power-of-two is chosen to represent the angle 2π and the modulo $\pi/2$ operation is implemented by removing the first two bits [82]. A further improvement of the complex phase estimation can be achieved by weighted signal averaging [83, 77].

Hoffmann [79, 80] employs a set of modulation-free position angles $\vartheta(k-N), \vartheta(k-N+1), \ldots, \vartheta(k+N)$ to obtain an estimated phase $\varphi(k)$ immediately, i. e. without function tables and complex calculations. An angle-based averaging equivalent to the barycenter algorithm [82] allows to obtain a sequence of K estimated phase values in parallel. This results in a sliding-window phase estimation process for a single estimation value $\varphi(k)$. It can be described as a real-valued scalar function of a real-valued input vector with $2N+1$ components. To obtain a sequence of K estimated phase angles in parallel, an K-dimensional vector function

$$[\varphi(k), \ldots, \varphi(k+K-1)] = \mathbf{\Phi}(\vartheta(k-N), \ldots, \vartheta(k+K+N-1)) \tag{3.210}$$

of an input vector with $2N+K$ components is required.

Fig. 3.55 left shows a suitable angle-based parallelized tree structure for $N = 2$ and $K = 4$. It is very hardware-efficient and imitates weighted averaging by multiple usage of center values. The partial tree for the component $\varphi(k+1)$ consists of the seven marked nodes. The nodes are basic cells that convert pairs of position angles α, β into average position angles μ. Basic cell functions $\mu(\alpha, \beta)$ are concatenated to form the phase estimator. Due to multiple usage of intermediate results, the vectorized function in Fig. 3.55 left requires only 17 basic cells instead of $K \cdot 7 = 28$.

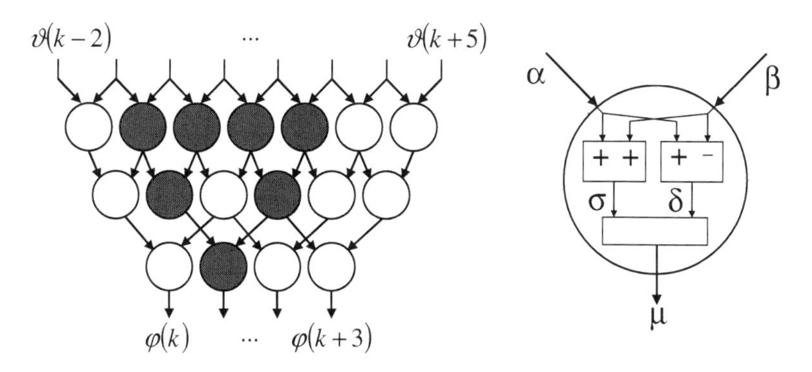

Fig. 3.55 Angle-based phase estimator structure (left) and internal structure of basic cell (right). © 2008 IEEE.

In order to imitate the behaviour of a Viterbi & Viterbi estimator with normalization [81, 77] the partial result from each basic cell has to be

$$\mu = \frac{1}{4}\left(\arg\left(e^{j4\alpha} + e^{j4\beta}\right)\bmod 2\pi\right). \tag{3.211}$$

The complex sum in can be rewritten as

$$e^{j4\alpha} + e^{j4\beta} = e^{j2\sigma} \cdot 2\cos(2\delta) \tag{3.212}$$

with the sum $\sigma = \alpha + \beta$ and the difference $\delta = \alpha - \beta$ of the input values. The argument of (3.212) depends on the exponent $j2\sigma$ and the sign of the cosine function because a negative sign of $\cos(2\delta)$ is equivalent to a rotation by π in the complex plane. Note that the cosine factor introduces a discontinuity of μ for $\delta \approx \pm\pi/4$. Taking the further operations of (3.211) into account, we obtain an alternative, computationally very simple calculation formula for μ based on σ and δ only,

$$\mu = \left(\frac{\sigma}{2} + \frac{\pi}{4}\left\lceil \frac{|\delta|}{\pi/4} \right\rceil\right) \bmod \frac{\pi}{2} \tag{3.213}$$

where $\lceil . \rceil$ means truncation to the next-lower integer.

Fig. 3.55 right shows the internal structure of the basic cell. The two input values are added to and subtracted from each other in parallel, and the final result μ is generated from the auxiliary quantities σ and δ according to (3.213), all within one digital signal processing unit (DSPU) clock cycle. The quantity δ is a directed phase increment that could be used for frequency estimation as described in [84]. The basic cell performs an interpolation of the phase track on the shortest possible path.

The estimation quality can be improved by selection, i. e. elimination of intermediate results from critical pairs α, β for which two possible paths have almost the same length. The acronym SML (Selective Maximum Likelihood) phase approximation refers to this feature.

The SML phase approximation was compared to a normalized block phase estimator with $K = 8$ and a normalized phase estimator with weighted averaging and $N = 5$ in a Monte-Carlo simulation. K is only important for the block phase estimator result, not for SML phase approximation or the weighted-normalized approach [84]. SMLPA with $N = 2$ as in Fig. 3.55 and a more complicated topology with $N = 4$ was employed. Asynchronous data recovery was also simulated as a reference. BER/OSNR curves were simulated with $4 \cdot 10^6$ random symbols for each BER value.

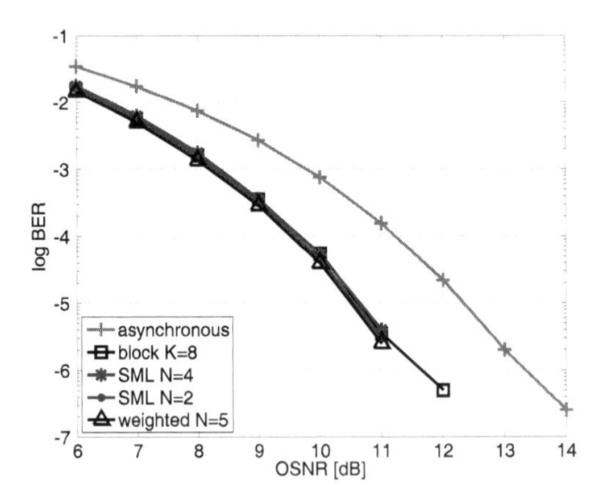

Fig. 3.56 BER/OSNR curves for different phase estimators for small phase noise. © 2008 IEEE.

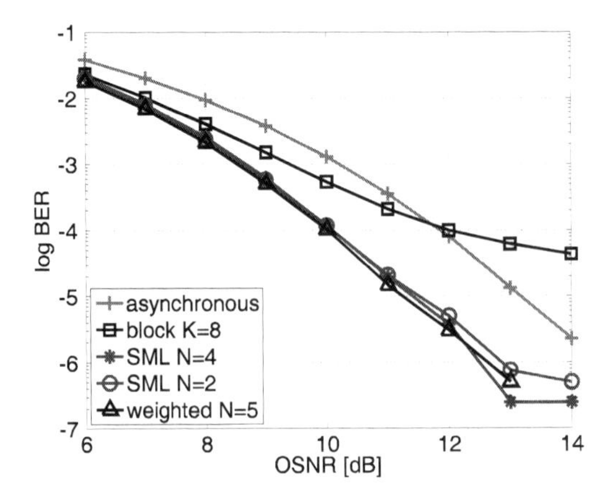

Fig. 3.57 BER/OSNR curves for different phase estimators for large phase noise. © 2008 IEEE.

Fig. 3.56 show the BER/OSNR results for a 28 GBaud system. DFB lasers with 1 MHz linewidth are assumed. With such a small linewidth, the block phase estimator performs almost as good as the other three phase estimators which track the phase drift.

Fig. 3.57 shows simulated BER/SNR curves for the same phase estimators and a 10 GBaud system with DFB lasers that have 1 MHz linewidth and a residual frequency mismatch of 20 MHz. While the block phase estimator is not able to cope with high phase noise, both SML phase approximation versions have results similar to the weighted averaging method. They fulfill the high phase noise requirements of 10 GBaud transmission systems with coarsely controlled DFB lasers. The angle-based phase estimation algorithm is a good replacement for the optimized moving average phase estimator. It has been used for the first realtime QPSK transmission [64], also with polarization multiplex and control [85].

In addition to the polarization transformation at the carrier frequency, polarization mode dispersion (PMD) and chromatic dispersion (CD) need also to be compensated. A surprising lead in this field has been taken by Nortel [86]. Another set of application-specific integrated circuits and its application in coherent QPSK transmission was presented in [87].

CD can as well be pre-compensated electronically in the transmitter that is equipped with an I&Q modulator [88]. In that case not even a coherent receiver is needed. Rather, for intensity modulation a standard direct-detection ASK receiver is sufficient.

CD and PMD can be electronically equalized with finite impulse response (FIR) filters in the time domain. Beside this, convolution in the time domain can be replaced by Fast Fourier Transform (FFT), multiplication in the frequency domain and inverse FFT (IFFT). Various equalizer configurations become thereby possible [89].

One possibility is it to perform the IFFT at the transmit end and the FFT at the receive end. In this way the symbols at the IFFT input can be understood as the amplitudes of orthogonal, narrowly spaced carriers. In order to maintain orthogonality in the presence of dispersion the temporal duration of each transmitted IFFT frame is cyclically extended by usually a few percent. This scheme, called orthogonal frequency division multiplex (OFDM), is well known in the wireless world but is also a serious candidate for high-performance optical communication [89–97].

3.3.6 Digital Coherent QAM Receiver

Polarization-multiplexed QPSK is that among the modulation formats which ideally work with no more than 18 photons/bit at a BER = 10^{-9} with the highest information rate, 4 bit/symbol. QPSK itself is a special case among the square quadrature amplitude modulation (QAM) schemes. We will extend and generalize the foregoing from QPSK to M-ary QAM schemes. M-QAM, in particular 16-QAM, is attractive for shorter transmission lengths where ultimate OSNR performance is not needed. Fig. 3.58 shows various QAM constellations with $q = 4$-fold angular symmetry, commonly referred to as square QAM [98, 99]. 4-QAM is the same as QPSK.

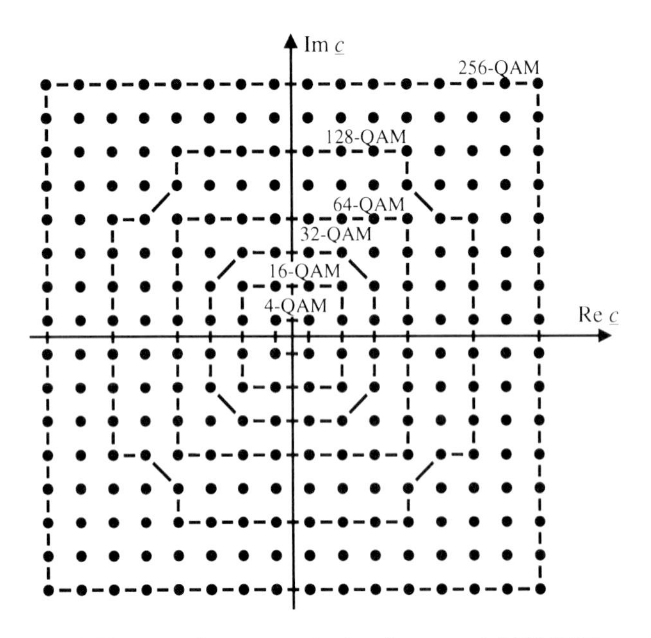

Fig. 3.58 Square QAM constellation diagrams. © 2009 IEEE.

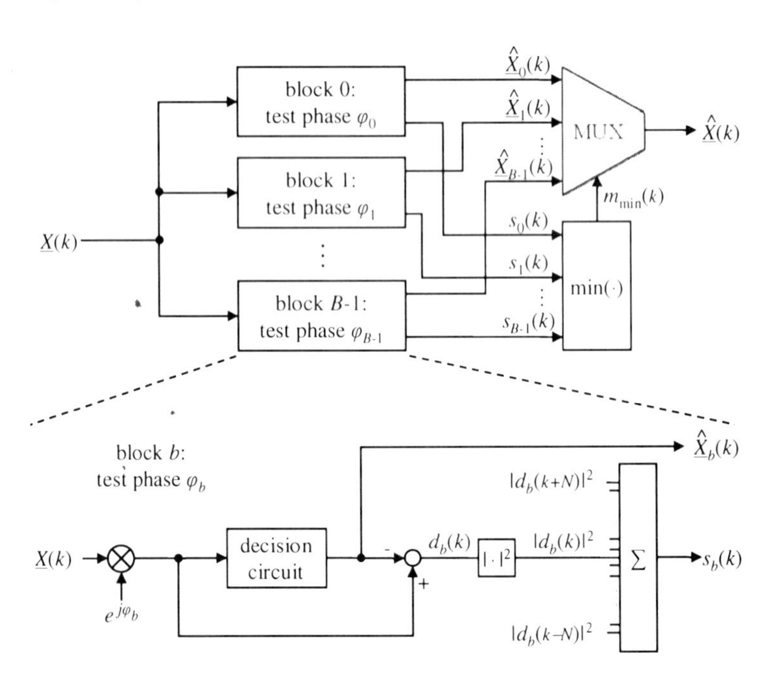

Fig. 3.59 Feedforward carrier recovery using B test phase values φ_b. © 2009 IEEE.

For true square M-QAM signals, the inphase and quadrature symbols (or optical field levels) can be written as

$$c = 2m + 1 - \sqrt{M}\,, \qquad m \in \left\{0, 1, \ldots, \sqrt{M} - 1\right\}. \tag{3.214}$$

The mean squared field is $\left\langle c^2 \right\rangle = \frac{M-1}{3}$. We divide by $\sqrt{2\left\langle c^2 \right\rangle}$ and write a complex QAM signal with unity mean squared magnitude,

$$\underline{c} = \sqrt{\frac{3}{2(M-1)}}\left(\left(2m_1 + 1 - \sqrt{M}\right) + j\left(2m_2 + 1 - \sqrt{M}\right)\right),$$

$$m_1, m_2 \in \left\{0, 1, \ldots, \sqrt{M} - 1\right\}, \qquad \left\langle \left|\underline{c}\right|^2 \right\rangle = 1\,. \tag{3.215}$$

For QPSK = 4-QAM this is identical with the $\underline{c}_p$ defined in Table 3.6. The modulation variables in the two quadratures are m_1, m_2.

A phase-noise tolerant, parallelizable feedforward carrier recovery for QAM signals, subsequently abbreviated as FCR-QAM, has been presented by Pfau [98, 99]. The setup is depicted in Fig. 3.59. The k-th input signal $\underline{X}(k)$ of the coherent receiver is sampled at the symbol rate, and perfect clock recovery and equalization are assumed. To recover the carrier phase the received signal $\underline{X}(k)$ is rotated by B test carrier phase angles φ_b with

$$\varphi_b = \frac{b}{B} \cdot \frac{2\pi}{q}\,, \qquad b \in \{0, 1, \ldots, B-1\}. \tag{3.216}$$

Then all rotated symbols are fed into a decision circuit and the squared distance

$$\left|d_b(k)\right|^2 = \left|\underline{X}(k)e^{j\varphi_b} - \left\lfloor \underline{X}(k)e^{j\varphi_b} \right\rfloor_D\right|^2 = \left|\underline{X}(k)e^{j\varphi_b} - \hat{X}_b(k)\right|^2 \tag{3.217}$$

to the closest constellation point is calculated in the complex plane, taking advantage of the rotational symmetry with angles $2\pi/q$. Here $\lfloor \cdot \rfloor_D$ means the rounding to the nearest of the symbols (3.215) as executed in the decision circuit. In order to remove noise, the distances of $2N+1$ consecutive test symbols rotated by the same carrier phase angle φ_b are summed up,

$$s_b(k) = \sum_{n=-N}^{N} \left|d_b(k-n)\right|^2. \tag{3.218}$$

The optimum value of N depends on the laser linewidth times symbol rate product. $N = 6 \ldots 10$ is a fairly good choice. After filtering the optimum phase angle is determined by searching the minimum sum of distance values. As the decoding was already executed in (3.217), the decoded output symbol $\hat{\underline{X}}(k)$ can be selected from the $\underline{\hat{X}}_b(k)$ by a switch controlled by the index $m_{\min}(k)$ of the minimum distance sum. Unless an error has occurred $\underline{\hat{X}}(k)$ is equal to the k-th transmitted symbol $\underline{c}(k)$ (3.215), but rotated by any of the q angles $2\pi \cdot (1,2,...,q-1)/q$.

In the case $q = 4$ with 4-fold ambiguity of the recovered phase, the first two bits which determine the quadrant of the complex plane should be differentially Gray-encoded. The differential encoding and decoding process is the same as for QPSK, see chapter 3.3.5. Decoding is described by

$$\begin{aligned} n_o(k) &= \left(n_r(k) - n_r(k-1) + n_j(k)\right) \bmod q \\ n_o(k), n_r(k), n_j(k) &\in \{0, 1, 2, \ldots, q-1\} \end{aligned} \tag{3.219}$$

where $n_o(k)$ is the differentially decoded quadrant number, $n_r(k)$ is the received quadrant number and $n_j(k)$ is the jump number. The only required modification of the decoding process compared to QPSK is that quadrant jumps are detected according to

$$n_j(k) = \begin{cases} 1 & B/2 < m_{\min}(k) - m_{\min}(k-1) \\ 0 & -B/2 \le m_{\min}(k) - m_{\min}(k-1) \le B/2 \\ q-1 & m_{\min}(k) - m_{\min}(k-1) < -B/2 \end{cases}. \tag{3.220}$$

For all other bits that determine the symbol within a quadrant of the complex plane normal Gray-coding is sufficient and no differential encoding/decoding is required. Fig. 3.60 illustrates the bit to symbol assignment including differential encoding/decoding exemplarily for square 16-QAM.

This FCR-QAM algorithm can also be applied to arbitrary QAM constellations. A q-fold rotational symmetry is already foreseen in (3.216), (3.219). If there is no symmetry then $q = 1$ holds. $\lceil \log_2 q \rceil$ bits must be differentially encoded/decoded, where $\lceil u \rceil$ is the smallest integer larger than or equal to u.

With polarization division multiplex, one may use two separate carrier recoveries. But due to the intrinsically low phase noise tolerance of QAM schemes it is advisable to implement a common carrier recovery for both polarizations. As a consequence, (3.218) is replaced by

$$s_b(k) = \sum_{p=1}^{2} \sum_{n=-N}^{N} \left| d_{p,b}(k-n) \right|^2 \tag{3.221}$$

where the added index p is the index of the polarization. Because twice as much data is available to determine the carrier phase angle, N can be halved, thereby increasing the phase noise tolerance by roughly a factor of 2.

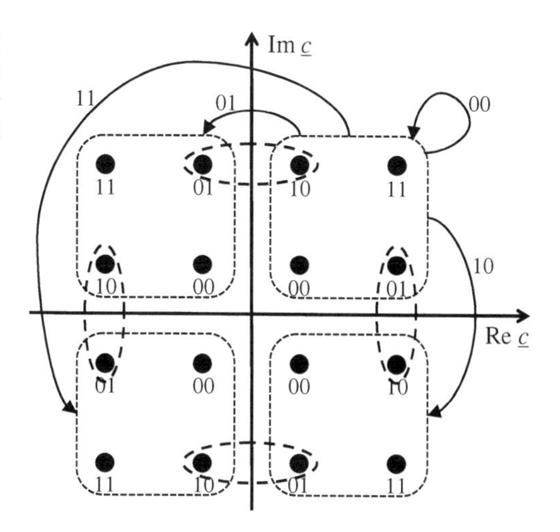

Fig. 3.60 16-QAM bit to symbol assignment: The dashed ellipses mark imperfect Gray coding of four symbol pairs due to differential quadrant encoding. © 2009 IEEE.

The rotation of a symbol in the complex plane normally requires a complex multiplication, consisting of four real-valued multiplications with subsequent summation. This would lead to a large number of multiplications to be executed, while achieving a sufficient resolution B for the carrier phase values φ_b. The hardware effort would therefore become prohibitive. Applying the CORDIC (coordinate rotation digital computer) algorithm [100, 101] can dramatically reduce the necessary hardware effort to calculate the B rotated test symbols. This algorithm can compute vector rotations simply by summation and shift operations. As for the calculation of the B rotated copies of the input vector intermediate results can be reused for different rotation angles, only $\sum_{b=1}^{\log_2 B} 2^{b+1}$ shift-and-add operations are required to generate the B test symbols. For example, to generate $B = 32$ rotated copies of Z_k the CORDIC algorithm requires only 124 shift-and-add operations instead of 124 real-valued multiplications and 62 summations.

To determine the closest constellation point $\hat{X}_b(k)$ the rotated symbols are fed into a decision circuit. The squared distance (3.217) at instant k can be rewritten as

$$|d_b(k)|^2 = \left(\mathrm{Re}\left(\underline{X}(k)e^{j\varphi_b}\right) - \mathrm{Re}\,\hat{X}_b(k)\right)^2 + \left(\mathrm{Im}\left(\underline{X}(k)e^{j\varphi_b}\right) - \mathrm{Im}\,\hat{\underline{X}}_b(k)\right)^2 \tag{3.222}$$

and requires to two multiplications and three additions/subtractions. However, the subtraction results are small in magnitude and the required result resolution is moderate. Therefore $|d_b(k)|^2$ is most efficiently determined by a look-up table or basic logic functions.

Highly parallelized systems allow for a very resourceful implementation of the summation of $2N+1$ consecutive values in (3.218), (3.221). The adders can be arranged in a binary tree structure where intermediate results from different modules are reused in neighbor modules.

The overall FCR-QAM hardware effort is on the order of B times higher than for QPSK. This may be viewed as dramatic since B is in the range of 16 to 64. But taking into account that the more bits/symbol are transmitted and that the hardware effort for electronic polarization control and PMD and CD compensation is also many times higher than that for QPSK carrier recovery it appears that the overall effort is reasonable. Furthermore it is possible to implement the FCR-QAM algorithm in two stages, which reduces the required hardware effort [102].

The FCR-QAM algorithm has been simulated for the constellations 4-QAM (QPSK), 16-QAM, 64-QAM and 256-QAM. These true square constellations are easiest to generate [103, 104] and are optimally immune against additive white Gaussian noise (AWGN) [105]. The transversal filter halfwidth is always set to $N=9$, and each data point is based on the simulation of 200,000 symbols. The results are compared against the theoretically achievable sensitivity [105]

$$\frac{E_S}{N_0}=\frac{M-1}{3}\left(\mathrm{Q}^{-1}\left(\frac{\log_2 M}{2}\left(1-\frac{1}{\sqrt{M}}\right)^{-1}\left(1-\sqrt{1-\mathrm{BER}}\right)\right)\right)^2 . \tag{3.223}$$

E_S/N_0 is the optical signal to noise ratio (OSNR), M is the number of constellation points, BER is the target bit error rate and $\mathrm{Q}(x)=\frac{1}{2}\mathrm{erfc}\left(\frac{x}{\sqrt{2}}\right)$ is the Q function. The inverse of (3.223) is

$$\mathrm{BER}=1-\left(1-\frac{2}{\log_2 M}\left(1-\frac{1}{\sqrt{M}}\right)\mathrm{Q}\left(\sqrt{\frac{3}{M-1}\frac{E_S}{N_0}}\right)\right)^2 . \tag{3.224}$$

A crucial quantity for FCR-QAM is the required number B of test phases φ_b. Fig. 3.61a shows the sensitivity penalty at the bit error rate 10^{-3} for 4-QAM and 16-QAM. FCR-QAM and a receiver with ideal carrier recovery were simulated with different resolutions for the carrier phase. Ideal carrier recovery means that the receiver knows the exact carrier phase (which is only realizable in simulation) and therefore the sensitivity penalty is only caused by differential quadrant encoding and quantization effects.

4-QAM attains a minimum penalty of 0.5 dB for the ideal receiver and 0.7 dB for FCR-QAM. The penalty difference of 0.2 dB is thus the implementation-induced penalty. For 16-QAM the minimum penalties decreases (0.4 dB for the ideal receiver, 0.6 dB for FCR-QAM), because only 2 out of 4 transmitted bits are differentially encoded. For all four receivers it can be seen that almost no additional penalty is induced due to the quantization of the carrier phase for $\log_2 B \geq 5$. Therefore in all following simulations for 4-QAM and 16-QAM B is set to 32.

Fig. 3.61b shows the same for 64-QAM and 256-QAM. The minimum penalty for 64-QAM is 0.3 dB with ideal carrier recovery and 0.5 dB with FCR-QAM. For 256-QAM the respective values are 0.35 dB and 0.55 dB. For both constellations the penalty due to the quantization of the carrier phase is tolerable only if $\log_2 B \geq 6$. The number of test phase values for 64-QAM and 256-QAM is therefore subsequently chosen as $B = 64$.

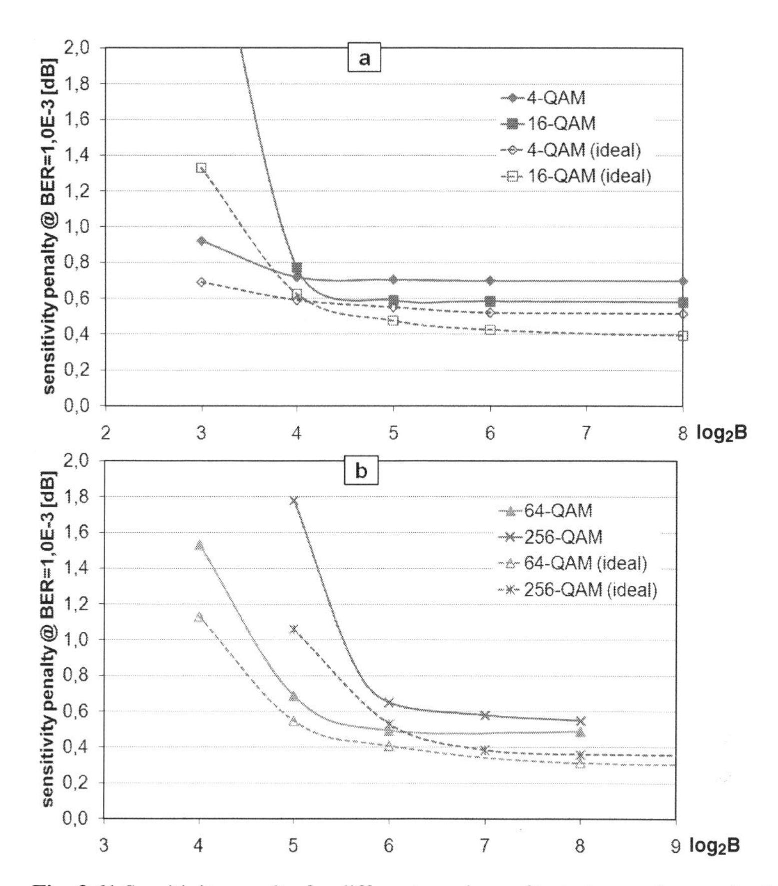

Fig. 3.61 Sensitivity penalty for different numbers of test phase values φ_b for (a) 4- and 16-QAM; (b) 64-QAM and 256-QAM. © 2009 IEEE.

FCR-QAM is advantageous because of its phase noise tolerance. Today's commercial transmission systems usually employ DFB lasers, because they are cost-efficient and have a small footprint. Their linewidth is in the range $100\,\text{kHz} < \Delta f_{\text{DFB}} < 10\,\text{MHz}$. The sum linewidth $\Delta f = 2\Delta f_{\text{DFB}}$ of signal and LO lasers is twice as much. Assuming a symbol rate of 20 Gbaud, linewidth times symbol duration products down to $10^{-5} < \Delta f \cdot T < 10^{-3}$ can be realized. Fig. 3.62 shows the sensitivity penalty of the FCR-QAM algorithm against the $\Delta f{\cdot}T$ product. Table 3.7 shows the maximum tolerable linewidth times symbol duration product for a sensitivity penalty of 1 dB at a BER of 10^{-3}. In a polarization multiplexed system with common carrier recovery these values can be approximately doubled.

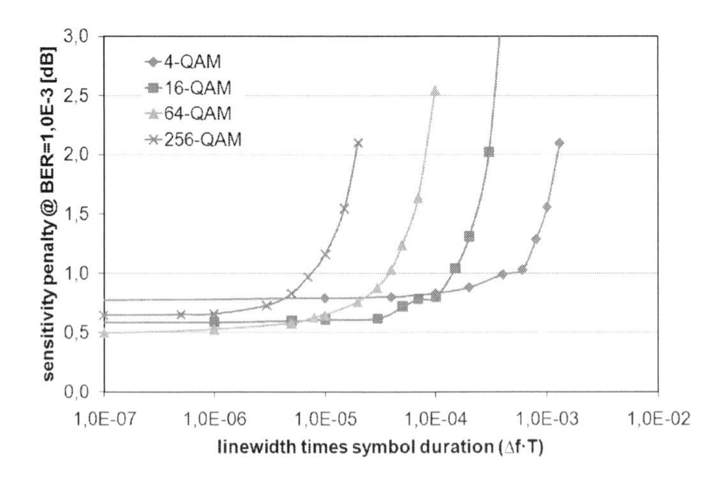

Fig. 3.62 Receiver tolerance against phase noise for different square QAM constellations. © 2009 IEEE.

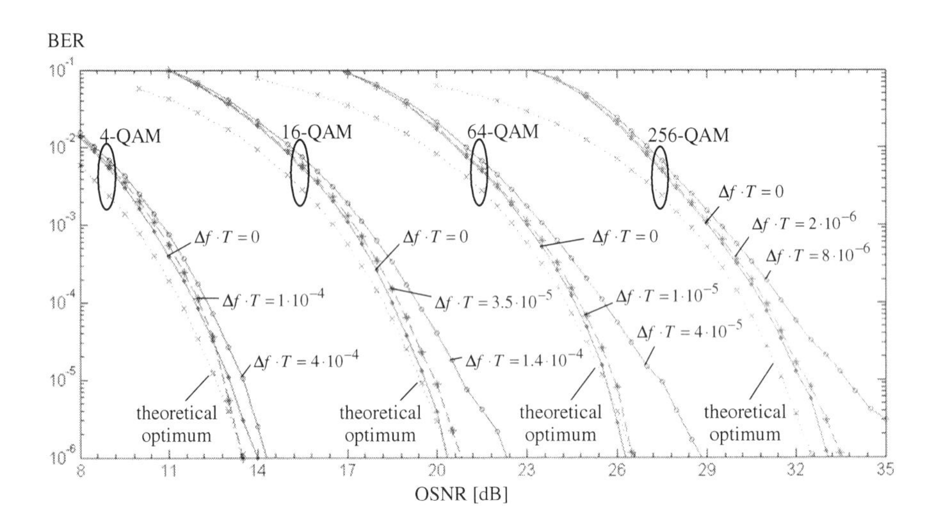

Fig. 3.63 Impact of different linewidth times symbol duration products on the receiver sensitivity of coherent QAM receivers. © 2009 IEEE.

Table 3.7 Linewidth requirements for feedforward carrier recovery with different square QAM constellations. © 2009 IEEE.

Constellation	Max. tolerable $\Delta f{\cdot}T$ for 1 dB penalty @ BER = 10^{-3}	Max. tolerable Δf_{DFB} for $1/T_S$ = 20 Gbaud
4-QAM	$4.1{\cdot}10^{-4}$	4.1 MHz
16-QAM	$1.4{\cdot}10^{-4}$	1.4 MHz
64-QAM	$4.0{\cdot}10^{-5}$	400 kHz
256-QAM	$8.0{\cdot}10^{-6}$	80 kHz

BER vs. OSNR has additionally been evaluated for selected values of $\Delta f{\cdot}T$ (Fig. 3.63). For BERs below 10^{-5} the results become inaccurate due to the low number of errors which occurred within the $2{\cdot}10^6$ symbols that were simulated per data point. The theoretical optimum is calculated by (3.224). A $\Delta f{\cdot}T_S$ value equal to 1/4 of that causing a ~1 dB penalty at BER = 10^{-3} gives excellent results at least down to BER = 10^{-6}.

For high OSNR the contribution of AWGN to phase noise can be considered to be also Gaussian with the variance $\left(2E_S/N_0\right)^{-1}$ [106]. Therefore the efficiency $e(N)$ of the phase estimator depending on the filter half width N is given by

$$e(N) = \frac{\frac{1}{2N+1}\left(2\frac{E_S}{N_0}\right)^{-1}}{\left\langle (\psi - \hat{\psi})^2 \right\rangle} \le 1 . \tag{3.225}$$

where the numerator is the Cramer-Rao lower bound [107], and the denominator is the mean squared error of the phase estimator output. Fig. 3.64 shows both the mean squared error together with the theoretical optimum given by the Cramer-Rao lower bound (top row) and the resulting estimator efficiency $e(N)$ (bottom row) for the different QAM constellations.

For higher QAM constellations the efficiency of the phase estimation reduces. This can be related to the fact that for lower OSNR other limiting factors like quantization and phase noise become more dominant, Note that for a 5 bit (4-QAM, 16-QAM) and 6 bit quantization (64-QAM, 256-QAM) of $\hat{\psi}$ the minimum mean squared errors are ~$2{\cdot}10^{-4}$ and ~$5{\cdot}10^{-5}$, respectively.

Fig. 3.64 also shows that the selected filter halfwidth N = 9 is always close to the optimum filter half width for a minimum mean squared error. In principle, receiver performance could be improved by optimizing N for each parameter set {OSNR, $\Delta f{\cdot}T_S$, M}.

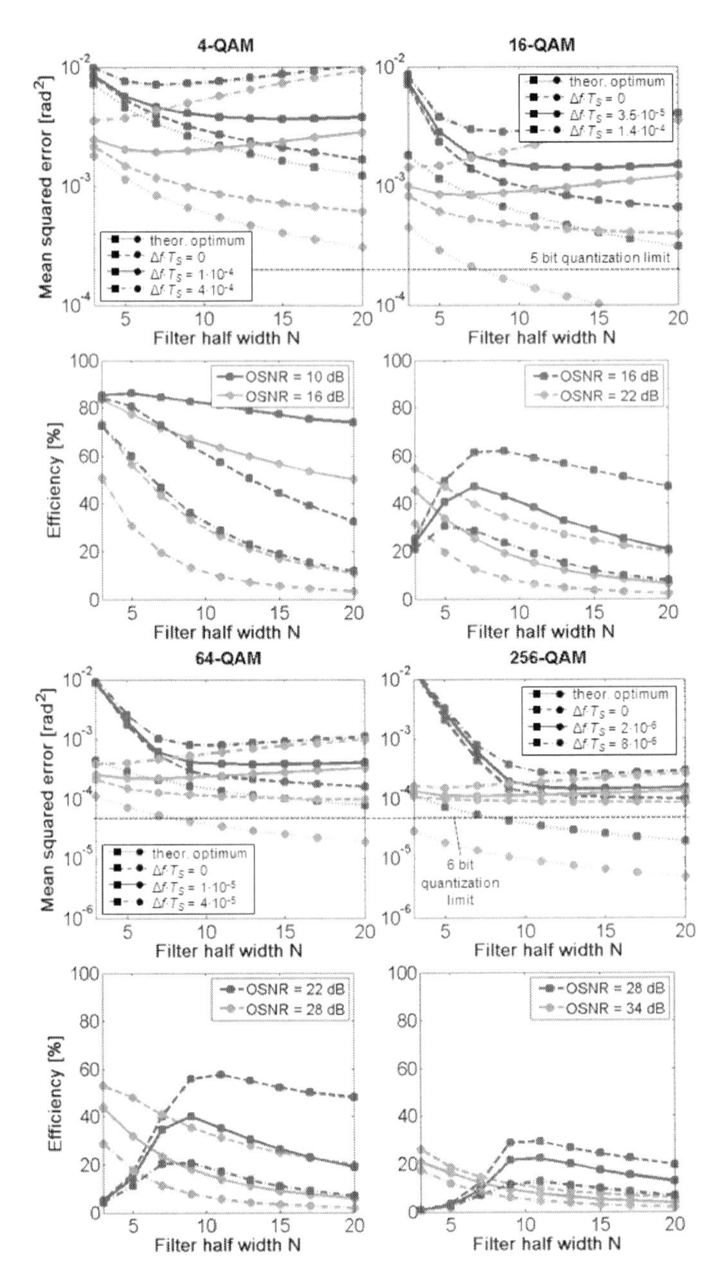

Fig. 3.64 Phase estimator mean squared error and efficiency $e(N)$ vs. filter half width N for different square QAM constellations. © 2009 IEEE.

Fig. 3.65 shows the effect of the analog-to-digital converter (ADC) resolution on receiver sensitivity. The necessary ADC resolution increases approximately by 1 bit if the number of constellation points is multiplied by 4. Table 3.8 summarizes the ADC requirements for a 100 Gb/s polarization multiplexed transmission system. Since commercial systems will also contain PMD and CD compensation, which necessitates oversampling, the values for $T_S/2$ sampling are also given. Tremendous progress is currently being made, and ADC resolution and speed targets are being met [108, 109].

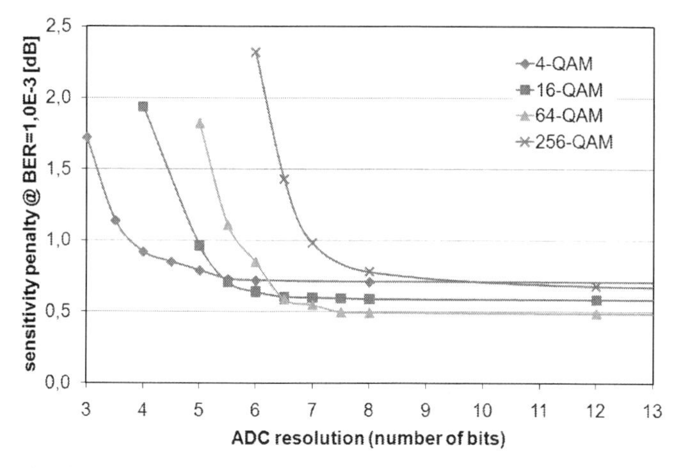

Fig. 3.65 Receiver sensitivity penalty vs. analog-to-digital converter resolution for different square QAM constellations. © 2009 IEEE.

Table 3.8 Analog-to-digital converter requirements for polarization-multiplexed 100 Gb/s transmission. © 2009 IEEE.

Constellation	ADC bandwidth	ADC sampling rate ($T_S/2$ sampling)	ADC effective number of bits
4-QAM	25 GHz	50 Gs/s	> 3.8
16-QAM	12.5 GHz	25 Gs/s	> 4.9
64-QAM	8.33 GHz	16.7 Gs/s	> 5.7
256-QAM	6.25 GHz	12.5 Gs/s	> 7.0

Fig. 3.66 shows the receiver sensitivity penalty against different resolutions of Re d_b and Im d_b. For all considered constellations a resolution ≥ 4 bits is sufficient. As for $|d_b|^2$ the penalty for a resolution ≥ 5 bits is tolerable (Fig. 3.67), and the square operations in (3.222) can be realized with a small look-up table (4 bit input, 4 bit output). The reason for the similar requirements in all QAM constellations is that the distance to the closest constellation point is independent of the number of constellation points. So, the hardware effort of FCR-QAM increases only moderately with the QAM order, which in turn determines overall spectral efficiency and hence the permissible cost frame.

To demonstrate the feasibility of optical high-order QAM transmission Figs. 3.68–3.71 depict the constellation diagrams at the receiver before and after carrier recovery for the different modulation formats and different optical signal-to-noise-ratios (OSNR). All constellation diagrams can be recovered using FCR-QAM for the different QAM modulation schemes and their respective linewidth times symbol rate products.

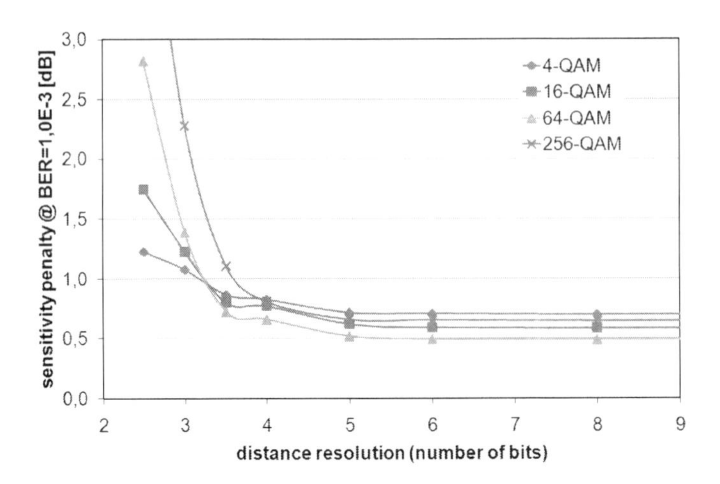

Fig. 3.66 Receiver sensitivity penalty vs. internal resolution for the distances Re d_m and Im d_m for different square QAM constellations. © 2009 IEEE.

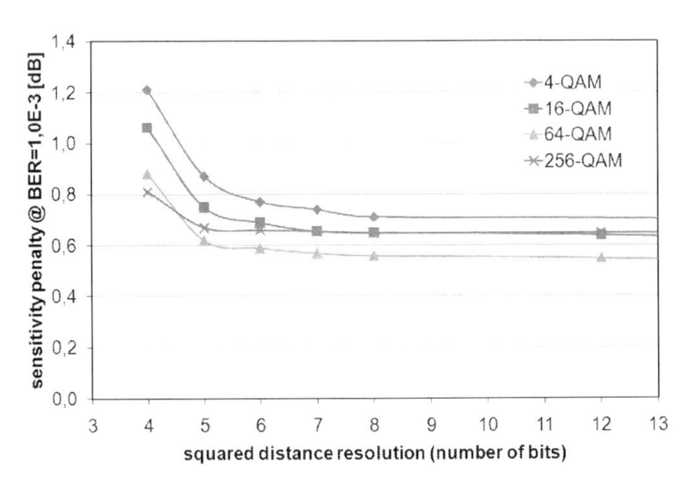

Fig. 3.67 Receiver sensitivity penalty vs. internal resolution for the squared distance $|d_b|^2$ for different square QAM constellations. © 2009 IEEE.

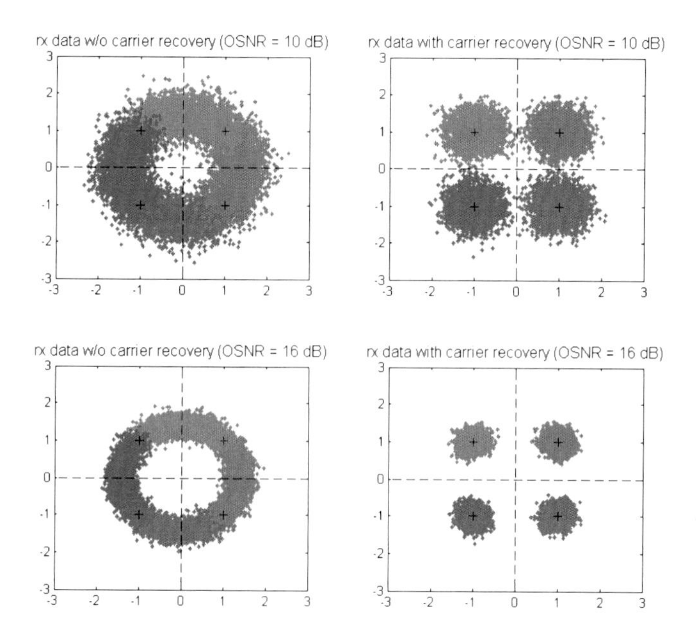

Fig. 3.68 4-QAM constellation diagrams before and after carrier recovery for $\Delta f{\cdot}T_S$ = $4.1{\cdot}10^{-4}$. © 2009 IEEE.

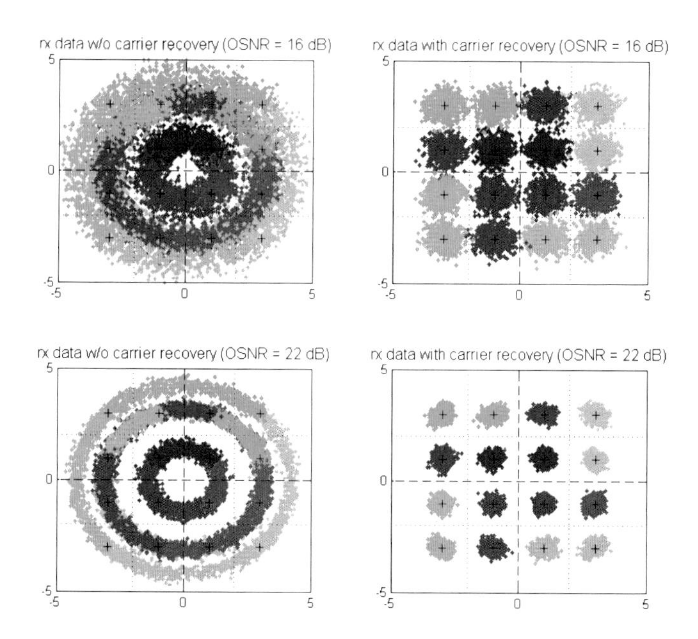

Fig. 3.69 16-QAM constellation diagrams before and after carrier recovery for $\Delta f{\cdot}T_S$ = $1.4{\cdot}10^{-4}$. © 2009 IEEE.

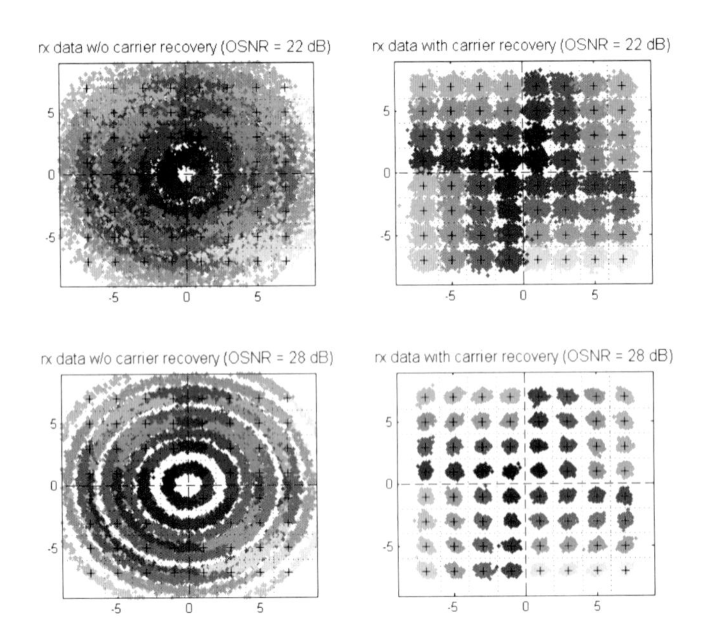

Fig. 3.70 64-QAM constellation diagrams before and after carrier recovery for $\Delta f \cdot T_S = 4.0 \cdot 10^{-5}$. © 2009 IEEE.

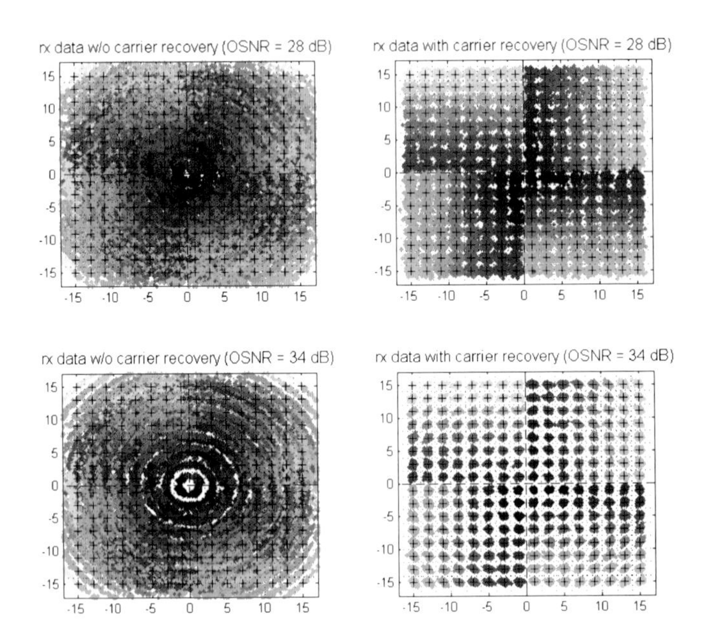

Fig. 3.71 256-QAM constellation diagrams before and after carrier recovery for $\Delta f \cdot T_S = 8.0 \cdot 10^{-6}$. © 2009 IEEE.

Let the transmitted symbol **c** and its recovered replica **r** (ideally equal to **c**, except for the time delay) have unit powers in each polarization ($\left|r_p\right|^2 = \left|c_p\right|^2 = 1$, $p = 1, 2$). The decision-directed electronic polarization control algorithm (3.200), (3.201) works also for polarization-multiplexed QAM.

Matters are more complicated for the constant-modulus polarization control algorithm (3.200), (3.204), which is non-data-aided. If it is used as is then the control gain g must be chosen so low that the deviations of $\left|\underline{X}_{1,2}\right|^2$ from 1 are averaged out effectively. This slows the control down considerably. In this context a CMA for 8-QAM signals [110] and a decision-directed CMA for QAM [111] have been published.

For adaptation of the CMA for QAM we modify (3.204) to become

$$\mathbf{T} = \begin{bmatrix} \Delta P_{1,\min} & 0 \\ 0 & \Delta P_{2,\min} \end{bmatrix} \mathbf{X}\mathbf{R}^+, \tag{3.226}$$

$$\Delta P_{p,\min} = \left[\Delta P_{p,h_p}\right]_{\min\limits_{\forall h_p}\left|\Delta P_{p,h_p}\right|} \qquad \Delta P_{p,h_p} = \hat{P}_{p,h_p} - \left|\underline{X}_p\right|^2. \tag{3.227}$$

$\Delta P_{p,h_p}$ is the power difference between the observed signal powers $\left|\underline{X}_p\right|^2$ in both polarizations $p = 1, 2$ and all expected values $\hat{P}_{p,h_p}$ ($h_p = 1, 2, \ldots, H$) of signal powers in case of zero polarization crosstalk. $\Delta P_{p,\min}$ is that value $\Delta P_{p,h_p}$ which has the smallest magnitude, hence most likely power difference. Once the algorithm has converged all power differences $\Delta P_{p,\min}$ would be zero in the absence of noise. The number H of distortion-free signal powers $\hat{P}_{p,h_p}$ equals the number of circles around the origin that are needed to touch all symbols in the chosen QAM constellation. The value is $H = 1, 3, 6, 9$ for 4-, 16-, 64- and 256-QAM, respectively.

The extension (3.205), (3.206) of the CMA for differential phase compensation is even more difficult to adapt for QAM. Similar as for FCR-QAM, a number of test phases

$$0 \le \varphi_c < 2\pi/q \tag{3.228}$$

is subtracted from the observed phase difference $\arg \underline{X}_1 - \arg \underline{X}_2$. Which test phases are needed depends on the signal powers $\hat{P}_{p,h_p}$ that have been estimated in the minimization process (3.227). Depending on the estimated powers $\hat{P}_{1,h_1}$,

$\hat{P}_{2,h_2}$ there are different numbers of test phases. Then one selects, for usage in (3.205),

$$\Delta\zeta = \left[\Delta\zeta_c\right]_{\substack{\min|\Delta\zeta_c| \\ \forall\,\varphi_c}} \qquad \Delta\zeta_c = \left(\left(\arg \underline{X}_1 - \arg \underline{X}_2 - \varphi_c + \varphi_f + \frac{\pi}{4}\right) \bmod \frac{\pi}{2}\right) - \frac{\pi}{4} \tag{3.229}$$

i.e., that value $\Delta\zeta$ among the various phase differences $\Delta\zeta_c$ found for the applicable set of test phases φ_c which has the lowest magnitude $|\Delta\zeta_c|$. For the time being let $\varphi_f = 0$. In order to improve accuracy in the presence of noise the phase difference can be weighted according to the available or estimated powers, for example using

$$\Delta\zeta = \left[\Delta\zeta_c\right]_{\substack{\min|\Delta\zeta_c| \\ \forall\,\varphi_c}} \hat{P}_{1,h_1} \hat{P}_{2,h_2} \,. \tag{3.230}$$

Consider 16-QAM as an example. We assume normalization with respect to the mean power. Then the expected powers in the individual polarization channels and the needed test phases are

$$\hat{P}_{p,h_p} = \begin{cases} 0.2 & \text{for } h_p = 1 \\ 1 & \text{for } h_p = 2 \\ 1.8 & \text{for } h_p = 3 \end{cases}, \qquad \varphi_c = \begin{cases} 0 & \text{for } h_1, h_2 \in \{1, 3\} \\ 0, \pm 0.64 & \text{for } h_1 = h_2 = 2 \\ \pm 0.46 & \text{otherwise} \end{cases} . \tag{3.231}$$

While the CMA with differential phase compensation can, in this configuration, track a once-acquired optimum, initial false locking is all the same possible. Similar to the p-fold phase ambiguity there exist different locking points, but only one of them yields a product **MJ** proportional to the unity matrix, while in the other cases there is a static phase shift between the two polarization channels. In order to quit a false optimum one can proceed similar as in FCR-QAM. One calculates $\Delta\zeta$ according to (3.229) not only for $\varphi_f = 0$ but for D equidistant phase offsets

$$\varphi_f = \frac{f}{F} \cdot \frac{2\pi}{q}, \qquad f \in \{0, 1, \ldots, F-1\}. \tag{3.232}$$

A good choice is $F = 8\sqrt{M}$; hence F may be equal to B specified in (3.216). The squares of the various $\Delta\zeta = \Delta\zeta(f,k)$ are added up over K subsequent symbols,

$$W_f = \sum_{k=1}^{K} \left(\Delta\zeta(f,k)\right)^2 . \tag{3.233}$$

One determines that integer f which corresponds to the smallest W_f. This indicates that a better optimum is available if one introduces a differential phase shift φ_f between the polarizations. So, after K symbols one sets

$$\mathbf{M} := \begin{bmatrix} e^{j\varphi_f/2} & 0 \\ 0 & e^{-j\varphi_f/2} \end{bmatrix} \mathbf{M} . \quad (3.234)$$

Thereafter the summation process (3.233) may start anew. But this is usually not needed because a single application of (3.234) usually yields the a differential phase very close to the optimum, which is subsequently improved and tracked as described above. As a consequence, the full set (3.229) obtained for all f and the summation (3.233) may be executed less frequently than the tracking calculation for $\varphi_f = 0$. This reduces the required hardware effort.

A more hardware-efficient way to avoid false locking is the following: After the decision circuits a framing information is detected which indicates whether data is being received correctly in both polarization channels. If not, then (3.234) is executed with a suitably chosen φ_f, or with all values given in (3.232), until data recovery is correct. This doesn't slow down normal control, since it occurs only at initial signal acquisition.

CMA-QAM and DPC-CMA-QAM were verified in short simulations without noise, for 4-, 16-, 64- and 256-QAM. Subsequently noise has been added, and standard CMA, CMA QAM and DPC-CMA-QAM have been compared for 16-QAM [76]. Fig. 3.72 shows the sensitivity degradation at BER = 10^{-3} as a function of control gain g, for a sum linewidth times symbol duration product of $\Delta f \cdot T = 2 \cdot 10^{-4}$. $1/g$ is proportional to the small-signal polarization control time constant, so a large g is advantageous. The total penalty reaches 2 dB for the standard CMA at $g \approx 2^{-6.5}$ while the CMA-QAM can control polarization about 3 times faster, with $g = 2^{-5}$. For both, the two carrier recoveries processed 19 symbols in parallel. The DPC-CMA-QAM with one joint carrier recovery for both polarizations processes only 9 temporal samples. This means that phase noise is better tolerated, and sensitivity is therefore increased by ~0.5 dB. Fig. 3.73 shows the same simulations, but at a doubled $\Delta f \cdot T = 4 \cdot 10^{-4}$. The sensitivity advantage of the common carrier recovery in the case of DPC-CMA-QAM is even more pronounced.

All QAM polarization control algorithms are compared in Fig. 3.74, again as a function of polarization control gain g [76]. Performance is worst for the original decision-directed algorithm (ODDA) [63] with $\mathbf{T} = (\mathbf{1} - \mathbf{Q})\mathbf{M}$ employed in (3.200) and $\mathbf{Q}$ given by (3.199). Penalty traces for CMA-QAM and DPC-CMA-QAM are fairly identical, due to $\Delta f \cdot T = 0$. The modified decision-directed algorithm (MDDA) (3.200), (3.201) performs best. At 0.5 dB penalty the MDDA is ~15 times faster than the standard CMA and ~4 times faster than the DPC-CMA-QAM.

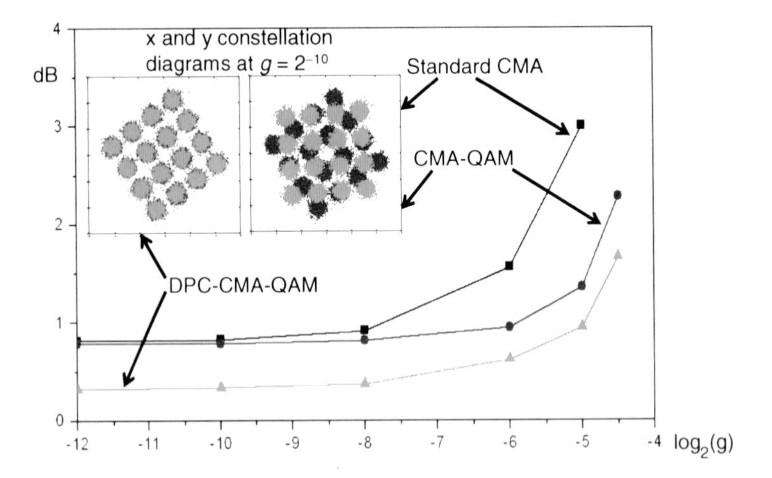

Fig. 3.72 Sensitivity penalty at BER = 10^{-3} of various CMAs applied to polarization-multiplexed 16-QAM signals vs. control gain g, for $\Delta f \cdot T_S = 2 \cdot 10^{-4}$, **MJ = 1**. Each data point corresponds to 200,000 symbols. Insets: Constellation diagrams, with phase difference between polarization channels compensated in the case of DPC-CMA-QAM. © 2010 IEEE.

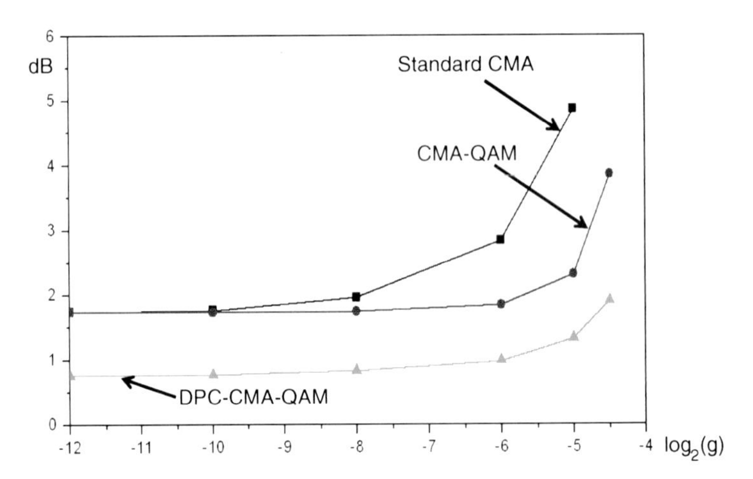

Fig. 3.73 As above, but for $\Delta f \cdot T_S = 4 \cdot 10^{-4}$. © 2010 IEEE.

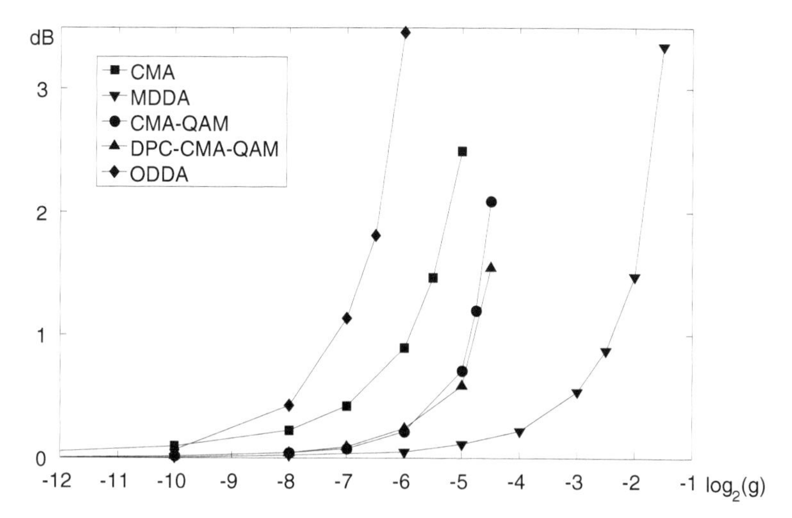

Fig. 3.74 Sensitivity penalty at BER = 10^{-3} of various 16-QAM polarization control algorithms vs. control gain g, for $\Delta f \cdot T_S = 0$, $\mathbf{MJ} = \mathbf{1}$. The MDDA outperforms all other algorithms. © 2010 IEEE.

References

1. Yariv, A.: Optical Electronics, 4th edn. Saunders College Publishing (1991)
2. Jeunhomme, L.B.: Single-mode fiber optics - principles and applications. Marcel Dekker Inc., New York (1983)
3. Gordon, J.P., Kogelnik, H.: PMD fundamentals: Polarization mode dispersion in optical fibers. PNAS 97(9), 4541–4550 (2000)
4. Elies, P., Le Jeune, B., Le Roy-Brehonnet, F., Cariou, J., Lotrian, J.: Experimental investigation of the speckle polarization for a polished aluminium sample. J. Phys. D: Appl. Phys. 30, 29–39 (1997)
5. Le Roy-Brehonnet, F., Le Jeune, B., Gerligand, P.Y., Cariou, J., Lotrian, J.: Analysis of depolarizing optical targets by Mueller matrix formalism. Pure Appl. Opt. 6, 385–404 (1997)
6. Cimini, L.J., Habbab, I.M.I., John, R.K., Saleh, A.A.M.: Preservation of polarization orthogonality through a linear optical system. Electronics Letters 23, 1365–1366 (1987)
7. Noé, R., Rehage, M., Harizi, C., Ricken, R.: Depolarizer based on acoustooptical TE-TM converters for suppression of polarization holeburning in long haul EDFA links. Electron. Lett. 30(18), 1500–1501 (1994)
8. Poole, C.D., Wagner, R.E.: Phenomenological approach to polarization dispersion in long single-mode fibers. Electron. Lett. 22(19), 1029–1030 (1986)
9. Peters, J., Dori, A., Kapron, F.: Bellcore's fiber measurement audit of existing cable plant for use with high bandwidth systems. In: NFOEC 1997, San Diego, September 21-25, vol. 2, pp. 19–30 (1997)
10. Noé, R., Sandel, D., Yoshida-Dierolf, M., Hinz, S., Mirvoda, V., Schöpflin, A., Glingener, C., Gottwald, E., Scheerer, C., Fischer, G., Weyrauch, T., Haase, W.: Polarization mode dispersion compensation at 10, 20 and 40 Gb/s with various optical equalizers. IEEE J. Lightwave Technology 17(9), 1602–1616 (1999)
11. Noé, R., Heidrich, H., Hoffmann, D.: Endless polarization control systems for coherent optics. IEEE J. Lightwave Techn. 6(7), 1199–1207 (1988)
12. Hidayat, A., Koch, B., Mirvoda, V., Zhang, H., Lichtinger, M., Sandel, D., Noé, R.: Optical 5 krad/s Endless Polarisation Tracking. Electronics Letters 44 (2008)
13. Koch, B., Hidayat, A., Zhang, H., Mirvoda, V., Lichtinger, M., Sandel, D., Noé, R.: Optical Endless Polarization Stabilization at 9 krad/s with FPGA-Based Controller. IEEE Photonics Technology Letters 20(12), 961–963 (2008)
14. Hidayat, A., Koch, B., Zhang, H., Mirvoda, V., Lichtinger, M., Sandel, D., Noé, R.: High-speed endless optical polarization stabilization using calibrated waveplates and field-programmable gate array-based digital controller. Optics Express 16(23), 18984–18991 (2008)
15. Koch, B., Hidayat, A., Mirvoda, V., Zhang, H., Sandel, D., Noé, R.: Fast Optical Polarisation Tracking Experiment with 2 Grad Trajectory Length. Electronics Letters 44(23), 1376–1378 (2008)

16. Noé, R., Koch, B., Mirvoda, V., Sandel, D.: Endless optical polarization control and PMD compensation. In: Proc. OFC 2010, San Diego, CA, USA, Paper OThJ1, March 21-25 (2010)
17. Noé, R., Koch, B., Mirvoda, V., Hidayat, A., Sandel, D.: 38 krad/s, 3.8 Grad, Broadband Endless Optical Polarization Tracking Using $LiNbO_3$ Device. IEEE Photonics Technology Letters 21(17), 1220–1222 (2009)
18. Noé, R., Sandel, D., Hinz, S., Yoshida-Dierolf, M., Mirvoda, V., Feise, G., Herrmann, H., Ricken, R., Sohler, W., Wehrmann, F., Glingener, C., Schöpflin, A., Färbert, A., Fischer, G.: Integrated optical $LiNbO_3$ distributed polarization mode dispersion equalizer in 20 Gbit/s transmission system. Electronics Letters 35(8), 652–654 (1999)
19. Heismann, F., Ulrich, R.: Integrated-optical single-sideband modulator and phase shifter. IEEE J. Quantum Electronics 18(4), 767–771 (1982)
20. Heismann, F., Whalen, M.S.: Broadband reset-free automatic polarization controller. Electron. Lett. 27(4), 377–379 (1991)
21. Johnson, M.: In-line fiber-optical polarization transformer. Applied Optics 18(9), 1288–1289 (1979)
22. Sandel, D., Mirvoda, V., Bhandare, S., Wüst, F., Noé, R.: Some enabling techniques for polarization mode dispersion compensation. J. Lightwave Techn. 21(5), 1198–1210 (2003)
23. Wüst, D., Sandel, V., Mirvoda, R.: Electrical slope steepness difference indicates higher-order PMD at 40Gbit/s. In: Optical Fiber Communication Conference (OFC 2003), Atlanta, GA, USA (March 2003)
24. Noé, R.: Combatting and equalizing the effects of PMD in 40 Gb/s systems and beyond. In: Proc. ECOC 2002, Kopenhagen, Denmark, vol. II, T4 (2002), Updated viewgraphs: http://ont.uni-paderborn.de/fileadmin/ont/publikationen/ecoc2002_noe_tut_add.pdf
25. Noe, R., Sandel, D.: In-service PMD monitoring and compensation. In: Ninth Optoelectronics and Communications Conference/Third International Conference on Optical Internet (OECC/COIN 2004), Invited Tutorial 14C2-1, Yokohama, Japan, July 12-16 (2004), http://ont.uni-paderborn.de/fileadmin/ont/publikationen/OECC2004_noe_tutorial_n12.pdf
26. Sharma, M., Ibe, H., Ozeki, T.: Optical circuits for equalizing group delay dispersion of optical fibers. IEEE J. Lightwave Technology 12, 1759–1765 (1994)
27. Takiguchi, K., Okamoto, K., Moriwaki, K.: Planar lightwave circuit dispersion equalizer. IEEE J. Lightwave Technology 14, 2003–2011 (1996)
28. Takiguchi, K., Kawanishi, S., Takara, H., Himeno, A., Hattori, K.: Dispersion slope equalizer for dispersion shifted fiber using a lattice-form programmable optical filter on a planar lightwave circuit. IEEE J. Lightwave Technology 16, 1647–1656 (1998)
29. Kudou, T., Shimizu, K., Harada, K., Ozeki, T.: Synthesis of grating lattice design. IEEE J. Lightwave Technology 17, 347–353 (1999)
30. Bohn, M., Horst, F., Offrein, B.J., Bona, G.L., Meissner, E., Rosenkranz, W.: Tunable Dispersion Compensation in a 40 Gb/s System using a Compact FIR Lattice Filter in SiON Technology. In: Proc. ECOC 2002, Kopenhagen, Denmark, vol. II., 4.2.3 (2002)
31. Noé, R., Gao, Z.: Mach-Zehnder lattice based tunable chromatic dispersion compensator design with simplified control and dispersion slope mitigation. In: Proc. European Conference on Integrated Optics (ECIO 2003), Prague, CZ, April, 2-4, WeA3.4, vol. 1, pp. 87–90 (2003)
32. Noé, R., Sandel, D., Mirvoda, V.: PMD in high-bit-rate systems and means for its mitigation. IEEE J. Selected Topics in Quantum Electronics 10, 341–355 (2005)

33. Heismann, F.: Jones matrix expansion for second-order polarization mode dispersion. In: Proc. ECOC-IOOC 2003, Th1.7.5 (2003)
34. Foschini, G.J., Jopson, R.M., Nelson, L.E., Kogelnik, H.: Statistics of polarization dependent chromatic fiber dispersion due to PMD. In: Proc. ECOC 1999, Nice, France, September 26-30, vol. II, pp. 56–59 (1999)
35. Shtaif, M., Mecozzi, A., Tur, M., Nagel, J.A.: A compensator for the effects of high-order polarization mode dispersion in optical fibers. IEEE Photonics Technology Letters 12(4), 434–436 (2000)
36. Möller, L.: Filter Synthesis for Broad-Band PMD Compensation in WDM Systems. IEEE Photonics Technol. Lett. 12(9), 1258–1260 (2000)
37. Harris, S.E., Ammann, E.O., Chang, I.C.: Optical Network Synthesis Using Birefringent Crystals. I. Synthesis of Lossless Networks of Equal-Length Crystals. J. Opt. Soc. Am. 54(10), 1267–1279 (1964)
38. Kogelnik, H., Nelson, L.E., Gordon, J.P.: Emulation and Inversion of Polarization-Mode Dispersion. IEEE J. Lightwave Technology 21(2), 482–495 (2003)
39. Eyal, A., Marshall, W.K., Tur, M., Yariv, A.: Representation of second-order polarization mode dispersion. Electron. Lett. 35(19), 1658–1659 (1999)
40. Eyal, A., Li, Y., Marshall, W.K., Yariv, A.: Statistical determination of the length dependence of high-order polarization mode dispersion. Optics Letters 25(12), 875–877 (2000)
41. Bergano, N.S., Kerfoot, F.W., Davidson, C.R.: Margin Measurements in Optical Amplifier Systems. IEEE Photonics Technol. Lett. 5, 304–306 (1993)
42. Sandel, D., Wüst, F., Mirvoda, V., Noé, R.: Standard (NRZ 1´40Gbit/s, 210km) and polarization multiplex (CS-RZ, 2´40Gbit/s, 212km) transmissions with PMD compensation. IEEE Photonics Technol. Lett. 14(8) (2002)
43. Noé, R., Sandel, D., Mirvoda, V., Wüst, F.: Polarization mode dispersion detected by arrival time measurement of polarization-scrambled light. IEEE J. Lightwave Techn. 20(2), 229–235 (2002)
44. Sandel, D., Mirvoda, V., Wüst, F., Noé, R., Hinz, S.: Signed online chromatic dispersion detection at 40Gb/s with a sub-ps/nm dynamic accuracy. 06.1.4. In: Proc. 28th European Conference on Optical Communication (ECOC 2002), Copenhagen, DK (2002)
45. Shimoda, K., Takahasi, H., Townes, C.H.: Fluctuations in amplifications of quanta with application to maser amplifiers. J. Phys. Soc. Japan 12, S.686–S.700 (1957)
46. Saleh, B.: Photoelectron Statistics. Springer, Heidelberg (1978)
47. Desurvire, E.: Erbium doped fiber amplifiers: Principles and Applications. Wiley, New York (1994)
48. Noé, R.: Coherent/WDM Optical Transmission Systems and Technologies. In: Proc. Fifth Optoelectronics Conference (OEC 1994), Makuhari Messe, Chiba, Japan, July 12-15, 14C4-1 (invited), pp. 188–189 (1994)
49. Haus, H.A.: The noise figure of optical amplifiers. IEEE Photon. Techn. Lett. 10, 1602–1604 (1998)
50. Noé, R.: Optical amplifier performance in digital optical communication systems. Electrical Engineering 83, 15–20 (2001)
51. Saplakoglu, G., Celik, M.: Output Photon Number Distribution of Erbium Doped Fiber Amplifiers. In: 3rd Topical Meeting of the Optical Society of America on Optical Amplifiers and their Applications, Santa Fe, New Mexico, USA, June 24-26 (1992)
52. Haus, H.A.: Private communication
53. Proakis, J.G.: Digital Communications, 2nd edn. McGraw-Hill, New York (1989)
54. Bhandare, S., Sandel, D., Milivojevic, B., Hidayat, A., Fauzi Abas Ismail, A., Zhang, H., Ibrahim, S., Wüst, F., Noé, R.: 5.94 Tbit/s, 1.49 bit/s/Hz (40×2×2×40 Gbit/s) RZ-DQPSK Polarization Division Multiplex C-Band Transmission over 324 km. IEEE Photonics Technology Letters 17, 914–916 (2005)

55. Noé, R.: Phase Noise Tolerant Synchronous QPSK/BPSK Baseband-Type Intradyne Receiver Concept with Feedforward Carrier Recovery. IEEE J. Lightwave Technology 23, 802–808 (2005)
56. Shpantzer, I., Achiam, Y., Cho, P.S., Greenblatt, A., Harston, G., Kaplan, A.: Optoelectronic Devices and Subsystems for Digital Coherent Optical Communication. In: Proc. IEEE-LEOS Summer Topicals 2008, MC1.3, Acapulco, Mexico, July 21-23 (2008)
57. Harston, G., Cho, P.S., Greenblatt, A.S., Kaplan, A., Achiam, Y., Shpantzer, I.: Active Control of an Optical 90° Hybrid for Coherent Detection. In: Proc. IEEE-LEOS Summer Topicals 2008, MC1.4, Acapulco, Mexico, July 21-23 (2008)
58. Leeb, W.R.: Realization of 90- and 180 degree hybrids for optical frequencies. Archiv fuer Elektronik und Uebertragungstechnik (ISSN 0001-1096) 37, 203–206 (1983)
59. Noé, R., Sessa, W.B., Welter, R., Kazovsky, L.G.: New FSK phase-diversity receiver in a 150 Mbit/s coherent optical transmission system. Electron. Lett. 24(9), 567–568 (1988)
60. Noé, R., Meissner, E., Borchert, B., Rodler, H.: Direct modulation 565 Mb/s PSK experiment with solitary SL-QW-DFB lasers and novel suppression of the phase transition periods in the carrier recovery. In: Proc. ECOC 1992, Th PD I.5, vol. 3, pp. 867–870 (1992)
61. Norimatsu, S., Iwashita, K.: Damping factor influence on linewidth requirements for optical PSK coherent detection systems. IEEE J. Lightwave Techn. 11(7), 1226–1233 (1993)
62. Derr, F.: Coherent optical QPSK intradyne system: Concept and digital receiver realization. J. Lightwave Techn. 10, 1290–1296 (1992)
63. Noé, R.: PLL-Free Synchronous QPSK Polarization Multiplex/Diversity Receiver Concept with Digital I&Q Baseband Processing. IEEE Photon. Technol. Lett. 17, 887–889 (2005)
64. Pfau, T., Hoffmann, S., Peveling, R., Bhandare, S., Ibrahim, S.K., Adamczyk, O., Porrmann, M., Noé, R., Achiam, Y.: First Real-Time Data Recovery for Synchronous QPSK Transmission with Standard DFB Lasers. IEEE Photonics Technology Letters 18, 1907–1909 (2006)
65. Pfau, T., Hoffmann, S., Peveling, R., Ibrahim, S., Adamczyk, O., Porrmann, M., Bhandare, S., Noé, R., Achiam, Y.: Synchronous QPSK Transmission at 1.6 Gbit/s with Standard DFB Lasers and Real-time Digital Receiver. Electronics Letters 18, 1175–1176 (2006)
66. Pfau, T., Hoffmann, S., Adamczyk, O., Peveling, R., Herath, V., Porrmann, M., Noé, R.: Coherent optical communication: Towards realtime systems at 40 Gbit/s and beyond. Opt. Express 16, 866–872 (2008), `http://www.opticsinfobase.org/abstract.cfm?URI=oe-16-2-866`
67. Noé, R.: Sensitivity comparison of coherent optical heterodyne, phase diversity, and polarization diversity receivers. J. Optical Communications 10, 11–18 (1989)
68. Noé, R., Meissner, E., Borchert, B., Rodler, H.: Direct modulation 565 Mb/s DPSK experiment with 62.3 dB loss span and endless polarization control. IEEE Photonics Technology Letters 4(10), 1151–1154 (1992)
69. Norimatsu, S., Iwashita, K.: The influence of cross-phase modulation on optical FDM PSK homodyne transmission systems. IEEE J. Lightwave Techn. 11(5/6), 795–804 (1993)
70. Noé, R.: Phase-noise tolerant feedforward carrier recovery concept for baseband-type synchronous QPSK/BPSK receiver. In: Proc. 3rd IASTED Int. Conf. on Wireless and Optical Communications, Banff, Canada, July 14–16 (2003) ISBN: 0-88 986-374-1, 197
71. Noé, R., Rodler, H., Ebberg, A., Gaukel, G., Noll, B., Wittmann, J., Auracher, F.: Comparison of polarization handling methods in coherent optical systems. IEEE J. Lightwave Techn. 9(10), 1353–1366 (1991)

72. Pfau, T., Peveling, R., Porte, H., Achiam, Y., Hoffmann, S., Ibrahim, S.K., Adamczyk, O., Bhandare, S., Sandel, D., Porrmann, M., Noé, R.: Coherent Digital Polarization Diversity Receiver for Real-Time Polarization-Multiplexed QPSK Transmission at 2.8 Gbit/s. IEEE Photonics Technology Letters 19(24), 1988–1990 (2007)
73. Godard, D.: Self-recovering equalization and carrier tracking in two-dimensional data communication systems. IEEE T. Commun. 28(11), 1867–1875 (1980)
74. Kikuchi, K.: Polarization-demultiplexing algorithm in the digital coherent receiver. In: Proc. IEEE/LEOS Summer Topical Meeting 2008, MC2.2, Acapulco, Mexico, July 21-23 (2008)
75. El-Darawy, M., Pfau, T., Hoffmann, S., Noé, R.: Differential Phase Compensated Constant Modulus Algorithm for Phase Noise Tolerant Coherent Optical Transmission. In: Proc. IEEE-LEOS 2009, TuC3.3, San Diego, USA, July 20-22 (2009)
76. Noé, R., Pfau, T., El-Darawy, M., Hoffmann, S.: Electronic Polarization Control Algorithms for Coherent Optical Transmission. IEEE Journal of Selected Topics in Quantum Electronics 16 (2010)
77. Tao, Z., Li, L., Isomura, A., Hoshida, T., Rasmussen, J.C.: Multiplier-free Phase Recovery for Optical Coherent Receivers. In: OFC/NFOEC 2008, OWT2 (2008)
78. Ly-Gagnon, D.-S., Tsukamoto, S., Katoh, K., Kikuchi, K.: Coherent Detction of Optical Quadrature Phase-Shift Keying Signals with Carrier Phase Estimation. IEEE JLT 24, 12–21 (2006)
79. Hoffmann, S., Peveling, R., Pfau, T., Adamczyk, O., Eickhoff, R., Noé, R.: Multiplier-Free Real-Time Phase Tracking for Coherent QPSK Receivers. IEEE Photonics Technology Letters 21(3), 137–139 (2009)
80. Hoffmann, S.: Hardwareeffiziente Echtzeit-Signalverarbeitung für synchronen QPSK-Empfang, Dissertation, Univ. Paderborn, Germany (2008)
81. Viterbi, A.J., Viterbi, A.N.: Nonlinear estimation of PSK-modulated carrier phase with application to burst digital transmission. IEEE Transactions on Information Theory 29(4), 543–551 (1983)
82. Boucheret, M.-L., Mortensen, I., Favaro, H., Belis, E.: A new algorithm for nonlinear estimation of PSK-modulated carrier phase, ECSC-3, 155–159 (November 1993)
83. van den Borne, D., Fludger, C.R.S., Duthel, T., Wuth, T., Schmidt, E.D., Schulien, C., Gottwald, E., Khoe, G.D., de Waardt, H.: Carrier phase estimation for coherent equalization of 43-Gb/s POLMUX-NRZ-DQPSK transmission with 10.7-Gb/s NRZ neighbours. In: ECOC 2007 (We 7.2.3), vol. 3, pp. 149–150 (2007)
84. Hoffmann, S., Bhandare, S., Pfau, T., Adamczyk, O., Wördehoff, C., Peveling, R., Porrmann, M., Noé, R.: Frequency and Phase Estimation for Coherent QPSK transmission With Unlocked DFB lasers. IEEE PTL 20(18), 1569–1571 (2008)
85. Pfau, T., Peveling, R., Samson, F., Romoth, J., Hoffmann, S., Bhandare, S., Ibrahim, S., Sandel, D., Adamczyk, O., Porrmann, M., Noé, R., Hauden, J., Grossard, N., Porte, H., Schlieder, D., Koslovsky, A., Benarush, Y., Achiam, Y.: Polarisation-Multiplexed 2.8 Gbit/s Synchronous QPSK Transmission with Real-Time Digital Polarization Tracking. In: Proc. ECOC 2007 (We 8.3.3), vol. 3, pp. 263–264 (2007)
86. Sun, H., Wu, K.-T., Roberts, K.: Real-time measurements of a 40 Gb/s coherent system. Optics Express 16, 873 (2008)
87. Herath, V., Peveling, R., Pfau, T., Adamczyk, O., Hoffmann, S., Wördehoff, C., Porrmann, M., Noé, R.: Chipset for a Coherent Polarization-Multiplexed QPSK Receiver. In: Proc. OFC 2009, OThE2, San Diego, CA, USA, March 22-26 (2009)
88. McNicol, J., O'Sullivan, M., Roberts, K., Comeau, A., McGhan, D., Strawczynski, L.: Electronic Domain Compensation of Optical Dispersion. Optical Fiber Communication Conference and Exposition and The National Fiber Optic Engineers Conference, Technical Digest (CD) (Optical Society of America, 2005), paper OThJ3 (2005)

89. Spinnler, B., Hauske, F.N., Kuschnerov, M.: Adaptive Equalizer Complexity in Coherent Optical Receivers. In: Proc. European Conference on Optical Communication (ECOC 2008), We.2.E.4, Brussels, Belgium, September 21-25 (2008)
90. Shieh, W., Athaudage, C.: Coherent optical orthogonal frequency division multiplexing. Electron. Lett. 42, 587–589 (2006)
91. Shieh, W., Yi, X., Tang, Y.: Transmission experiment of multi-gigabit coherent optical OFDM systems over 1000 km SSMF fiber. Electron. Lett. 43, 183–185 (2007)
92. Lowery, J.: Amplified-spontaneous noise limit of optical OFDM lightwave systems. Opt. Express 16, 860–865 (2008)
93. Shieh, W., Bao, H., Tang, Y.: Coherent optical OFDM: theory and design. Opt. Express 16, 841–859 (2008)
94. Lowery, J., Wang, S., Premaratne, M.: Calculation of power limit due to fiber nonlinearity in optical OFDM systems. Opt. Express 15, 13282–13287 (2007)
95. Lowery, J.: Fiber nonlinearity pre- and post-compensation for long-haul optical links using OFDM. Opt. Express 15, 12965–12970 (2007)
96. Yang, Q., Kaneda, N., Liu, X., Shieh, W.: Demonstration of Frequency-Domain Averaging Based Channel Estimation for 40-Gb/s CO-OFDM with High PMD. IEEE Photonics Technology Letters 21 (2009)
97. Nazarathy, M., Khurgin, J., Weidenfeld, R., Meiman, Y., Cho, P., Noé, R., Shpantzer, I., Karagodsky, V.: Phased-array cancellation of nonlinear FWM in coherent OFDM dispersive multi-span links. Opt. Express 16, 15777–15810 (2008)
98. Pfau, T., Hoffmann, S., Noé, R.: Hardware-Efficient Coherent Digital Receiver Concept with Feedforward Carrier Recovery for M-QAM Constellations. IEEE J. Lightwave Technology 27, 989–999 (2009)
99. Pfau, T.: Development and real-time implementation of digital signal processing algorithms for coherent optical receivers, Dissertation, Univ. Paderborn, Germany (2009)
100. Volder, J.: The CORDIC Trigonometric Computing Technique. IRE Transactions on Electronic Computers EC-8(3), 330–334 (1959)
101. Walther, J.: A Unified Algorithm for Elementary Functions. In: Proc. Spring Joint Computer Conference, Atlantic City, NJ, USA, May 18-20, pp. 379–385 (1971)
102. Pfau, T., Noé, R.: Algorithms for Optical QAM Detection. In: Proc. IEEE-LEOS 2009, MC3.2, San Diego, USA, July 20-22 (2009)
103. Sakamoto, T., Chiba, A., Kawanishi, T.: 50-Gb/s 16 QAM by a quad-parallel Mach-Zehnder modulator. In: Proc. ECOC 2007, PD 2.8, Berlin, Germany, September 16-20 (2007)
104. Doerr, C., Winzer, P., Zhang, L., Buhl, L., Sauer, N.: Monolithic InP 16-QAM Modulator. In: Proc. OFC/NFOEC 2008, PDP20, San Diego, CA, USA, February 24-28 (2008)
105. Webb, W., Hanzo, L.: Modern quadrature amplitude modulation. Pentech Pr., London (1994)
106. Tretter, S.: Estimating the Frequency of a Noisy Sinusoid by Linear Regression. IEEE Trans. Inf. Theory 31(6), 832–835 (1985)
107. Cramer, H.: Mathematical Methods of Statistics. Princeton University Press, Princeton (1999)
108. Ellermeyer, T., Mullrich, J., Rupeter, J., Langenhagen, H., Bielik, A.: DA and AD Converters for 25 GS/s and above. In: Proc. SUM 2008, TuC3.1, Acapulco, Mexico, July 21-23 (2008)
109. `http://www.fujitsu.com/emea/news/pr/fme_20090127.html`
110. Zhou, X., Yu, J., Magill, P.: Cascaded two-modulus algorithm for blind polarization demultiplexing of 114-Gb/s PDM-8-QAM optical signals. In: Proc. OFC 2009, OWG3, San Diego (2009)
111. Louchet, H., Kuzmin, K., Richter, A.: Improved DSP algorithms for coherent 16-QAM transmission. In: Proc. ECOC 2008, TU.1.E.6, Brussels, September 21-25 (2008)

Index